NCCER

MW01490006

Instrumentation Level Four

PEARSON

Prentice
Hall

Upper Saddle River, New Jersey
Columbus, Ohio

NCCER

President: Don Whyte
Director of Curriculum Revision and Development: Daniele Dixon
Instrumentation Project Manager: Deborah Padgett
Production Manager: Debie Ness
Quality Assurance Coordinator: Jessica Martin
Editor: Cristina Escobar
Desktop Publishers: Laura Parker and Rachel Ivines

The NCCER would like to acknowledge the contract service provider for this curriculum:
Topaz Publications, Liverpool, New York.

This information is general in nature and intended for training purposes only. Actual performance of activities described in this manual requires compliance with all applicable operating, service, maintenance, and safety procedures under the direction of qualified personnel. References in this manual to patented or proprietary devices do not constitute a recommendation of their use.

20 19 18 17 16 15 14 13
ISBN 0-13-108922-6

Preface

This volume was developed by the National Center for Construction Education and Research (NCCER) in response to the training needs of the construction, maintenance, and pipeline industries. It is one of many in NCCER's *Contren® Learning Series*. The program, covering training for close to 40 construction and maintenance areas, and including skills assessments, safety training, and management education, was developed over a period of years by industry and education specialists.

NCCER also maintains a National Registry that provides transcripts, certificates, and wallet cards to individuals who have successfully completed modules of NCCER's *Contren® Learning Series*, when the training program is delivered by an NCCER Accredited training Sponsor.

The NCCER is a not-for-profit 501(c)(3) education foundation established in 1995 by the world's largest and most progressive construction companies and national construction associations. It was founded to address the severe workforce shortage facing the industry and to develop a standardized training process and curricula. Today, NCCER is supported by hundreds of leading construction and maintenance companies, manufacturers, and national associations, including the following partnering organizations:

PARTNERING ASSOCIATIONS

- American Fire Sprinkler Association
- American Petroleum Institute
- American Society for Training & Development
- American Welding Society
- Associated Builders & Contractors, Inc.
- Association for Career and Technical Education
- Associated General Contractors of America
- Carolinas AGC, Inc.
- Carolinas Electrical Contractors Association
- Citizens Democracy Corps
- Construction Industry Institute
- Construction Users Roundtable

- Design-Build Institute of America
- Merit Contractors Association of Canada
- Metal Building Manufacturers Association
- National Association of Minority Contractors
- National Association of State Supervisors for Trade and Industrial Education
- National Association of Women in Construction
- National Insulation Association
- National Ready Mixed Concrete Association
- National Systems Contractors Association
- National Utility Contractors Association
- National Vocational Technical Honor Society
- North American Crane Bureau
- Painting & Decorating Contractors of America
- Plumbing-Heating-Cooling Contractors National Association
- Portland Cement Association
- SkillsUSA
- Steel Erectors Association of America
- Texas Gulf Coast Chapter ABC
- U.S. Army Corps of Engineers
- University of Florida
- Women Construction Owners & Executives, USA

Some features of NCCER's *Contren® Learning Series* are:

- An industry-proven record of success
- Curricula developed by the industry for the industry
- National standardization providing portability of learned job skills and educational credits
- Credentials for individuals through NCCER's National Registry
- Compliance with Apprenticeship, Training, Employer, and Labor Services (ATELS) requirements for related classroom training (CFR 29:29)
- Well-illustrated, up-to-date, and practical information

Acknowledgments

This curriculum was revised as a result of the farsightedness and leadership of the following sponsors:

Austin Industrial
Cianbro Corporation
Lee College

This curriculum would not exist were it not for the dedication and unselfish energy of those volunteers who served on the Authoring Team. A sincere thanks is extended to:

Gordon Hobbs
Glenn "Turkey" Pratt
Jonathan Sacks
Doug Smith
Richard Tunstall

Contents

12401-03 Digital Logic Circuits 1.i

12402-03 Instrument Calibration and Configuration 2.i

12403-03 Performing Loop Checks 3.i

12404-03 Troubleshooting and Commissioning a Loop 4.i

12405-03 Tuning Loops 5.i

12406-03 Programmable Logic Controllers 6.i

12407-03 Distributed Control Systems 7.i

12408-03 Analyzers 8.i

Instrumentation Level Four . . . . .Index

Digital Logic Circuits

COURSE MAP

This course map shows all of the modules in the fourth level of the Instrumentation curriculum. The suggested training order begins at the bottom and proceeds up. Skill levels increase as you advance on the course map. The local Training Program Sponsor may adjust the training order.

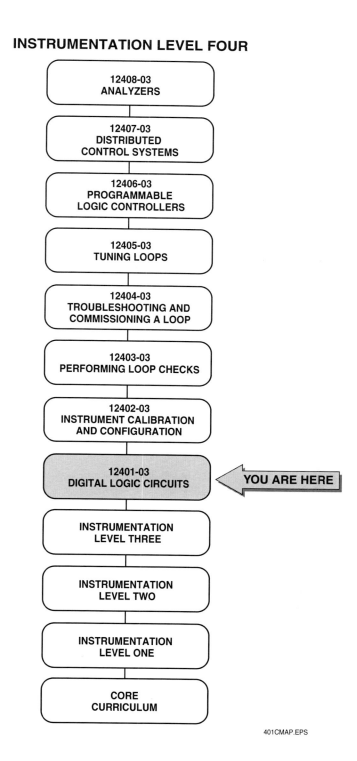

INSTRUMENTATION LEVEL FOUR

12408-03
ANALYZERS

12407-03
DISTRIBUTED
CONTROL SYSTEMS

12406-03
PROGRAMMABLE
LOGIC CONTROLLERS

12405-03
TUNING LOOPS

12404-03
TROUBLESHOOTING AND
COMMISSIONING A LOOP

12403-03
PERFORMING LOOP CHECKS

12402-03
INSTRUMENT CALIBRATION
AND CONFIGURATION

12401-03
DIGITAL LOGIC CIRCUITS ◁ YOU ARE HERE

INSTRUMENTATION
LEVEL THREE

INSTRUMENTATION
LEVEL TWO

INSTRUMENTATION
LEVEL ONE

CORE
CURRICULUM

401CMAP.EPS

1.0.0 INTRODUCTION 1.1

  1.1.0 AND Gate 1.1

  1.1.1 Multiple Input AND Gate 1.2

  1.1.2 AND Timing Diagram 1.4

  1.2.0 OR Gate 1.4

  1.2.1 OR Gate Pulsed Input Timing Diagram 1.4

  1.2.2 Applying AND/OR Gates to Furnace Burner Logic 1.5

  1.3.0 Amplifier 1.5

  1.4.0 Inverter 1.5

  1.5.0 NAND Gate 1.5

  1.6.0 NOR Gate 1.8

  1.7.0 Exclusive OR Gate 1.8

  1.8.0 Combination Logic 1.10

2.0.0 BASIC FLIP-FLOPS AND LATCHES 1.10

  2.1.0 RS NOR Latch 1.10

  2.2.0 RS NAND Latch 1.12

  2.3.0 Clocked RS Latch 1.13

  2.4.0 Data (D) Latch 1.14

  2.5.0 D Flip-Flop 1.15

  2.6.0 JK Master-Slave Flip-Flop 1.16

  2.7.0 Toggle (T) Flip-Flop 1.18

3.0.0 SHIFT REGISTERS 1.18

  3.1.0 Basic 1.18

  3.2.0 Serial In–Serial Out 1.22

  3.3.0 Serial In–Parallel Out 1.23

  3.4.0 Parallel In–Serial Out 1.25

  3.5.0 Parallel In–Parallel Out 1.28

  3.6.0 Universal 1.28

  3.6.1 Shift Left 1.28

  3.6.2 Shift Right 1.30

  3.6.3 Receive Parallel Inputs 1.30

  3.6.4 Disable Shift Register 1.30

4.0.0 COUNTERS 1.30

  4.1.0 Four-Bit Binary 1.31

  4.2.0 Up 1.32

  4.3.0 Down 1.33

  4.4.0 Up/Down 1.34

  4.5.0 Synchronous 1.36

  4.6.0 Ripple Carry 1.37

  4.7.0 Binary Coded Decimal 1.37

      4.8.0   Ring and Johnson 1.38

      4.9.0   Programmable 1.40

5.0.0 ARITHMETIC ELEMENTS 1.42

      5.1.0   Half Adder 1.42

      5.2.0   Full Adder 1.42

6.0.0 DECODERS 1.42

7.0.0 ANSI/ASQC STANDARDS 1.43

SUMMARY 1.44

REVIEW QUESTIONS 1.45

GLOSSARY 1.47

REFERENCES 1.49

Figures

Figure 1   AND gate 1.2

Figure 2   Two-level, three-input AND gate 1.3

Figure 3   Three-level, four-input AND gate 1.3

Figure 4   Timing diagram for an AND gate 1.4

Figure 5   OR gate 1.4

Figure 6   Timing diagram for an OR gate 1.5

Figure 7   Applying OR gates to furnace burner logic 1.6

Figure 8   Amplifier symbol and truth table 1.6

Figure 9   Inverter symbol and truth table 1.6

Figure 10   NAND gate 1.7

Figure 11   Two forms of NAND gate symbols 1.7

Figure 12   Timing diagram for a NAND gate 1.8

Figure 13   NOR gate 1.8

Figure 14   Two forms of a NOR gate 1.9

Figure 15   Timing diagram for a NOR gate 1.9

Figure 16   Exclusive OR symbol and truth table 1.9

Figure 17   Exclusive OR logic circuit 1.9

Figure 18   Timing diagram for an exclusive OR gate 1.9

Figure 19   Combination logic circuit 1.10

Figure 20   Basic memory element 1.10

Figure 21   RS latch 1.11

Figure 22   RS latch symbol and timing diagram 1.11

Figure 23   RS latch NAND gates 1.12

Figure 24   Clocked RS latch 1.13

Figure 25   D latch circuit and timing diagram 1.14

Figure 26   D flip-flop 1.15

Figure 27   JK flip-flop 1.16

Figure 28   Master-slave JK flip-flop 1.17

Figure 29   JK master-slave flip-flop timing diagram 1.17

Figure 30   T flip-flop and associated waveforms 1.18

Figure 31   Basic data movements in registers 1.19

Figure 32   Basic shift register circuit 1.19

Figure 33   Methods of shifting data with a shift register 1.20

Figure 34   Weighted coding of flip-flops in a shift register 1.21

Figure 35   Serial in–serial out register .1.22

Figure 36   Data serially entered into a register 1.23

Figure 37   Data serially shifted out of a register 1.24

Figure 38   Shift left shift register 1.25

Figure 39   Timing diagram and shift table
             for shift left register 1.25

Figure 40   Serial in–parallel out register 1.26

Figure 41   Parallel in–serial out shift register 1.27

Figure 42   Parallel in–serial out register and waveforms 1.28

Figure 43   Parallel in–parallel out shift register 1.29

Figure 44   Universal shift register 1.29

Figure 45   Symbol for a four-bit binary counter 1.31

Figure 46   Four-bit binary counter circuit 1.31

Figure 47   Up counter circuit and timing diagram 1.33

Figure 48   Down counter circuit and timing diagram 1.34

Figure 49   Up/down counter 1.35

Figure 50   Up/down counter timing diagram 1.35

Figure 51   Synchronous counter 1.36

Figure 52   Ripple carry counter 1.37

Figure 53   BCD counter and truth table 1.38

Figure 54   Ring counter with preload 1.39

Figure 55   Self-correcting ring counter .1.39

Figure 56   Johnson counter 1.40

Figure 57   Self-correcting Johnson counter 1.40

Figure 58   Programmable counter with preset inputs 1.41

Figure 59   Programmable counter whose final count
             can be preselected 1.41

Figure 60   Half adder 1.42

Figure 61   Full adder 1.43

Figure 62   BCD-to-decimal decoder 1.43

Tables

Table 1   Truth Table for the AND Gate .1.2
Table 2   Three-Input AND Gate 1.3
Table 3   Truth Table for the OR Gate .1.5
Table 4   RS Latch Truth Table for NOR Gates 1.12
Table 5   D Latch Truth Table 1.14
Table 6   JK Flip-Flop Truth Table 1.16
Table 7   Binary Number Code 1.32
Table 8   Binary Counter Truth Table 1.32
Table 9   Down Counter Truth Table .1.34
Table 10   Counter Sequence 1.36

Digital Logic Circuits

Objectives

When you have completed this module, you will be able to do the following:

1. Identify the different gates and circuits in digital logic.
2. Describe the truth tables and timing diagrams for various digital gates.
3. Describe the operation of different digital flip-flops.
4. Describe the operation of shift registers.
5. Describe the operation of counters.
6. State the purpose of the American National Standards Institute (ANSI) Q90–Q93 standards.

Prerequisites

Before you begin this module, it is recommended that you successfully complete the following: Core Curriculum; Instrumentation Levels One through Three.

Required Trainee Materials

1. Pencil and paper
2. Appropriate personal protective equipment

1.0.0 ◆ INTRODUCTION

Digital logic uses a set or combination of electronic circuit elements that perform a function by manipulating a discrete electrical signal. This module discusses the circuitry associated with digital signals. In digital logic there are eight basic elements, as follows:

- AND gate
- OR gate
- Amplifier
- Inverter
- NAND gate
- NOR gate
- Exclusive OR gate
- Combination logic

What they do is very simple, but it is essential that you understand them. Interconnecting a number of these gates into circuits allows them to perform various increasingly complex functions, such as addition, multiplication, or division of any two numbers, counting, keeping the time of day, and even running a whole computer.

1.1.0 AND Gate

The AND gate is a device whose output is a logic 1 if both of its inputs are a logic 1. If only one input is a 1 with the other a logic 0, the output will be a 0. The numbers 1 and 0 are used to represent the two different levels of logic. Forms you may encounter that are used to describe logic levels of 1 and 0 are as follows:

- On and Off
- H (high) and L (low)
- T (true) and F (false)
- Yes and No

The AND gate is represented by the symbol shown in *Figure 1*, where the two inputs are on the left of the symbol, marked A and B, and the output is on the right of the symbol, marked C. These inputs and outputs may be marked E, F, and G; X, Y, and Z; or any other combination of letters.

To visualize the AND gate, use the light bulb circuit in *Figure 1* as an analogy. In this circuit, both switches A and B must be closed for the light

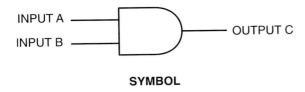

SYMBOL

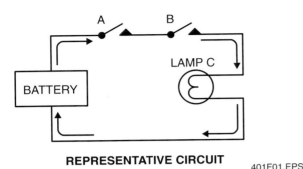

REPRESENTATIVE CIRCUIT

401F01.EPS

Figure 1 ◆ AND gate.

bulb to be on. If only one of the switches is closed, the light bulb will be off. The two switches are therefore analogous to the AND gate inputs A and B, while the light bulb corresponds to the output.

The various combinations of input states of an AND gate and its response to these inputs can be expressed in a table. *Table 1* shows the input and output combinations and is called a **truth table**, which is an important tool in digital logic. You will frequently use one to represent the operation of many kinds of digital circuits. Referring to *Table 1* to determine the input-output relationships for the AND gate.

The two columns on the left show the states of the inputs to the AND gate and the column on the right shows the corresponding output. If you relate this table to the lamp circuit, a 0 represents the lamp off or open switch condition, and a 1 represents the lamp on or closed switch condition. If you read horizontally along the lines of this table, you will see the response of the output (or lamp) to all combinations of the inputs. The number of possible states for the AND gate truth table is determined by the number of inputs. The AND

gate has two inputs so there is a possible combination of four states (2^2). If the AND gate contained three inputs there would be eight possible states (2^3).

Another method of representing the operation of an AND gate is called the Boolean logic equation or simply the logic equation. The AND statement represents the combination of variables by logic multiplication. The symbol used to represent the AND function is [•]. The AND function in *Figure 1* can be written as follows:

A and B = C

A × B = C (not frequently used)

A • B = C

(A)(B) = C

AB = C

Referring again to the lamp circuit, the full logic equation that relates the output C to the inputs A and B is C = A • B. It reads C equals A and B. This equation means that both A and B must be logic 1 if C is to be logic 1.

Propagation delay is a very important characteristic of logic circuits because it limits the speed (frequency) at which they can operate. The terms *low speed* and *high speed*, when applied to logic circuits, refer to the propagation delays: the shorter the propagation delay, the higher the speed of the circuit.

A propagation delay of a gate is basically the time interval between the application of an input pulse and the occurrence of the resulting output pulse. There are two propagation delays associated with a logic gate: the delay time from the **positive-going** edge of the input pulse to the **negative-going** edge of the output pulse is called the turn-on delay (tpHL); and the delay time from the negative-going edge of the input pulse to the positive-going edge of the output pulse is called the turn-off delay (tpLH). In many circuits, the delays are not equal, and the larger of the two is the worst case propagation delay. If several gates are arranged in series, their propagation delays are additive.

1.1.1 Multiple Input AND Gate

Figure 2A shows a multiple input AND gate. This logic circuit consists of a two-level, three-input AND gate. **Transistor-transistor logic (TTL)** propagation delay is 10 **nanoseconds (ns)**. Propagation delay can then be calculated as follows:

Propagation delay = (typically) 10ns × number of levels

For the multiple input AND gate shown:

Propagation delay = (10ns)(2) = 20ns

Table 1 Truth Table for the AND Gate

INPUTS		OUTPUT
A	B	C
0	0	0
0	1	0
1	0	0
1	1	1

401T01.EPS

INSTRUMENTATION LEVEL FOUR — TRAINEE MODULE 12401-03

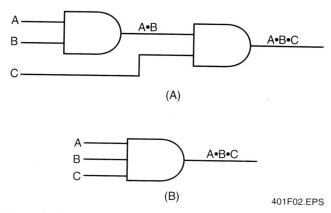

(A)

(B)

401F02.EPS

Figure 2 ◆ Two-level, three-input AND gate.

Table 2 Three-Input AND Gate

INPUTS			
A	**B**	**C**	**OUTPUT**
0	0	0	0
0	0	1	0
0	1	0	0
0	1	1	0
1	0	0	0
1	0	1	0
1	1	0	0
1	1	1	1

401T02.EPS

To turn on the first gate you need a high input (1) at both A and B, but to turn on the second gate you need high inputs (1) at A, B, and C—a total of three high inputs. These two gates can be considered as one logic circuit and drawn, as shown in *Figure 2B*, as a single AND gate with three separate inputs. The truth table is constructed exactly as presented earlier, except now there are three inputs to deal with instead of just two and eight possible combinations of inputs instead of just four. Possible combinations are $2^3 = 2 \times 2 \times 2 = 8$. The truth table for this multiple input AND gate is shown in *Table 2*.

Next, consider a three-level, four-input AND gate, as shown in *Figure 3A*.

Propagation delay = (10ns) × number of levels

Propagation delay = (10ns)(3) = 30ns

Again, you can analyze this logic circuit to determine what conditions are necessary to obtain a high (1) output from this three-level, four-input AND gate. To turn on the first gate, you need a high input (1) at both A and B. This condition will provide a high input to the second gate. This high input along with one high input from C will provide a high output from the second gate, as shown in *Figure 3A*. The output from the second gate provides an input to the third gate along with input D. Both of these inputs must be high to obtain a high output. So you can see that the inputs must be high throughout the entire circuit if the output is going to be high. Therefore, the three-level AND gate circuit can be reduced to a single AND gate with four inputs, as shown in *Figure 3B*.

There are 16 possible combinations for this four-input AND gate. Possible combinations are $2^4 = 2 \times 2 \times 2 \times 2 = 16$. You can write up a truth table for this gate to confirm that, as was explained, the output will be high only when all the inputs (A, B, C, and D) are high. Any combination other than this will produce a low output.

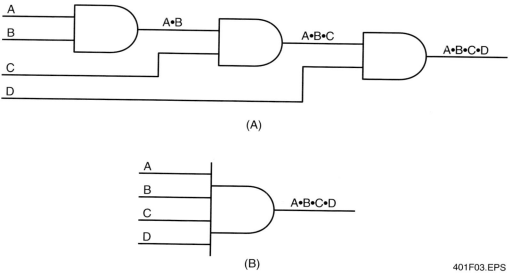

(A)

(B)

401F03.EPS

Figure 3 ◆ Three-level, four-input AND gate.

1.1.2 AND Timing Diagram

When a logic gate is performing a useful function in a circuit, its inputs can change and its output will react to these changes according to the truth tables for that gate. It is often useful to create a chart of these logic states as they change over time. This is known as a **timing diagram**.

In *Figure 4*, the operation of an AND gate is shown in a timing diagram for an arbitrary sequence of high and low logic signals applied to the A and B inputs.

Input A, input B, and output C of the AND gate are each represented by a continuous line that proceeds from left to right as time progresses. Each of these lines has two possible levels that correspond to whether the particular input or output is logic 1 (high) or logic 0 (low) at a particular time. Note that output C corresponds to the level predicted by the truth table and logic equation of the two-input AND gate presented earlier. A and B are both at a high level only during the time indicated on the bottom line representing output C.

The main purpose of a timing diagram is to show what the conditions in a logic circuit are at any one particular time. By using timing lines, it is possible to oversee all inputs and outputs simultaneously. If any input or output line is displayed on an oscilloscope screen, it would appear very much like the timing diagram shown in *Figure 4*.

1.2.0 OR Gate

The OR gate is a device whose output is a logic 1 if either or both of its inputs are a logic 1. The OR gate is shown by the symbol in *Figure 5* with the two inputs A and B again on the left and the output C on the right.

To visualize the OR gate, use the light bulb circuit with the switches connected in parallel rather than in series. The bulb can now be turned on by closing either switch A, switch B, or both.

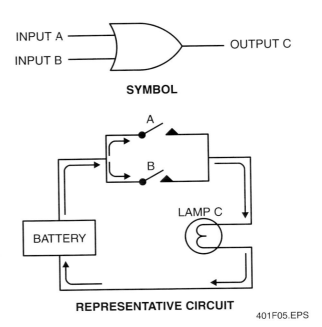

SYMBOL

REPRESENTATIVE CIRCUIT

401F05.EPS

Figure 5 ◆ OR gate.

The truth table for the OR gate is shown in *Table 3*, and the logic equation is $C = A + B$. This equation reads $C = A$ or B. Note the differences between the truth tables and the logic equations of the OR and AND gates. The symbol for the OR function is (+), but this should not be confused with the plus sign of mathematics; the Boolean symbols have functions that are unique to Boolean algebra.

1.2.1 OR Gate Pulsed Input Timing Diagram

Study the operation of an OR gate with pulsed inputs. Keep in mind what you have learned about its logical operation, and look at it in the form of a timing diagram. Again, the important thing in analysis of gate operation with pulsed waveforms is the relationship of all the waveforms to each other, as shown in *Figure 6*.

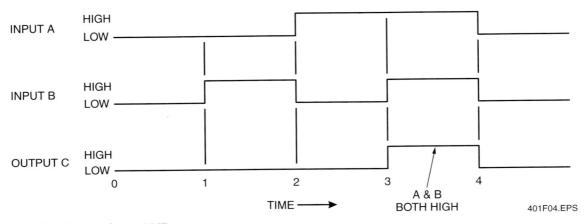

Figure 4 ◆ Timing diagram for an AND gate.

INSTRUMENTATION LEVEL FOUR — TRAINEE MODULE 12401-03

Table 3 Truth Table for the OR Gate

| INPUTS | | OUTPUT |
A	B	C
0	0	0
0	1	1
1	0	1
1	1	1

401T03.EPS

The inputs A and B are both high during interval T1, making the output high. During interval T2, A is low, but, because input B is high, the output is high. Both inputs are low during interval T3, giving a low output during this time. During T4, the output is high because input A is high. In this illustration, you are simply applying the truth table operation of the OR gate to each of the intervals.

1.2.2 Applying AND/OR Gates to Furnace Burner Logic

The majority of explosions in the burner systems of furnaces can be attributed to uncontrolled ignition of combustible mixtures that have accumulated in the furnace or the related exhaust ducting systems. The National Fire Protection Association (NFPA) developed Standard 86, which lists required furnace safeguards as well as the protection these safeguards provide. A sample of safeguards required includes low-pressure fuel detectors, high-pressure fuel detectors, flame detectors, and airflow detectors.

In most furnace designs, these safeguards function as inputs and outputs to a digital controller whose functions include controlling and monitoring the fuel and combustion conditions of the furnace. The controller normally sets the main fuel valve at a minimum setting for ignition and verifies that the flame has remained lit once ignition has been verified. If, for any reason, any one of these conditions cannot be verified, the controller is designed to close a fuel safety valve and shut down the burner's operation. *Figure 7* shows how

OR gates may be used in series logic to control the safety valve; however, normally AND gates are also incorporated because the actual digital circuitry within one of these burner systems contains many more interlocks and safety features to control the burner operation. This is just an example of how these digital devices are incorporated.

1.3.0 Amplifier

The amplifier (*Figure 8*) is a device whose output assumes the high state if and only if the input assumes the high state. They are used to restore logic signal levels or to convert logic signals into usable power to drive solenoids, contactors, lamps, and other electrical devices. The Boolean equation for the amplifier is A = B.

1.4.0 Inverter

The simplest element of digital logic is the inverter (*Figure 9*), which is different from the AND and OR gates in that it has only a single input. As a result, it does not perform a decision-making function dependent on a combination of inputs. Instead, the inverter simply converts a logic 1 at its input to a logic 0 at its output and, conversely, a logic 0 to a logic 1. The inverter can be represented by either of the symbols shown, and its logic equation, $C = \overline{A}$, reads C equals A not (or not A).

The inverter, also known as the NOT function, produces an output that is always opposite of the input. Inverters may also be used to perform the same function as described for the amplifier.

1.5.0 NAND Gate

Another type of logic gate often found in digital circuits is the NAND gate (NOT AND) gate. As the name implies, a NAND gate is an AND gate with an inverter on its output. The truth table and symbol for a two-input NAND gate are shown in *Figure 10*.

The truth table shows that the output of a NAND gate assumes the 0 state if and only if all inputs assume the 1 state. In comparing the NAND gate

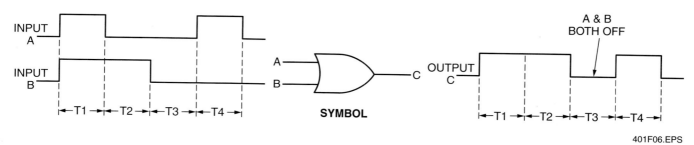

Figure 6 ◆ Timing diagram for an OR gate.

401F06.EPS

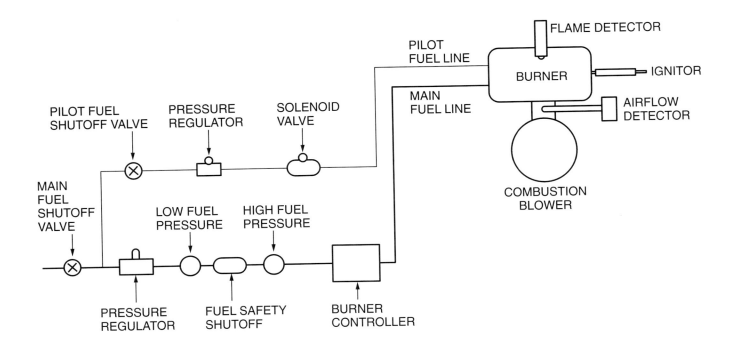

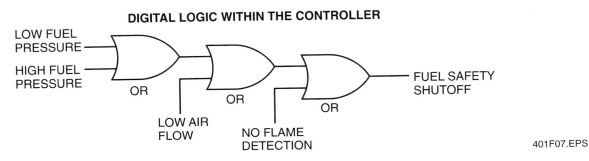

Figure 7 ◆ Applying OR gates to furnace burner logic.

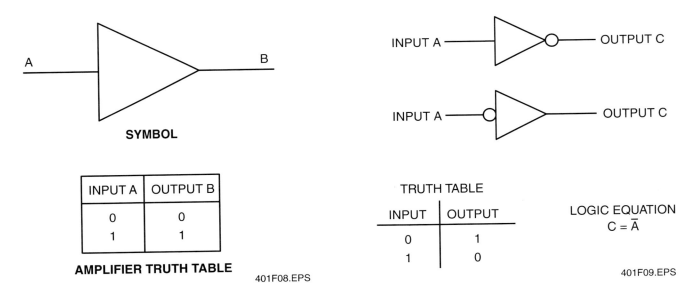

Figure 8 ◆ Amplifier symbol and truth table.

Figure 9 ◆ Inverter symbol and truth table.

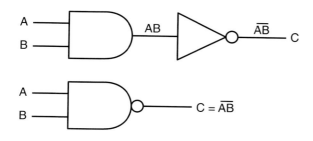

INPUT		OUTPUT
A	B	C
0	0	1
0	1	1
1	0	1
1	1	0

401F10.EPS

Figure 10 ◆ NAND gate.

truth table to that of the AND gate, you will see that in each case the outputs of the NAND are opposite those of the AND for the same input conditions. This is represented in the logic equation by placing a NOT symbol (a bar) over A • B, hence $\overline{A \cdot B}$. As shown, the inversion in the NAND gate is represented by a small circle at the output of the device.

If you look at the truth table for the NAND gate shown in *Figure 11*, it is apparent that by inverting (changing 1s to 0s and 0s to 1s) all the inputs to the NAND gate, you have exactly the truth table of the OR gate.

In other words, if the inputs (A, B) applied to the NAND are already individually inverted ($\overline{A}$, $\overline{B}$), it will perform the OR function. If these inputs are not inverted with the output inverted, it will perform as an AND. This dual function is represented by having two symbols for the NAND.

Now look at the pulsed operation of the NAND gate. The input and output relationships are shown in the timing diagram in *Figure 12*. This shows the pulses for a NAND with two inputs (A, B) and one output (C).

Remember from the truth table for a NAND gate that any time all the inputs are high, the output will be low, and this is the only time a low output occurs. This can be seen in the timing diagram because C is low only when both A and B are high.

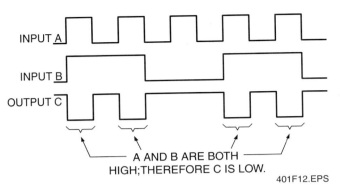

Figure 12 ◆ Timing diagram for a NAND gate.

1.6.0 NOR Gate

In addition to the NAND gate, the other logic gate often found in digital circuits is the NOR gate. *Figure 13* shows the NOR gate composition along with the symbol and truth table.

As you can see, the NOR gate is an OR gate with an inverter on its output. When comparing the NOR gate truth table to that of the OR gate, you will see that, in each case, the outputs of the NOR gate are opposite those of the OR for the same input conditions. This is represented in the logic equation by placing a NOT symbol over A + B, hence $\overline{A+B}$. As shown, the inversion in the NOR gate is represented by a small circle at the output of the device. Also from the truth table, you can see that the output of a NOR gate assumes the 0 state when any input assumes the 1 state.

By inspecting the NOR truth table in *Figure 14*, you can see that inverting all the inputs in the NOR truth table yields the AND truth table. NOR can, therefore, not only be used as an OR with an inverter on the output but also as an AND if the

inputs (A, B) are already in the individual inverted form ($\overline{A}$, $\overline{B}$). The two symbols for the NOR are shown in *Figure 14*. Note again that the small circle indicates the inversion function.

Now look at the pulsed operation of the NOR gate. The input and output relationships are shown in the timing diagram in *Figure 15*. This shows the pulses for a two-input (A, B) NOR gate with the output pulse (C) and verifies the truth table for the NOR gate. Whenever an input, A or B, is high, the output C, is low.

1.7.0 Exclusive OR Gate

There is one more gate that needs to be considered: the exclusive OR gate (XOR), shown in *Figure 16*. The exclusive OR gate has a logic high output when either of its inputs is high but not when both are high.

The exclusive OR functions exactly like the OR gate with one exception: when the OR receives two high inputs, the output is also high; however, when the exclusive OR receives two high inputs, the output is low. The exclusive OR gate is quite useful because its output is high only when the inputs are different. The logic equation for this gate introduces a new symbol, $\wedge$, called exclusive OR. Therefore, the logic equation for the gate in *Figure 17* is C = A $\wedge$ B.

The exclusive OR has been presented thus far as a single logic circuit with two inputs and one output. The exclusive OR gate is actually composed of many individual logic gates. *Figure 17* shows the exclusive OR logic circuit.

The output expression for this circuit is C = A$\overline{B}$ + $\overline{A}$B. This logic circuit can be used to verify the truth table previously given for the exclusive OR gate.

Now look at the pulsed operation of the exclusive OR. The input and output relationships are shown in the timing diagram of *Figure 18*. This shows the pulses for the two inputs (A and B) and the output pulse (C) and verifies the truth table for the exclusive OR gate. There is an output only when one of the inputs is high.

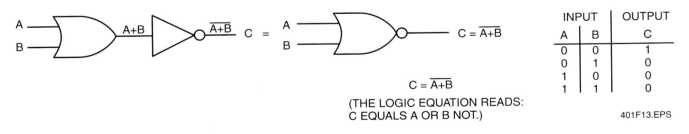

Figure 13 ◆ NOR gate.

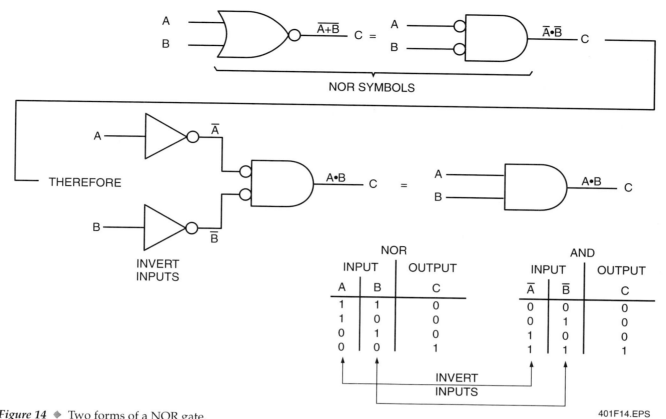

Figure 14 ◆ Two forms of a NOR gate.

401F14.EPS

NOR			AND		
INPUT		OUTPUT	INPUT		OUTPUT
A	B	C	$\overline{A}$	$\overline{B}$	C
1	1	0	0	0	0
1	0	0	0	1	0
0	1	0	1	0	0
0	0	1	1	1	1

INVERT INPUTS

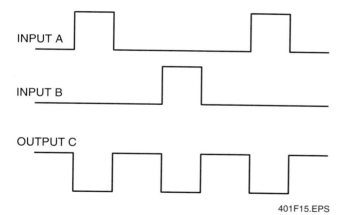

Figure 15 ◆ Timing diagram for a NOR gate.

401F15.EPS

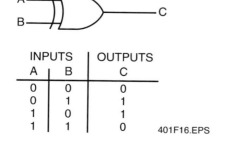

Figure 16 ◆ Exclusive OR symbol and truth table.

INPUTS		OUTPUTS
A	B	C
0	0	0
0	1	1
1	0	1
1	1	0

401F16.EPS

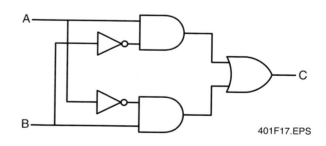

Figure 17 ◆ Exclusive OR logic circuit.

401F17.EPS

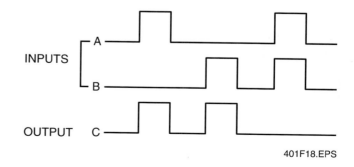

Figure 18 ◆ Timing diagram for an exclusive OR gate.

401F18.EPS

1.8.0 Combination Logic

Figure 19 shows a combination logic circuit consisting of two AND gates and one OR gate. Each of the three gates has two input variables as indicated, which can be either a high (1) or a low (0). Because there are four input variables, there are sixteen possible combinations of the input variables ($2^4 = 16$). To illustrate an analysis procedure, assign one of the sixteen possible input combinations and see what the corresponding output value is.

First, make each input variable a low and examine the output of each gate in the network in order to arrive at the final output, Y. If the inputs to gate G1 are both low, the output of gate G1 is low. Also, the output of gate G2 is low because its inputs are low. As a result of the lows on the outputs of gates G1 and G2, both inputs to gate G3 are low, and therefore its output is low. You have determined that the output function of the logic circuit of *Figure 19* is low when all of its inputs are low. You can verify each of the 16 possible conditions on the logic diagram.

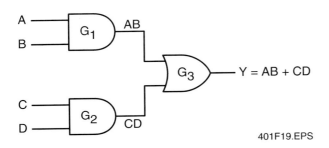

Figure 19 ◆ Combination logic circuit.

2.0.0 ◆ BASIC FLIP-FLOPS AND LATCHES

Basic logic devices discussed thus far require continuous input signals (0 or 1) to perform a function. The second basic group of digital logic elements contains the memory elements. Even though memory elements are constructed from AND and OR gates, the performance characteristics are quite different. In actual practice, the role of memory elements in digital logic circuits is so fundamentally important that they are regarded as an independent group. The basic characteristics and some applications of these elements (called **flip-flops** and latches) will be studied in the following sections.

The same AND and OR gates that have decision-making capability can also be interconnected to provide memory. That is, they can remember if a signal of logic 1 or 0 level has been connected to their inputs and make this fact available at the outputs. The output of a memory circuit is therefore determined by the past inputs as well as the present inputs, whereas AND and OR gates simply make decisions based on their present inputs. It is very interesting that decision-making and memory capabilities are so closely related, because by simply considering the basic characteristics of either gate, it is hard to imagine how it could remember a history of inputs. For purposes of developing this concept, only a very limited memory capability will be considered at first. An elementary memory element constructed from a single OR gate is shown in *Figure 20*.

In analyzing *Figure 20*, assume that the output Q and input A are initially at logic 0, and therefore the B input (which feeds back from output Q) is also at logic 0. If the signal at A is changed to a logic 1, the output Q will go to logic 1, as well as input B. However, if input A now returns to logic 0, the output will not change because input B will still be at logic 1 and will keep the output at logic 1. Thus, the OR gate remembers that it received a logic 1 level, and the only way to erase that fact from this memory is to physically disconnect the wire between Q and B and place B at logic 0.

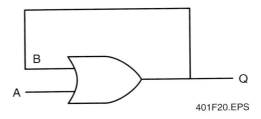

Figure 20 ◆ Basic memory element.

2.1.0 RS NOR Latch

If the memory in *Figure 20* only has to remember one event (one logic 1) in its lifetime and does not need to be erased and reused for another event, it is adequate. However, in real situations, that is not the case, and instead, a memory is needed that can be erased or reset when it no longer needs to remember an event. A memory is normally used to store information (data) for some period of time and then discard it. To implement this type of memory requires two NOR gates. This circuit is shown in *Figure 21*.

This circuit is very important and will be analyzed in considerable detail before proceeding to develop other circuits from it. It is called a reset-set (RS) flip-flop or RS latch and is the most basic form of the class of circuits called flip-flops. For the time being, because they are similar, the words flip-flop and latch will be used interchangeably.

The reason for difficulty in analyzing the RS latch lies in the fact that the outputs of the circuit are connected back to the inputs. Thus, any signal that is input to the circuit goes through the circuit and always returns to the input lines. As a result, the input signal has multiple effects. This method of connecting an output of a circuit back to its own input is called feedback. A feedback connection is essential in giving a logic circuit its memory capability.

To analyze the RS latch, begin by raising the set input S to a logic 1 level while holding the reset input R to a logic 0, and see what happens. Remember, a NOR gate output is a logic 0 whenever one or both of its inputs are a logic 1. First of all, $\overline{Q}$ output goes to the logic 0 state. But logic 0 at $\overline{Q}$ is connected to the lower gate and now the lower gate has two logic 0 inputs: its output Q goes to logic 1. The logic 1 at Q is connected back to the input of the upper gate and the upper gate now has two logic 1 inputs. Being a NOR gate, it needs only one high input to maintain its output in the logic 0 state. Therefore, the logic 1 at S can now become a logic 0, and the $\overline{Q}$ output will not change. The original input has been traced through the circuit in a figure-eight pattern. If the logic 1 at S goes to a logic 0, the figure-eight pattern can be traced again and all conditions still remain the same. Note that the S input was used to activate (or set) this circuit. As a result, the gates have latched themselves in certain states, and as long as the R input remains a logic 0, nothing will change. Thus, this device is called a latch.

To reset the circuit, raise R to a logic 1. Tracing a figure 8 beginning at R will yield the same results as before, except now Q becomes a logic 0 and $\overline{Q}$ becomes a logic 1 (you should trace and verify this fact). The circuit has been reset and its memory erased.

The RS latch has the property that whenever the set input is raised to a logic 1, the Q output will become latched in a Q = 1 state and, whenever the reset input is raised to a logic 1, Q will become a logic 0. If you alternately raise one input and then the other, Q and $\overline{Q}$ will each alternate between logic 1 and logic 0 states.

There is a final condition that should be considered: both inputs R and S being high simultaneously. Because any one high input to the NOR gates will cause the output to be logic 0, both outputs will be logic 0 under these conditions. This is a special state of the RS latch, which, as you will learn later, should be avoided. However, it does not change the basic function of the circuit, which is that of remembering a logic 1 or 0 at its inputs.

For simplicity, the RS NOR latch is shown as a rectangle with labeled inputs and outputs, as in *Figure 22*. This representation is standard for all flip-flops and shows the flip-flop as a simple memory element whose essential purpose is to store a logic 1 or 0.

This stored logic level is always available at the Q output of the flip-flop and its complement (opposite) is present on the $\overline{Q}$ output. That is, if Q is a logic 1, $\overline{Q}$ is a logic 0 and, conversely, if Q is a logic 0, $\overline{Q}$ is a logic 1. The stored logic level can be removed by storing another logic level in its place. If a logic 1 is stored in the flip-flop (Q = 1), the flip-flop is referred to as being set; if a logic 0 is stored (Q = 0), the flip-flop is referred to as being reset.

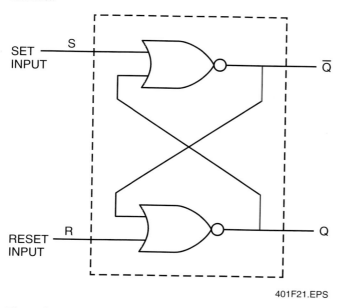

Figure 21 ◆ RS latch.

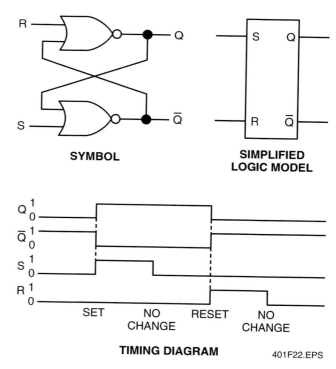

Figure 22 ◆ RS latch symbol and timing diagram.

There are other control inputs that can be added to the standard flip-flop. These will be discussed in the paragraphs that follow.

Using the knowledge of how a NOR gate operates, you can develop the truth table for the RS latch. *Table 4* is the truth table for the NOR gate RS latch.

This truth table can be interpreted in the following manner, based upon the logic diagram and NOR truth tables.

In the first condition:

S = 0

R = 1 makes Q go to 0

Q and S are now both equal to 0; $\overline{Q}$ = 1

$\overline{Q}$ and R are both 1; Q = 0

R = 1 is the activating or reset condition that forces the flip-flop into the condition Q = 0 and $\overline{Q}$ = 1

In the second condition:

If S = 1 and R = 0, the S = 1 will make $\overline{Q}$ go to 0; S = 1 is the set condition that forces the flip-flop into the condition Q = 1, $\overline{Q}$ = 0

In the third condition:

If R is returned to 0 (S still at 0), Q = 1 will keep $\overline{Q}$ at 0 and there is no change in the output level

In the fourth condition:

If both S and R = 1, both Q and $\overline{Q}$ will go to 0; this condition is not allowed

2.2.0 RS NAND Latch

The basic RS latch circuit can also be constructed from NAND gates, as shown in *Figure 23*. The analysis of this circuit is very similar to that of the NOR gate circuit, so it is not essential to repeat it in as much detail. However, because logic designers use both types of circuits, you should readily recognize them and know the differences.

The two types of RS latches are the same in their ability to store a logic level. The NAND gate implementation of the latch is set and reset using logic 0s, instead of logic 1s, as indicated by the inversion bars on the $\overline{R}$ and $\overline{S}$ inputs in *Figure 23*. To store a logic 1 at Q (logic 0 at $\overline{Q}$), the condition $\overline{S}$ = 0 and $\overline{R}$ = 1 must be at the inputs. This condition is remembered when the inputs return to $\overline{R}$ = 1 and $\overline{S}$ = 1. Conversely, a logic 0 is stored at Q (logic 1 at $\overline{Q}$) when the inputs are $\overline{S}$ = 1 and $\overline{R}$ = 0.

The input condition to avoid in the NAND gate RS latch is $\overline{R}$ = 0 and $\overline{S}$ = 0 simultaneously. Just as R = 1 and S = 1 in the NOR gate version of the latch, the condition $\overline{R}$ = 0 and $\overline{S}$ = 0 in the NAND gate version creates an undefined output state.

The most significant reason for avoiding the $\overline{R}$ = 0, $\overline{S}$ = 0 input state in the NAND gate latch is the possibility of what is called a **race condition**. The race condition can be created by the following sequence of input signals. Assume that both inputs of the RS latch shown in *Figure 23* are at logic 0, and therefore the outputs are both at logic 1. If both inputs are simultaneously raised to a logic 1 state, both NAND gates have two logic 1 inputs, and they must change their output states to logic 0.

But the logic 0 outputs remove one logic 1 input from each NAND gate and now both gates must switch back again to logic 1 outputs. These outputs establish two logic 1 inputs to each NAND gate and are back where they started: the outputs switch again to logic 0.

As long as the two gates change state at exactly the same time, this sequence continues indefinitely. Both gates continuously switch back and forth between logic 0 and logic 1 outputs as fast as they can.

Table 4 RS Latch Truth Table for NOR Gates

INPUTS		OUTPUTS	
R	S	Q	$\overline{Q}$
1	0	0	1 (RESET)
0	1	1	0 (SET)
0	0	NO CHANGE (PREVIOUS STATE IS MAINTAINED)	
1	1	NOT ALLOWED (Q AND $\overline{Q}$ = 0)	

401T04.EPS

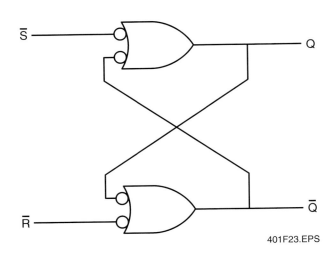

Figure 23 ◆ RS latch NAND gates.

401F23.EPS

Instead of continuing indefinitely, however, one of the gates will switch from one state to the other just a little faster than the other and cause the flip-flop to latch. The gates literally race each other to change states, and whichever changes its output first prevents the other gate from changing. In this situation, the final output cannot be predicted, and therefore this sequence of input states should be avoided.

The NOR gate latch can also be trapped in a race condition if two logic 1 inputs are followed by two logic 0 inputs. This possibility was not analyzed earlier, but to understand the RS latch fully, you should study the race condition in that circuit also.

2.3.0 Clocked RS Latch

Before other flip-flops are described, three basic control signals common to most flip-flops will be considered. These are the clock, preset, and clear signals.

In *Figure 24*, two gates are connected to the inputs of the latch, and a clock signal is connected so that it can enable or disable both gates simultaneously.

The enable and disable gates prevent the R and S inputs from causing a change in the state of the flip-flop while the clock is low. When the clock is raised to a logic 1, any logic 1 signal on the R or S inputs is gated in (allowed to enter). Then, the clock can go to logic 0 again to disable the input gates and keep other signals from entering the latch. The clock signal thus creates what may be called a window. Unless this window is open, the state of the flip-flop cannot be changed by the R and S inputs. Consider how such a window may be useful in a calculator circuit that uses several flip-flops for the storage of numbers. To perform addition, a number is first entered by depressing keys on the calculator keyboard. Then, a clock signal connected to all flip-flops is raised to a logic 1 level, the numbers are gated into the flip-flops, and the flip-flops are disabled again by placing the clock in the logic 0 state. After that, the clock signal is maintained at logic 0 so that the flip-flops are prevented from receiving additional numbers while calculations are being carried out by other circuits.

Thus, a clock signal can be used to clock or gate data into both inputs of the RS latch. This is analogous to opening and closing our eyes to control the entry of information through our visual senses into the brain and its memory.

Another fundamental purpose of a clock signal is to synchronize. In the above calculator example, storage of data required several flip-flops. The same clock signal was used to enable and disable the data inputs to all flip-flops, thus entering the data into the flip-flops synchronously. Using the analogy of our eyes, the same signal in the nervous system causes both of our eyes to blink synchronously. This concept of synchronization is very important and will be considered in detail later in this topic.

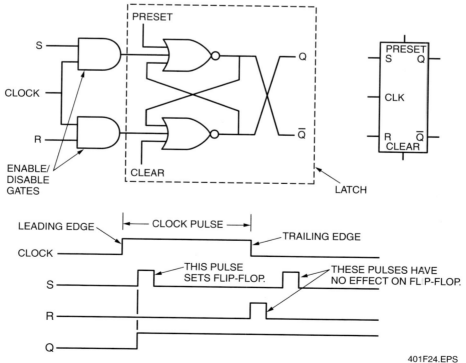

Figure 24 ◆ Clocked RS latch.

Preset and clear are inputs used to set or reset a flip-flop without involving the data and clock inputs. In other words, preset and clear can be used to set and reset the flip-flop when the clock signal is low or 0. Therefore, it is said that the preset and clear inputs are used to set and reset the flip-flop **asynchronously**.

To understand how the preset and clear inputs are used, consider that when power is first applied to a circuit, it is not known whether any given flip-flop is in the 1 or 0 state (set or reset). The preset or clear inputs can then be used to initialize each flip-flop to a known state and ensure that all following logic sequences will proceed correctly. This operation is very similar to clearing a calculator before starting a new calculation.

2.4.0 Data (D) Latch

The data (D) latch is shown in *Figure 25*. Because an inverter is used to produce the R input, the inputs to the NOR gates will always be complementary: when one is logic 1, the other must be logic 0. In this manner, the race condition is avoided. Thus, the D latch offers a simple solution to the race problem of the RS latch and is a very commonly used circuit.

Note that the D latch has only one data input. The latch is set (Q = 1) by clocking in a logic 1 and reset (Q = 0) by clocking in a logic 0.

To use the D latch correctly, the clock signal is raised, the desired high or low input is applied to the data line, and then the clock signal is removed, before allowing the data line input signal to change. As soon as the clock signal is removed, the circuit is latched and the data input line can change in any manner without affecting the outputs. This is illustrated in the timing diagram in *Figure 25*. Note that, alternatively, the D input can be applied prior to raising the clock signal. From the timing diagram shown you can develop a truth table (*Table 5*) for the D latch.

Table 5 D Latch Truth Table

INPUTS		OUTPUTS	
D	CLK	Q	Q̄
LOW	ABSENT	NO CHANGE	
LOW	PRESENT	LOW	HIGH
HIGH	ABSENT	NO CHANGE	
HIGH	PRESENT	HIGH	LOW

401T05.EPS

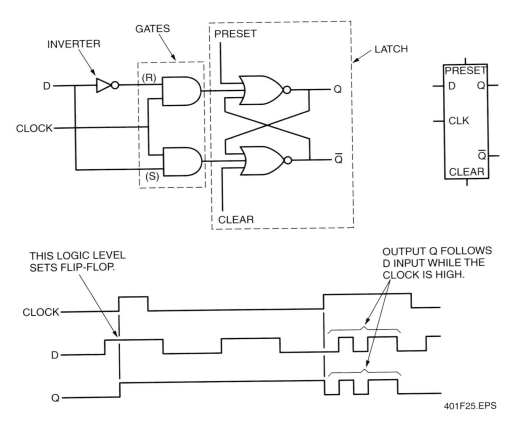

Figure 25 ◆ D latch circuit and timing diagram.

2.5.0 D Flip-Flop

The D flip-flop in *Figure 26* looks considerably more complicated than the D latch, but functionally the two are closely related. Both have a single data input and a logic 1 or 0 signal is used to set or reset them.

The difference between the D flip-flop and the D latch is associated with a new way of using the clock signal for gating in data. As you analyze the circuit, you will notice that any change in status of the output latch takes place only at the instant when the clock signal changes from a logic low to a logic high level and at no other time. In other words, the essential feature of this flip-flop is that it samples data present at the input only with the rising edge of the clock pulse, as shown in the timing diagram of *Figure 26*. This is known as edge triggering and is characteristic of most flip-flops, such as the D flip-flop and the JK flip-flop.

Many times in digital circuits it is desirable to sample the data inputs at a unique point in time. This type of sampling technique virtually elimi-

nates the possibility of an input data change during the time input data is transferred to the output. That inputs can be sampled at a specific point in time is one of the advantages of edge triggering. An analysis of the D flip-flop also reveals that the circuit consists of two interconnected input latches, gates U1/U2 and U3/U4; and an output latch, gates U5/U6. The input latches are interconnected so that when the clock signal goes from a low to a high level (the leading edge of the clock pulse), it causes the input latches to latch in complementary states; that is, one always supplies a logic 1 to the output latch and the other a logic 0. Which way they latch is determined by the state of the data line at the leading edge of the clock pulse. Once the clock is high, it holds both input latches in their existing states, and the data line cannot have any further effect. When the clock goes low again, both input latches supply a logic 1 to the output latch, and the data line can only affect the status of gates U1 and U4.

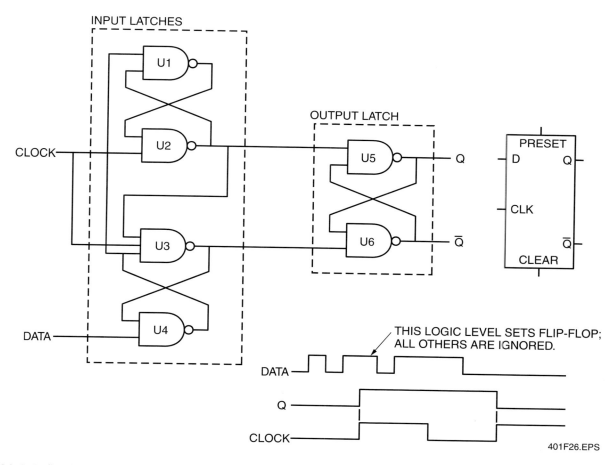

Figure 26 ◆ D flip-flop.

2.6.0 JK Master-Slave Flip-Flop

The JK flip-flop, shown in *Figure 27*, is the most commonly used flip-flop and the most versatile and sophisticated. Like the RS latch, it has two data inputs, but it has none of the shortcomings of the RS latch. It cannot have an undefined output, and its latches are not subject to the race condition. Like the D flip-flop, it is clocked and is always edge triggered; however, most versions of the JK flip-flop are controlled by the trailing edge instead of the leading edge of the clock.

The JK flip-flop works as follows. First, if one of its inputs is a logic 1 and the other a logic 0, it will be set or reset by the clock edge in exactly the same manner as the RS latch. Second, if both of its inputs are logic 0 when the clock edge occurs, it will simply remain in the same state as before the clock edge. Third, if both of its inputs are held at logic 1, the flip-flop will change states when the clock edge occurs. That is, if it was set before the clock pulse, it will be reset afterwards, and vice versa. This switching of a state to its complement is called toggling.

To summarize the behavior of a JK flip-flop in a truth table (*Table 6*), it is convenient to expand the conventions for truth tables. This will permit you to specify whether the output being referred to is before or after the clock edge.

The time before the clock edge will be called t_n and the time after will be called t_{n+1}. Likewise, the state of the Q output before the clock edge will be denoted as Q_n and the state after as Q_{n+1}. With these conventions in mind, look at the truth table for the JK flip-flop. Note that if the J and K inputs are both 0 before the clock edge, the Q_{n+1} output after the clock will be Q_n—the same as before. If J and K both are at logic 1, the Q_{n+1} output will be $\overline{Q_n}$—the opposite of Q_n.

Table 6 JK Flip-Flop Truth Table

INPUTS (AT t_n)		OUTPUTS (AT t_{n+1})
J	K	Q_{n+1}
0	0	Q_n
0	1	0
1	0	1
1	1	$\overline{Q_n}$

401T06.EPS

This method of designating the time before and after the clock edge applies equally well to any clocked flip-flop and allows you to make a simpler, more generalized truth table for the JK flip-flop.

There are two basic types of JK flip-flops. You should learn to recognize the circuits of the two basic types and know the differences in their operating characteristics. The first type can be recognized as an RS latch whose outputs are cross-connected back to the inputs and ANDed with the J and K inputs. One typical version of this circuit was shown in *Figure 27*, together with a simplified logic model of it. This circuit is triggered by the negative going (trailing) edge of the clock pulse. One significant detail in its operation is that the J and K input signals appear at the outputs of gates U1 and U2 after the clock signal is raised to a logic high. If the status of the J and K lines is changed at this time, the changes will be reflected at the outputs of U1 and U2. The cross-connection of the outputs back to the input gates determines the actual input conditions to R and S when J = 1 and K = 1. This causes the toggling action that occurs under these input conditions and prevents the undefined output or race condition.

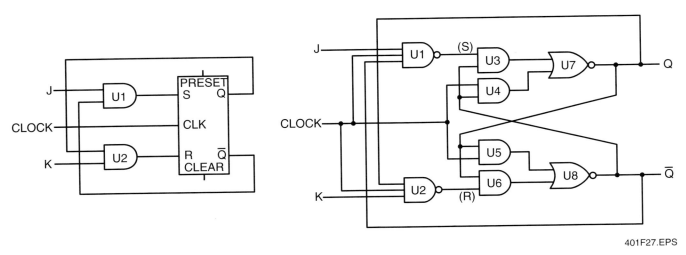

Figure 27 ◆ JK flip-flop.

The second type of JK flip-flop is shown in *Figure 28* and is known as a master-slave JK. The master-slave principle is not limited to JK flip-flops alone but is any arrangement in which the data is input into one RS latch and subsequently transferred to a second. In this type JK circuit, the leading edge of the clock pulse enables the input gates to the first latch (master) while at the same time isolating it from the second (slave). On the falling edge of the clock pulse, the data is transferred to the slave latch (to the output).

In order to aid in the understanding of the functions of the JK flip-flop, a timing chart is shown in *Figure 29*. This chart relates specifically to the TTL 7476 flip-flops. Across the top of the chart are clock pulses t_1 through t_{10}. These are followed by the other inputs and the outputs.

At time zero (t_0), the clear is active (a low level is present), and the preset is not active (a high level is present). Therefore, the flip-flop must be initially in the reset state because these two inputs have priority over all others. The result is that the set output (Q) is low level and the reset output ($\overline{Q}$) is high level.

At time t_1, both J and K are inactive (both are low level). Therefore, the flip-flop is instructed to remain in its present state (remember). Because the flip-flop is in the reset state, it will remain there.

Before t_2, K becomes active (high), and J is inactive (low). The flip-flop is now instructed to reset. However, because it is already in the reset state, it will remain there. Q and $\overline{Q}$ do not change.

At t_3, J is active (high), and K is inactive (low). This instructs the flip-flop to set. As a result, Q becomes high level, and $\overline{Q}$ becomes low level.

At t_4, the input variables J and K are reversed. J is inactive and K is active. The instruction is to reset. Q becomes low level, while $\overline{Q}$ becomes high level.

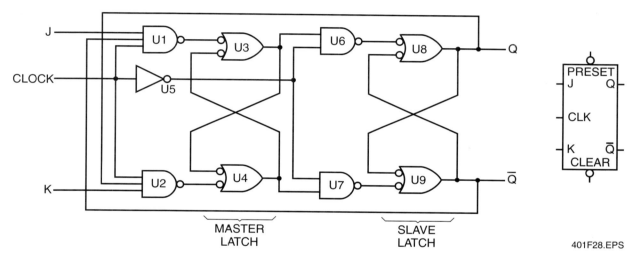

Figure 28 ◆ Master-slave JK flip-flop.

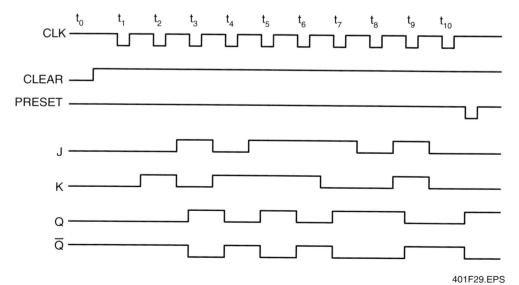

Figure 29 ◆ JK master-slave flip-flop timing diagram.

At t_5, both J and K are active. This is the instruction to change state (complement). Because the flip-flop is in the reset state, it will set.

At t_6, J and K are still both active. The effect is that the flip-flop again changes state (resets) with the clock pulse.

At t_7, J is active and K is inactive. This is another instruction to set. Q goes high, while $\overline{Q}$ goes low.

At t_8, the flip-flop is again instructed to remember. There is no change in the output.

Clock pulse t_9 causes the flip-flop to change state because both J and K are active. Q goes low, and $\overline{Q}$ goes high.

The state of the flip-flop remains unchanged at t_{10} because both J and K are inactive. However, following t_{10}, the preset input becomes active, which causes the flip-flop to set.

Remember that the J and K inputs are used to tell the flip-flop what to do, and the CLK input is used to tell it when to do it. Keep in mind, however, that the preset and clear inputs do not require a clock pulse.

2.7.0 Toggle (T) Flip-Flop

The toggle (T) flip-flop may be regarded as a simplified version of the JK. It is similar because it toggles with every clock pulse, exactly as the JK flip-flop does when both of its inputs are at logic 1 (J = K = 1). The T flip-flop is usually not manufactured by itself but, instead, is used in such circuits as counters and dividers. *Figure 30* shows a T flip-flop with its associated input and output waveforms. The T flip-flop, with no inverter on the input, changes states on the leading edge, and with an inverter, it changes states on the trailing edge.

The output of a T flip-flop is cross-connected back to the input so that the flip-flop is conditioned to its opposite state. For example, if the flip-flop is in the reset state, the S input will have a high level present, and the R input will be low level. This conditions the flip-flop to set with the next input pulse. The flip-flop will then set and

become conditioned to reset. Referring to *Figure 30*, two complete input cycles are required for each output cycle, resulting in a frequency divider of 2:1. Two such flip-flops connected in series will result in a frequency divider of 4:1.

3.0.0 ◆ SHIFT REGISTERS

Shift registers are very important in applications involving the storage and transfer of data in a digital system. In general, a register is used solely for storing and shifting data (1 and 0) entered into it from an external source, and it possesses no characteristic internal sequence of states.

The flip-flop is the basic element of shift registers, and the storage capability of a register is one of its two basic functional characteristics that makes it an important type of memory device.

3.1.0 Basic

The storage capacity of a register is the number of bits (1s and 0s) of digital data it can retain. Each stage of a shift register represents one bit of storage capacity; therefore, the number of stages in a register determines its total storage capacity.

Registers are commonly used to temporarily store data within a digital system. Registers are implemented with a flip-flop for each stage, or they use the inherent capacitance in **MOSFET** devices for the storage elements. The shift capability of a register permits the movement of data stored from stage to stage within the register or into or out of the register upon application of clock pulses. *Figure 31* symbolically shows the types of data movement in shift register operations. The blocks represent arbitrary registers, and the arrows indicate direction and type of data movement.

A shift register is a simple application of flip-flops. The shift register is one of the most basic logic circuits, and various forms of it are used as building blocks in many other logic circuits. A

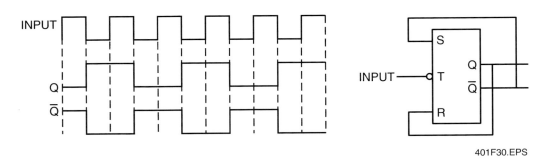

401F30.EPS

Figure 30 ◆ T flip-flop and associated waveforms.

shift register consists of a series of flip-flops interconnected so that the output of one flip-flop becomes the input of another, as shown in *Figure 32*. D flip-flops have been chosen for the illustration in *Figure 32*, but JK flip-flops could have been used equally well.

All the flip-flops in the shift register have a common clock signal connection and are all set or reset synchronously. In this discussion, the clock is a series of symmetric pulses, as shown in *Figure 32*, and the flip-flops are all reset initially. As the flip-flops are clocked, a logic 1 is supplied to the D input of the first flip-flop, FF1, just long enough to be sampled by the leading edge of one clock pulse. With the clock pulse, the 1 is stored in FF1 and appears at its output a very short time after the leading edge of the clock pulse. This small propagation delay, usually in the 10s of nanoseconds, is

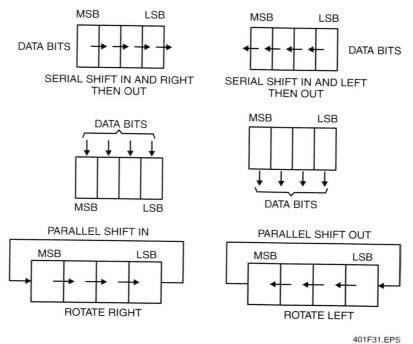

401F31.EPS

Figure 31 ◆ Basic data movements in registers.

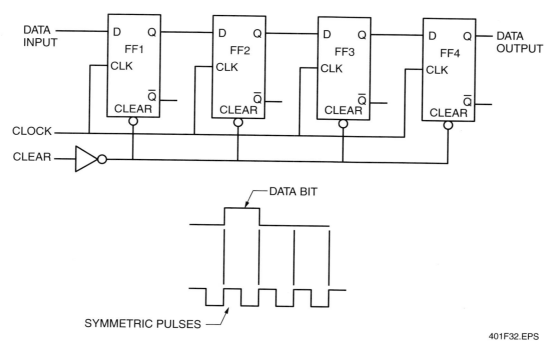

401F32.EPS

Figure 32 ◆ Basic shift register circuit.

important to note. On the next clock pulse, FF2 will receive the 1 from the output of FF1 and will be set. Meanwhile, FF1 will sample its input again at the leading edge of the next clock pulse and, with a 0 at the input, will load a logic 0. On the third clock pulse, the 1 bit will move into FF3 from FF2 and, on the fourth clock pulse, into FF4.

It can be seen that the logic 1 is simply being shifted or stepped through the flip-flops by the clock signal. This single logic 1 is regarded as an entity and, as it is moved in the shift register, it is called a bit or data bit. Similarly, the logic 0s in the shift register are called bits; their logic state is just as important in defining the contents of the shift register. If the shift register receives more than four clock pulses, the logic 1 will be shifted out of FF4, and all flip-flops will contain logic 0 again.

A shift register may be used to store a number in binary form. The number of bits in the register determines the size of the number that can be stored. For example, a four bit register can hold a four bit binary number. The number can then range from 0000 (decimal 0) to 1111 (decimal 7). An 8 bit register can then hold an 8 bit binary number from 0000 0000 (decimal 0) up to 1111 1111 (decimal 255). In a serial shift register, each bit is loaded one at a time. If the number is loaded starting with the most significant bit (MSB) and ending with the least significant bit (LSB), then the register is a shift-left register. If the LSB is loaded first and the MSB is loaded

last, the register is a shift-right register. This is shown in *Figure 33*. Note that the leading edge is the only point in time at which the flip-flops will sample input data.

If the shift register is first cleared (all flip-flops reset) and the four data bits shown in the timing diagram of *Figure 33* are shifted through the register, the flip-flops will have the following states at the end of four clock pulses:

FF1 = 0   FF2 = 1   FF3 = 1   FF4 = 1

This can be written in an abbreviated manner, simply as 0111. If, instead, you had stored the number 2, it would be represented by 0011 and the number 4 by 1111. With these four flip-flops, five numbers can be stored: 0, 1, 2, 3, and 4.

At this point, it is appropriate to define several terms related to the shift register and describe how it is used. First, because the bits were loaded one after another and the shift register shifted them from one flip-flop to another, the sequence is referred to as loading serial data, and the circuit is called a four-bit serially loaded shift register. The alternative to serial loading of the shift register is parallel loading. In this case, a separate line is connected to the preset input of each of the four flip-flops, as shown in the top of *Figure 33*. All the data bits are loaded simultaneously by setting the appropriate flip-flops to logic 1 through the preset input. Because this loading of the flip-flops takes

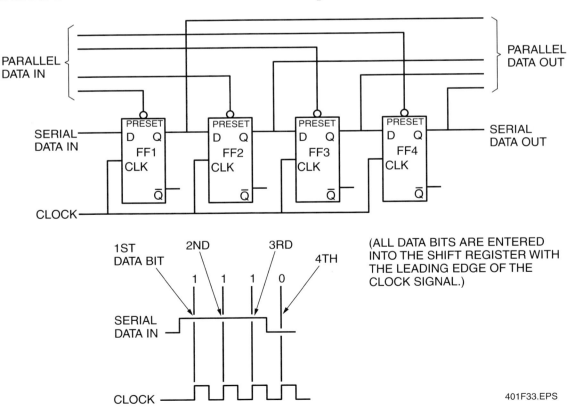

Figure 33 ◆ Methods of shifting data with a shift register.

place without the use of a synchronizing clock signal, it is said to be asynchronous. The data on the four preset inputs is called parallel data.

Just as the data bits can be loaded into the shift register in a parallel or serial manner, they can also be read out all at once or one at a time. A parallel readout is performed by simultaneously sampling data at the outputs of each of the flip-flops (see *Figure 33*), and a serial readout is performed by shifting data through the flip-flops and sampling at the output of FF4.

If the data is loaded serially and read out in parallel, the shift register is functioning as a serial-to-parallel converter. If the data is loaded in parallel and shifted out serially, the shift register is a parallel-to-serial converter.

Four bits can be stored in a four flip-flop shift register like the one illustrated in *Figure 33*. To store a number larger than 4, the shift register would have to be expanded by adding more flip-flops. Even as compact as integrated circuits are, this solution would cease to be practical as soon as large numbers, like 1,000,000, are considered. Therefore, another method of representing numbers is needed. The method to be discussed is called weighted coding.

Weighted coding consists of assigning different values or weights to each flip-flop. For example, if a five-bit shift register is used, as shown in *Figure 34*, the flip-flop on the right is assigned the weight of 1. The next flip-flop to the left is assigned the weight of 2, and the remaining three are weighted 4, 8, and 16.

When any one of the flip-flops is set, it represents the weight assigned to it. If more than one flip-flop is set, the individual values, or weights, are added to determine what number is stored. With this simple code, the range of numbers that can be represented by five flip-flops is greatly expanded, compared to the same five flip-flops used without weighted coding. The five-bit shift register can store any number up to 31, or if 0 is included, 32 different numbers in weighted code, compared to only six numbers (including 0) in the non-weighted code.

You can see how the weighted code works by storing some sample numbers. To store the number 1, the bit sequence 00001 is shifted into the shift register so that the flip-flop representing the weight of 1 is set. To store a 2, the serial bits 00010 are loaded, and only the flip-flop representing the number 2 is set. To store 3, 00011 is loaded to set the 2 flip-flop and the 1 flip-flop—the sum of the two weighted bits equals 3. To store 4, 00100 is entered. To store 5, 00101 is entered, and so on. The maximum number that can be represented with five flip-flops using this code is 11111, or $16 + 8 + 4 + 2 + 1 = 31$.

To store a number higher than 31, the shift register in *Figure 34* is simply extended by one more flip-flop and the weight of the new flip-flop is twice that of the largest already there, which in this case is $2 \times 16 = 32$. In this manner, the capacity can be increased very rapidly because the weight of each succeeding flip-flop is doubled: 64, 128, 256, 512, 1024, and so on. Thus, a 10-bit shift register can store 1024 different numbers. To find out how many numbers a shift register can store, there is a shorthand method: count the number of flip-flops in the register and raise 2 to that power. For example, a 6-bit register can store 64 different numbers (0 through 63), because $2^6 = 2 \times 2 \times 2 \times 2 \times 2 \times 2 = 64$.

Two important terms you should know are the *most significant bit (MSB)* and *least significant bit (LSB)*. The logic 1 or 0 bit stored in the flip-flop with the smallest weight (the flip-flop marked 1) is the least significant bit. Conversely, the bit stored in the flip-flop with the highest weight is the most significant. These terms are not limited to shift registers alone: they are used to identify the orientation of bits in any bit stream. For example, in the bit stream that was used for the number 5, the terms are applied as follows:

Figure 34 ◆ Weighted coding of flip-flops in a shift register.

3.2.0 Serial In–Serial Out

Serial in–serial out shift registers accept digital data serially; that is, one bit at a time on one line. They also produce the stored information on output in serial form. First look at the serial entry of data into a typical shift register with the aid of *Figure 35*, which shows a four-bit device implemented with RS flip-flops.

With four stages, this register can store up to four bits of digital data; its storage capacity is four bits. The following will illustrate the entry of the four-bit binary number 1010 into the register, beginning with the right-most bit. The 0 is put onto the data input line, making $S_A = 0$ and $R_A = 1$. When the first clock is applied, FFA is reset, thus storing the 0. Next, the 1 is applied to the data input, making $S_A = 1$ and $R_A = 0$. $S_B = 0$ and $R_B = 1$ because they are connected to the QA and $\overline{Q_A}$ outputs, respectively. When the second clock pulse occurs, the 1 on the data input is shifted into FFA because FFA sets, and the 0 that was in FFA is shifted into FFB. The next 0 in the binary number is now put onto the data input line and a clock pulse is applied. The 0 is entered into FFA, the 1 stored in FFA is shifted into FFB, and the 0 stored in FFB is shifted into FFC. Examination of the S and R inputs of each of the flip-flops will verify this operation. The last bit in the binary number, a 1, is now applied to the data input, and a clock pulse is applied to the CK line. This time the 1 is entered into FFA, the 0 stored in FFA is shifted into FFB, the 1 stored in FFB is shifted into FFC, and the 0 stored in FFC is shifted into FFD. This completes the serial entry of the four-bit number into the shift register, where it can be retained for any length of time. *Figure 36* illustrates each step in the shifting of the four bits into the register.

To get the data out of the register, it must be shifted out serially and taken off the QD output. After CK_4 in the data entry operation described above, the right-most 0 in the number appears on the Q_D output. When clock pulse CK_5 is applied, the second bit appears on the Q_D output. CK_6 shifts the third bit to the output, CK_7 shifts the fourth bit to the output, and CK_8 clears the register, as shown in *Figure 37*. Notice that while the original four bits are being shifted out, a new four-bit number can be shifted in.

Figure 38 shows a shift register that can be used to shift bits to the left with each successive clock pulse. This register consists of four JK flip-flops. Any one of the flip-flops may produce the output or receive the input in relationship to the other flip-flops.

To determine the operation of this register, you must first determine which flip-flop is used to receive the input data and which flip-flop is used to produce the output. The register in *Figure 38* provides an input on flip-flop (FFD), and the output is taken from Q of FFA. These input and output connections show that this register is also a serial in–serial out register. In this case, A through D may be wired differently in different cases.

With four stages, this register can store up to four bits of digital data. The design of this register is such that whatever condition is initially in FFD will appear in FFC after one clock pulse, and so on.

To analyze this shift left register, a timing diagram and shift table will be helpful. The timing diagram and shift table in *Figure 39* shows the output of each flip-flop with reference to the clock pulse.

Initially all the flip-flops are at the 0 state. Once a data input of logic 1 is received, J = 1 and K = 0 for FFD. At the same time, the first clock pulse is applied to all the flip-flops. On the trailing edge of the clock pulse, FFD stores the 1. With this 1 at the output of Q_D, now FFC has J = 1 and K = 0. With clock pulse 2, the 1 that was in FFD is shifted to FFC. Referring to the timing diagram, you can see that the data input pulse is removed after the first clock pulse, so now FFD is reset and stores the 0. The 1 stored in FFC is shifted to FFB with the third clock pulse, and FFC resets to 0. On the fourth clock pulse, there is another logic 1 input to FFD, and this 1 is stored in FFD along with the shift of the 1 stored in FFB to FFA. Now this register contains the bits 1001.

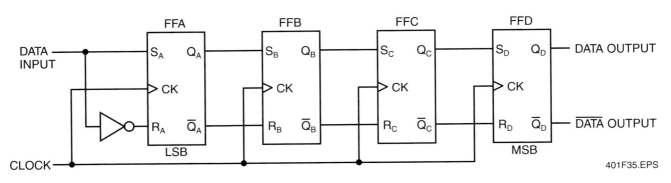

Figure 35 ◆ Serial in–serial out register.

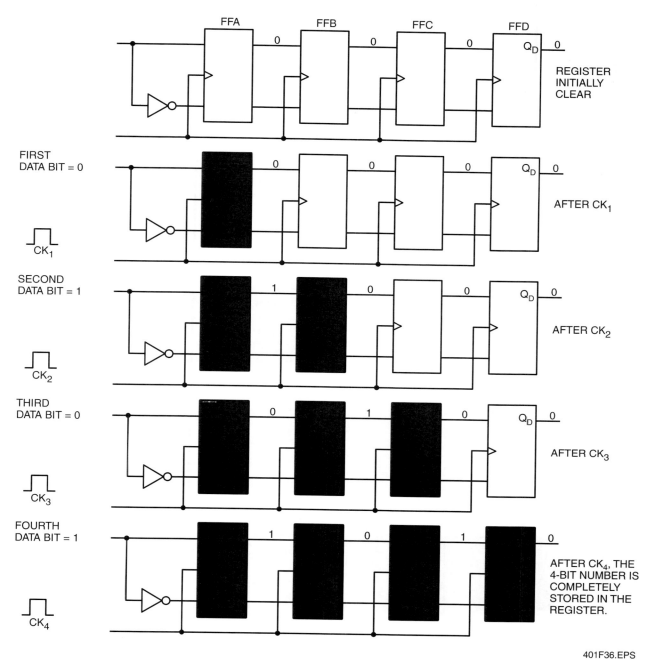

Figure 36 ◆ Data serially entered into a register.

You have analyzed this shift left register through four clock pulses. With no more logic 1 inputs, each of the following clock pulses will cause the data to shift left. This continues until the eighth clock pulse, when all the flip-flops are storing 0s.

The serial in–serial out shift register can be used for temporary data storage. Also, this type of register can be used as a buffer register. The term *buffering* refers to changing data transmission rates. Buffering implies that serial data is loaded into the register at one rate and then shifted out at another rate.

Because it takes one shift pulse to move data 1 bit, data arrives at the next bit position delayed by one pulse. Pulses are delayed as they propagate through the flip-flops. A desired delay can be built by selecting the number of flip-flops needed in a register to obtain the desired delay.

3.3.0 Serial In–Parallel Out

Data is entered serially into a serial in–parallel out shift register in the same manner as discussed earlier. The difference is the way in which the data is taken out of the register; in the parallel output

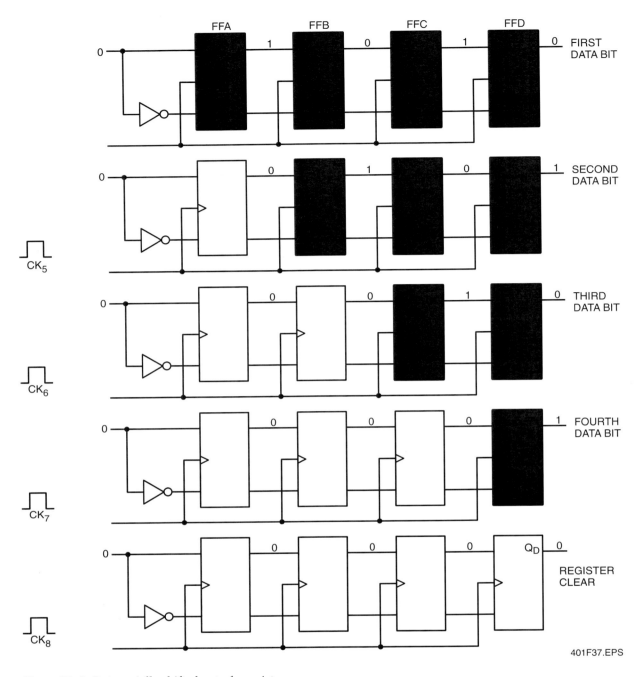

Figure 37 ◆ Data serially shifted out of a register.

register, the output of each stage is available at one time. Once the data is stored, each bit appears on its respective output line, and all bits are available simultaneously rather than on a bit-by-bit basis as with the serial output. *Figure 40* shows a four-bit serial in–parallel out register, its equivalent logic block symbol, and the associated waveform.

This register consists of four flip-flops. The output is fed to the input of the next flip-flop, with the exception of the last flip-flop.

Assume the register initially contains all 1s. With the waveforms shown in *Figure 40*, you can analyze the serial in–parallel out register operation and determine the register contents after four clock pulses.

The discussion of this register operation will be focused around the clock pulses, for the clock determines when the register will operate. Initially the register contains all 1s, and register operation is as follows:

- *Clock 1* – The data input is a 0 state. FFA resets on the leading edge of the first clock pulse. The register now contains 0111 in storage.

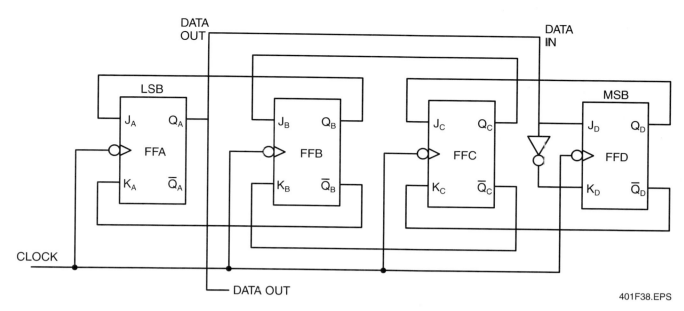

Figure 38 ◆ Shift left shift register.

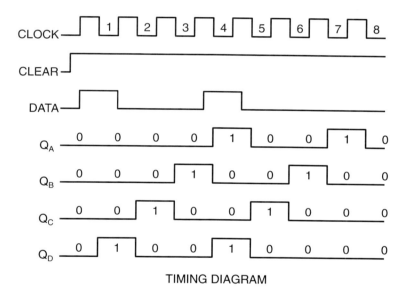

TIMING DIAGRAM

SHIFT PULSE	Q_A	Q_B	Q_C	Q_D
X	0	0	0	0
1	0	0	0	1
2	0	0	1	0
3	0	1	0	0
4	1	0	0	1
5	0	0	1	0
6	0	1	0	0
7	1	0	0	0
8	0	0	0	0

SHIFT TABLE

401F39.EPS

Figure 39 ◆ Timing diagram and shift table for shift left register.

- *Clock 2* – Now the data input is a 1 state. FFA sets and stores 1, and the previously stored 0 is shifted to FFB. FFC and FFD are still storing 1s. The register now contains 1011 in storage.
- *Clock 3* – The data input is still a logic 1. FFA stores this input and shifts its previously stored 1 to FFB. FFB shifts its 0 to FFC. FFD is still storing a 1. The register now contains 1101 in storage.
- *Clock 4* – The data input is changed to logic 0. FFA resets and stores the 0 while shifting its previously stored 1 to FFB. The shift of bits continues, and the final contents of the register after the fourth clock pulse is 0110.

A common application of the serial in–parallel out shift register is to convert serial data from a serial data bus to parallel data for transmission on a parallel data bus.

3.4.0 Parallel In–Serial Out

For a register with parallel data inputs, the bits are entered simultaneously into their respective stages on parallel lines rather than on a bit-by-bit basis on one line as with serial data inputs. The serial output is executed as previously described, once the data is completely stored in the register.

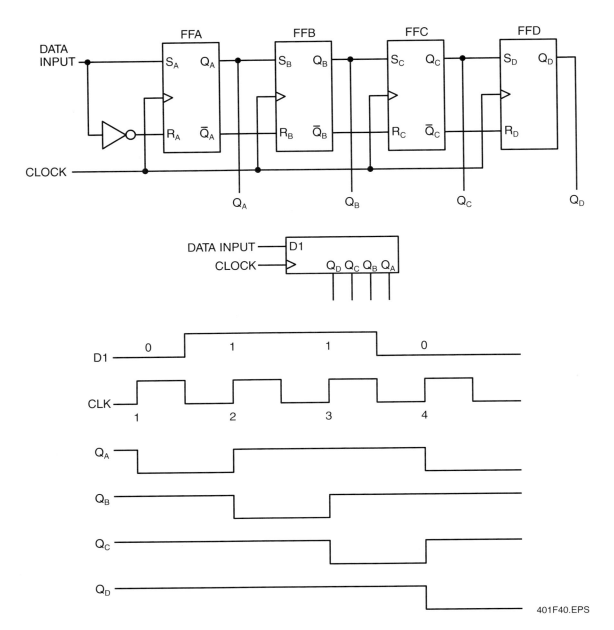

Figure 40 ◆ Serial in–parallel out register.

Refer to *Figure 41*. In this circuit, a serial data output is obtained at Q. The JK flip-flops that are utilized have direct set (S) and reset (R) inputs. By manufacturer's design, if the set input is high and the reset input is low, the flip-flop is placed in the reset state. When the direct set and reset inputs are both high, then the J and K inputs determine the state of the flip-flop.

If a low appears at the enable input for parallel data, any parallel data at A, B and C will not be loaded into the register. In other words, the parallel inputs do not have an effect on the register. Serial data could be applied to the serial data input, and a serial data output would be obtained from the Q output of FFC.

When parallel data appears at A, B, and C, it is loaded into the register by applying a high at the enable input for parallel data. For example, as shown at the bottom of *Figure 41*, suppose that a 1 data bit appears at A, a 0 data bit appears at B, and a 1 data bit appears at C.

At the A input, the high data bit is inverted and applied to gate D, which NANDs the enable pulse and the inverted data bit. Gate D, therefore, NANDs a low (inverted data bit) and a high (enable pulse). The NANDed output of gate D is high. Gate E NANDs the data bit (high) and the enable pulse (high). The NANDed output of gate E is low. Consequently, a high is applied to the reset input, and a low is applied to the set input.

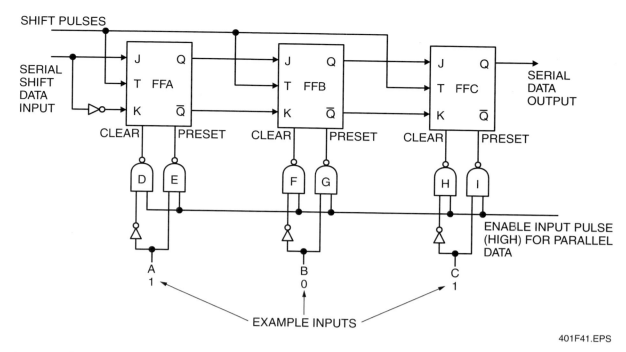

Figure 41 ◆ Parallel in–serial out shift register.

This will force FFA into the set state, and its Q output will be high.

At the B input, the low data bit is inverted and applied to gate F, which NANDs the inverted data bit and the enable pulse. Consequently, gate F NANDs a high (inverted data bit) and a high (enable pulse). The NANDed output of gate F is low. Gate G NANDs the data bit (low) and the enable pulse (high). The NANDed output of gate F is high. With a high applied to the set input and a low applied to its reset input, FFB will be forced into the reset state, and its Q output, like the data bit, will be low.

The high data bit at input C will be loaded into FFC in exactly the same manner that the high data bit at input A was loaded into FFA.

As a result of the parallel inputs, the register has been preset. The Q output of FFA is high because its data bit was high. The Q output of FFB is low because its data bit was low. The Q output of FFC is high because its data bit was high. The enable pulse is removed, and the parallel inputs no longer affect the register.

If three shift pulses are now applied to the register, the data that was loaded into the register will be serially shifted out of the register at the data output, which is the Q output of FFC. If the Q outputs of these flip-flops are sensed, a parallel output can also be obtained from this register.

Figure 42 shows a four-bit parallel in–serial out shift register. Notice that there are four data input lines—D_A, D_B, D_C, and D_D—and a parallel enable line, PE, that allows four bits of data to be entered in parallel into the register. When the PE line is high, gates G_1 through G_4 are enabled, allowing each data bit to be applied to the S input of its respective flip-flop and the data bit complement to be applied to the R input. When a clock pulse is applied, the flip-flops with a 1 data bit will set and those with a 0 data bit will reset, thereby storing the data word on one clock pulse.

When the PE line is low, parallel data input gates G_1 through G_4 are disabled, and shift enable gates G_5 through G_7 are enabled, allowing the data bits to shift from one stage to the next. The OR gates allow either the normal shifting operation to be carried out or the parallel data entry operation to be accomplished, depending on which AND gates are enabled by the level on the PE line.

Assume that this four-bit register is initially cleared and the bits 1010 are to be entered as shown in *Figure 42*. Referring to the waveforms shown, you can determine register operation.

On clock pulses 1, 2, and 3, the same parallel data are respectively loaded into the register, keeping Q_D at 0. On clock pulse 4, the 1 from Q_C is shifted onto Q_D; on clock pulse 5, the 0 is shifted onto Q_D; on clock pulse 6, the next 1 is shifted onto Q_D; and on clock pulses 7 and 8, all data bits have been shifted out, and only 0s remain in the register because no new data has been entered.

A common application for the parallel in–serial out register is to convert parallel data to serial data for transmission on a serial transmission line.

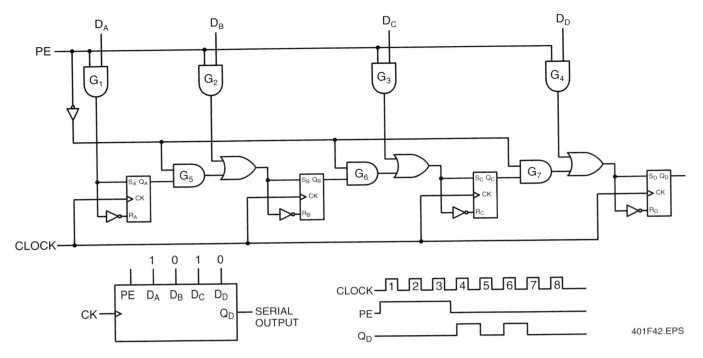

Figure 42 ◆ Parallel in–serial out register and waveforms.

3.5.0 Parallel In–Parallel Out

Parallel entry of data was described earlier and parallel output of data was also previously discussed. The parallel in–parallel out register employs both methods. Immediately following the simultaneous entry of all data bits, the bits appear on the parallel outputs. This type of register is shown in *Figure 43*.

The flip-flops for this register are **trailing edge triggered**. The load signal should occur when input data on the D lines is stable. Loading a flip-flop and reading the output from that flip-flop should not occur simultaneously. Each flip-flop in this register can store 1 bit of data indefinitely.

A common application of the parallel in–parallel out register is temporary storage of data.

3.6.0 Universal

The universal shift register (*Figure 44*) is extremely versatile. It can shift data in either direction, load data either serially or in parallel, and can output data either serially or in parallel.

Data can be loaded in this register by one of the three following means:

- Shift right serial input
- Shift left serial input
- Parallel inputs

There are two types of output available with this type of register: parallel outputs are available from Q of each individual flip-flop; and a serial output is provided from Q_4. There are two inputs for mode control: S_0 and S_1. The mode control inputs determine the mode of operation by determining the type of input that will cause register operation.

Now use *Figure 44* to analyze the conditions needed to cause the register to shift data different directions and to accept parallel inputs.

3.6.1 Shift Left

S_1 = Low (0)

S_0 = High (1)

Shift left serial input = High (1)

These mode control inputs provide the register the ability to shift bits to the left. With $S_1 = 0$ and $S_0 = 1$, AND gate 12 is the only AND gate that will produce a 1 output. This output from AND gate 12 causes a 0 output from NOR gate 4. This 0 output is applied to the R input of FF4. Also, the 0 from the output of NOR gate 4 is inverted, and a 1 is applied to the S input. This causes FF4 to set, and a 1 is stored. Each bit applied to the shift left serial input will cause a shift of bits in the left direction.

INSTRUMENTATION LEVEL FOUR — TRAINEE MODULE 12401-03

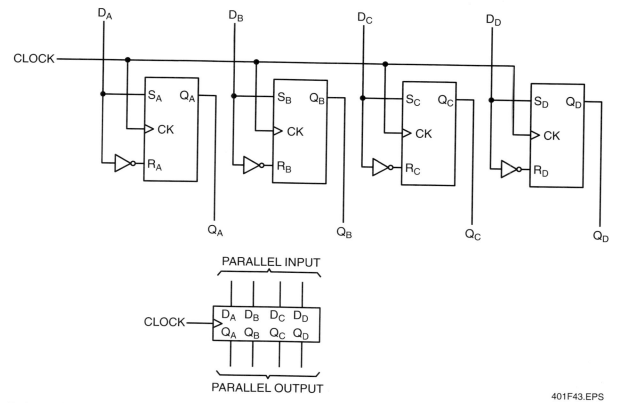

Figure 43 ◆ Parallel in–parallel out shift register.

401F43.EPS

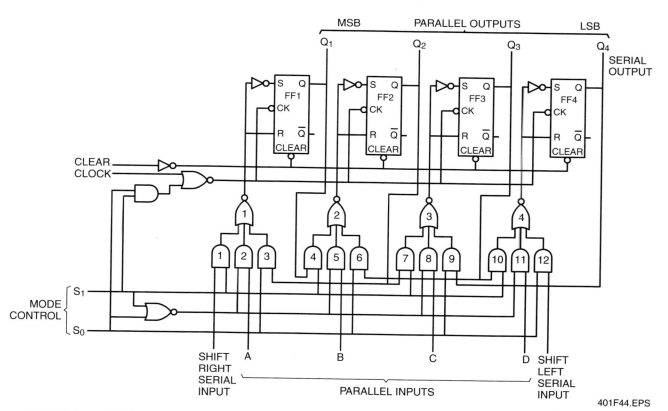

Figure 44 ◆ Universal shift register.

401F44.EPS

3.6.2 Shift Right

S_1 = High (1)

S_0 = Low (0)

Shift right serial input = High (1)

These mode control inputs provide the register the ability to shift bits to the right. With $S_1 = 1$ and $S_0 = 0$, AND gate 1 is the only AND gate that receives two 1 inputs; therefore, the output of AND gate 1 is a logic 1. This logic 1 output from AND gate 1 causes NOR gate 1 to produce a logic 0 output. This output from the NOR gate is applied to the R input of FF1. This logic 0 is inverted by an inverter, and a logic 1 is applied to the S input of FF1. Assuming all flip-flops are initially reset, FF1 sets and stores a 1. Each successive input from the shift right serial input will cause the bits stored in the register to shift bits in the right direction.

3.6.3 Receive Parallel Inputs

S_1 = Low (0)

S_0 = Low (0)

Parallel inputs—four bits of data applied

These conditions enable the register to receive parallel inputs and inhibit serial inputs from either the left or right inputs. With $S_1 = 0$ and $S_0 = 0$, AND gates 2, 5, 8, and 11 all receive an input of logic 1. Assuming all four parallel inputs are loading a 1, AND gates 2, 5, 8, and 11 will produce a logic 1 output. This output characteristic causes all the NOR gates to produce an output that is a logic 0. This logic 0 is applied to the R input of each of the four flip-flops. The 0 output from the four NOR gates is inverted again to a logic 1 and applied to the S input of the four flip-flops. With S = 1 and R = 0 in each flip-flop, each flip-flop will set and store a logic 1. This stored condition matches the data that was input from the parallel inputs.

3.6.4 Disable Shift Register

S_1 = High (1)

S_0 = High (1)

With S_1 and $S_0 = 1$, the shift register is disabled. Regardless of what information is loaded, the shift register does nothing. With S_1 and $S_0 = 1$, the clock pulse is inhibited, and therefore the flip-flops in the register do not set or reset. Because of its versatile construction, this shift register can be used in any application the previous shift register could be used in.

4.0.0 ◆ COUNTERS

When counting a sequence of numbers, two basic factors are involved: you must know where you are in the sequence at any given time, and you must know the next number in the sequence. A digital counter is a logic circuit that progresses through a sequence of numbers or states when activated by a clock pulse input. The output of a counter indicates the binary number obtained within the counter at any given time. Counters are all characterized by a storage or memory capability because they must be able to retain their present state until the clock forces a change to the next state in the sequence.

The true binary counter circuits differ from all forms of shift registers described in that their flip-flops are interconnected in a different manner. The object of a counter circuit is to output data in a specific pattern or to maximize the number of different states that can be obtained with a given number of flip-flops.

Most counters output data coded in 8421, 2421, Excess 3, or some other common binary code, but by designing interconnecting logic, any arbitrary output pattern can be obtained.

Counters are usually used as basic building blocks in other logic circuits. They are used for counting, sequencing of equipment or process operations, frequency measurement and division, arithmetic manipulation, time interval measurement, and many other purposes.

There are many different variations of counter circuits. All are made of either JK, T, RS, or D flip-flops and can be classified into two very basic types: asynchronous (also known as serial or ripple counters) or synchronous (also known as parallel counters).

In synchronous counters, all flip-flops change state simultaneously; in asynchronous counters one flip-flop changes state, and this change triggers a second flip-flop, which then may trigger a third, then a fourth, and so on.

Within each of the two basic categories, a counter can be designed that counts to any desired binary number before it starts the count sequence over again. The number of consecutive states through which a particular counter sequences before repeating is called its modulus. Counters with a modulus of 2, 4, 8, 16, or some other number that is a power of 2 are the easiest to construct; however, counters with a modulus of 6 or 10 are also common.

Counters may also be classified according to the weighted code in which they count (for example, the 8421 binary code or Excess 3). In addition, a counter may either count up, count down, or may do both, depending on the logic level at a mode control input.

4.1.0 Four-Bit Binary

Another common application of flip-flops is in binary counters. These counters can be used whenever it is desired to monitor the number of occurrences of a certain event. For example, a counter could be used to record the number of cars that pass a traffic sensor or the number of seconds for a person to run 100 yards. A four-bit binary counter is represented by the block in *Figure 45*.

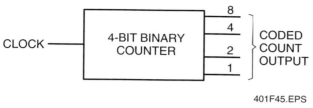

Figure 45 ◆ Symbol for a four-bit binary counter.

The counter receives a clock signal and outputs a binary coded count on four signal lines. The flip-flops and gates that make up a circuit inside the block are interconnected to advance the binary coded count output by one every time a clock pulse is received. The output of the counter consists of the Q outputs of four flip-flops. The sequence followed by the counter is for numbers from zero through fifteen. The counter needs only four flip-flops, representing columns 1, 2, 4, and 8.

To determine how the counter outputs the binary code, analyze the circuit in *Figure 46* with respect to the requirements of the code in *Table 7*. Notice that the circuit consists of four flip-flops in series, which are interconnected either directly or through AND gates in a way similar to a shift register. Because the clock is connected in parallel to all flip-flops of the counter, their output states change state simultaneously. As a result, the counter is called a synchronous counter.

The pattern of alternate 1s and 0s in the 1s column can be recognized as the output of a JK flip-flop that is toggling. The first JK flip-flop FF1 in *Figure 46* is, therefore, made to toggle by placing both J and K at logic 1. The 2s column of *Table 7* shows a similar alternating pattern of 1s and 0s, except that they change back and forth every two clock pulses. In comparing the sequence of logic states in the 2s column with the sequence in the 1s column, it can be seen that when the 1s column is a logic 1, the next count requires that the 2s column change its logic state. This is implemented by connecting the inputs of FF2 to the Q output of FF1 and allowing this signal to toggle FF2.

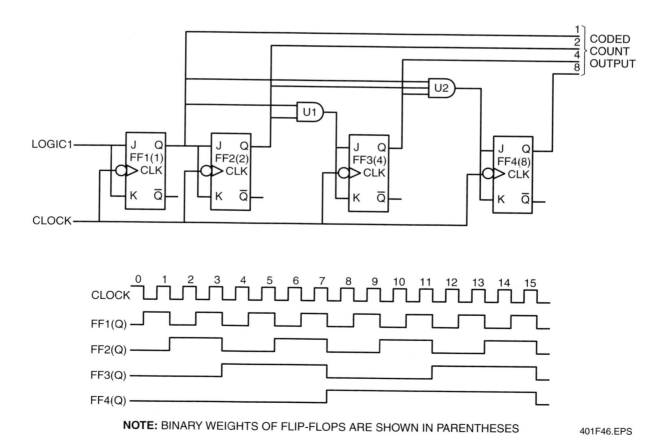

NOTE: BINARY WEIGHTS OF FLIP-FLOPS ARE SHOWN IN PARENTHESES

401F46.EPS

Figure 46 ◆ Four-bit binary counter circuit.

Table 7 Binary Number Code

NUMBER	CODED BITS:	8	4	2	1
0		0	0	0	0
1		0	0	0	1
2		0	0	1	0
3		0	0	1	1
4		0	1	0	0
5		0	1	0	1
6		0	1	1	0
7		0	1	1	1
8		1	0	0	0
9		1	0	0	1
10		1	0	1	0
11		1	0	1	1
12		1	1	0	0
13		1	1	0	1
14		1	1	1	0
15		1	1	1	1

401T07.EPS

Table 8 Binary Counter Truth Table

CLK	FF1	FF2	FF3	FF4
1	1	0	0	0
2	0	1	0	0
3	1	1	0	0
4	0	0	1	0
5	1	0	1	0
↓	↓	↓	↓	↓
15	1	1	1	1

401T08.EPS

The above relationship can also be seen in the timing diagram of *Figure 46* where FF1 is a logic 1 just before FF2 changes state. With the logic 1 from FF1 present only on alternate clock pulses, FF2 toggles only on alternate clock pulses.

If you analyze FF3, you will notice that for FF3 to toggle, it is necessary for both FF1 and FF2 to be in the logic 1 state because the 1s and 2s columns are 1 at the count just before the 4s column changes state. The outputs of FF1 and FF2 are therefore ANDed together to provide the toggling signal to J and K inputs of FF3.

A similar analysis holds true for FF4, or any other flip-flops, if they were added to extend the counter: the J and K input to any flip-flop consists of the ANDed outputs of all the previous flip-flops.

If you were to develop a truth table for this four-bit binary counter (*Table 8*) it would be similar to the table given for the binary number codes. The only difference is that the truth table labels the binary output according to each flip-flop instead of the binary coded output.

4.2.0 Up

A binary ripple (asynchronous) counter is the most basic of all the counters. A series of JK flip-flops can be connected to count up in binary code, as shown in *Figure 47*.

The circuit in *Figure 47* has one major characteristic that classifies it as asynchronous and allows you to identify any circuit of this type. The output of the first flip-flop, FF1, triggers the second flip-flop, FF2, at its clock input; FF2 output triggers FF3; FF3 output, in turn, triggers FF4. Thus, the effect of a clock pulse introduced at the input of FF1 will propagate from one flip-flop to another until it reaches the last in the series. Hence, it has the names ripple counter or series counter.

To analyze the asynchronous up counter, note further that each JK flip-flop has its J and K inputs at logic 1. This causes the flip-flop to toggle (change state) each time a clock pulse is received. Because the output of one flip-flop is connected to the clock input of the next flip-flop, each succeeding flip-flop changes state half as often as the previous flip-flop. In other words, the clock input of a flip-flop must go from low to high and back to low again for its output to change state once. From the timing diagram in *Figure 47*, you can see that this divide-by-two action of each flip-flop creates state changes that conform to the 8421 binary code shown earlier in *Table 7*. Using the timing diagram in *Figure 47*, you can analyze this counter's operation.

Before clock pulses are applied, it is assumed that the flip-flops will be reset; therefore, the Q outputs of all the flip-flops would be low. On the trailing edge of the first clock pulse, FF1 will be triggered into the set state. A transition from a low to a high appears at the CLK input of FF2. This is the leading edge of a pulse; consequently, FF2 does not respond and remains in the reset state. On the trailing edge of the second clock pulse, FF1 is triggered into the reset state. This produces a

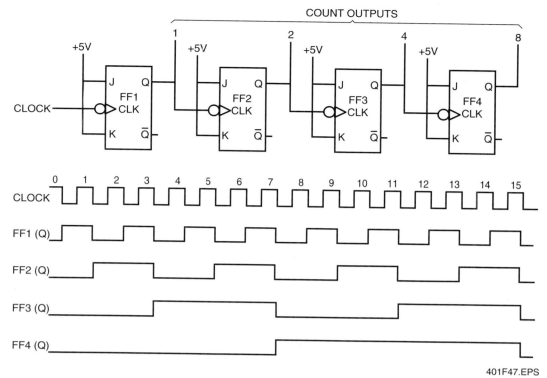

Figure 47 ◆ Up counter circuit and timing diagram.

transition from a high to a low on the CLK input of FF2 that is the trailing edge of the pulse. Consequently, FF2 is triggered into the set state. On the trailing edge of the third clock pulse, FF1 is triggered into the set state. This is the leading edge of a pulse at the CLK input of FF2, so FF2 remains in its previous state. On the fourth clock pulse, FF1 is reset, which resets FF2; this is the trailing edge for the input to FF3, and FF3 sets. This cycle continues until the 15th clock pulse, at which time the counter contains a binary 1111. This counter will count to fifteen and then the cycle repeats. The problem with this type of counter is the propagation delay. The clock pulse is applied to the input of only the least significant flip-flop, and all other flip-flops depend upon the previous flip-flop for a trigger pulse. This limits the counting speed. Later, you will see another type of counter that will eliminate some of the propagation delay.

4.3.0 Down

The asynchronous down counter (*Figure 48*) is the same in principle as the up counter except the $\overline{Q}$ output (instead of Q) of each flip-flop causes the following flip-flop to toggle, which reverses the code sequence.

The rules for determining the required number of flip-flops are the same as those previously utilized for the up counter. For example, if the highest number that will be counted is a binary 1111, a flip-flop will be required for each bit of the binary number. It should be noted that the $\overline{Q}$ sides of the FFs are used to trigger the next flip-flop. This is the primary difference between the up counter and the down counter.

For the timing diagram and counter shown in *Figure 48*, assume all the flip-flops are initially set and the counter contents are 1111. This counter will be analyzed with reference to clock pulses.

- *CLK 1 0111* – FF1 resets, and $\overline{Q}$ of FF1 goes to a logic 1 and is applied to the CLK input of FF2. The trailing edge of a pulse will trigger the flip-flops; therefore, FF2 remains unchanged.

- *CLK 2 1011* – FF1 sets, and $\overline{Q}$ of FF1 goes to a logic 0, which triggers FF2. FF2 resets, and $\overline{Q}$ of FF2 is applied to the CLK of FF3. FF3 remains unchanged but now has a logic 1 applied to its CLK input.

- *CLK 3 0011* – FF1 resets, and the CLK input of FF2 goes high. FF2, FF3, and FF4 are unaffected.

- *CLK 4 1101* – FF1 sets, and this triggers FF2 to the set state also. FF2 now causes the CLK input of FF3 to go from high to low, and FF3 resets.

- *CLK 5 0101* – FF1 resets and causes the CLK input of FF2 to go high. All flip-flops except FF1 remain unchanged.

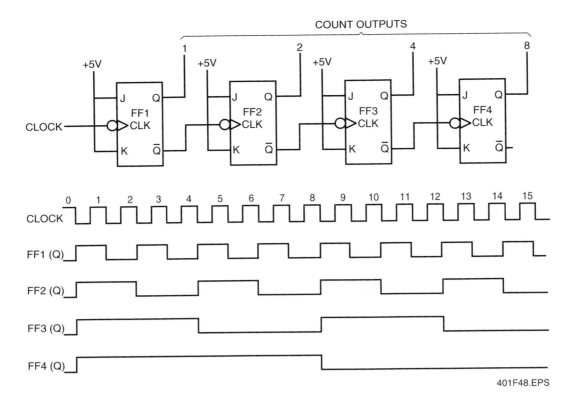

COUNT OUTPUTS

Figure 48 ◆ Down counter circuit and timing diagram.

The timing diagram for this down counter shows the operation of the counter through CLK pulse 15. The operation has been analyzed through CLK 5 but is the same throughout. The trailing edge of the clock pulse triggers the flip-flops just as in the up counter. This counter can count from 0–15 just as the up counter, but because of the clock connections, the counter counts down. The timing diagram for the down counter can aid in determining the truth table for a down counter. (See *Table 9*.)

Table 9 Down Counter Truth Table

CLK	FF1	FF2	FF3	FF4
0	1	1	1	1
1	0	1	1	1
2	1	0	1	1
3	0	0	1	1
4	1	1	0	1
5	0	1	0	1
6	1	0	0	1
↓	↓	↓	↓	↓
15	0	0	0	0
16	1	1	1	1

401T09.EPS

4.4.0 Up/Down

The up counter and down counter can be combined into one up/down counter, whose mode of counting is selected by a control signal through some logic gates. The circuit in *Figure 49* shows how an OR gate can be used to connect either the Q or the $\overline{Q}$ output of any flip-flop to the clock input of the following flip-flop.

Also included in the up/down counter is a count enable feature, which consists of connecting the J and K signals to a switchable control signal instead of simply to a logic 1. If this count enable signal is at a logic 0 level, all J and K inputs are at logic 0, and the flip-flops will not change state when clocked. That is, the counter remains at the last count before the count enable signal was removed.

You can use a timing diagram to analyze the up/down counter operation. Remember the count enable signal must be at a logic 1 or the flip-flops will not change. *Figure 50* shows the timing diagram for an up/down counter. The counter starts with all flip-flops initially reset and a high input on count up/down input.

Both a count up and a count down operation of this up/down counter will be analyzed with relation to clock pulses. Remember that JK flip-flops only trigger on a transition from a high to a low on the CLK input.

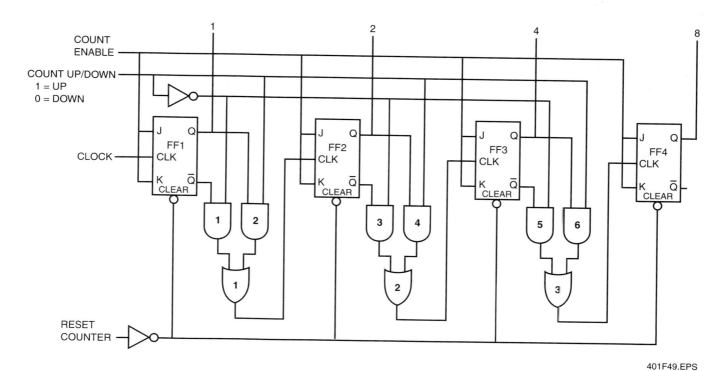

Figure 49 ◆ Up/down counter.

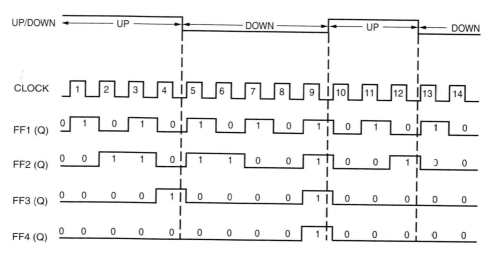

Figure 50 ◆ Up/down counter timing diagram.

- *CLK 1 1000* – FF1 sets, which causes a high output from OR gate 1, which is applied to the CLK input of FF2.
- *CLK 2 0100* – FF1 resets, which causes a low output from OR gate 1, which triggers FF2, and FF2 sets. When FF2 sets, it causes the output of OR gate 2 to produce a high input to the CLK input of FF3.
- *CLK 3 1100* – FF1 sets, which causes a high input to FF2, but all flip-flops except FF1 remain unchanged.
- *CLK 4 0010* – FF1 resets, which triggers FF2 to also reset. When FF2 resets, OR gate 2 causes a

high to low transition on the CLK input of FF3, and FF3 sets. FF4 remains unchanged.
- *CLK 5 1100* – Now the count up/down inputs are changed to a logic 0 to indicate the counter will now count down. Now with the count up/down change, OR gates 1 and 2 produce a high input to the CLK input of FF2 and FF3 respectively. When clock pulse 5 is applied, FF1 will set, which causes FF2 to set. When FF2 sets, FF3 will reset.
- *CLK 6 0100* – FF1 resets, and the remaining flip-flops stay unchanged.

This concludes the up/down counter operation through six clock pulses. The timing diagram can be used to trace the counter operation during up count and down count operations. From the waveforms you can see the counter sequence in *Table 10*.

4.5.0 Synchronous

A synchronous counter is based on the same JK (or type T) flip-flop circuit as an asynchronous counter, except all flip-flops are clocked by a common clock signal and, therefore, all change states synchronously. The J and K inputs of any flip-flop are connected to the Q outputs of all previous flip-flops in the counter chain through an AND gate, as shown in *Figure 51*.

Any one flip-flop will toggle when the AND gate at its input causes the J and K inputs to be a logic 1, and that occurs only when all previous flip-flops in the counter chain are in the 1 state. Use *Figure 51* to analyze circuit operation of this synchronous counter.

Assume that all flip-flops are initially reset. All of the AND gates are receiving logic 0 inputs. When the first clock pulse is applied, FF1 will set and provide a logic 1 input to J and K of FF2. On the second clock pulse, FF1 will reset, and FF2 will set. Referring to *Figure 51*, you can see that any of the flip-flops, except FF1 and FF2, will only produce a logic 1 when their respective AND gate provides a high input to the J and K inputs.

This counter counts exactly like the up counter presented earlier except for the difference in the propagation delay. Notice that because all flip-flops are clocked and change states at the same time, the total delay, regardless of how many flip-flops there are, is that of just one flip-flop. If the total propagation delay time of a JK flip-flop and the AND gate interconnecting its output to another flip-flop is 35ns, the clock pulses can occur at a maximum frequency of approximately 30MHz in a synchronous counter. This compares to a 10MHz maximum frequency for the four-bit asynchronous counter using the same flip-flop propagation delay figures.

Table 10 Counter Sequence

FF4	FF3	FF2	FF1	
0	0	0	0	
0	0	0	1	
0	0	1	0	UP
0	0	1	1	
0	1	0	0	
0	0	1	1	
0	0	1	0	
0	0	0	1	DOWN
0	0	0	0	
1	1	1	1	
0	0	0	0	
0	0	0	1	UP
0	0	1	0	
0	0	0	1	DOWN
0	0	0	0	

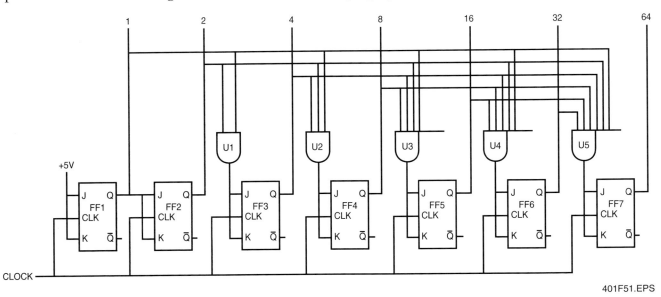

Figure 51 ◆ Synchronous counter.

401F51.EPS

Another desirable feature of the synchronous counter is that all output lines change simultaneously. Thus, there are no intermediate states when the counter outputs are incorrect as the counter advances from one state to another.

The synchronous counter does, of course, have limitations. First, it requires more logic gating to implement and thus is more complex and costly than a comparable asynchronous counter. Second, notice that the last AND gate has to have six input lines or even more if a count enable signal is used. If the counter were to be expanded any further, the number of inputs to the subsequent AND gates would become a practical limitation.

4.6.0 Ripple Carry

The counter shown in *Figure 52* is a simplified version of the straight synchronous counter and is often referred to as a ripple carry counter. This counter is still synchronous in that all flip-flops change state at the same time, but the connection between the J and K inputs of any flip-flop and the Q outputs of all previous flip-flops is made through AND gates that are in series rather than in parallel. As a result, the propagation delay of the AND gates is cumulative, and the frequency of operation is reduced somewhat when compared to synchronous counters. As the counter is extended much beyond four flip-flops, the delay becomes proportionately greater, and its speed advantage over asynchronous counters diminishes. Although this counter is a slower device than a purely synchronous counter, because the outputs change states simultaneously and the circuit is much simpler, it is an attractive compromise between asynchronous and synchronous counters.

4.7.0 Binary Coded Decimal

A binary coded decimal (BCD) counter (*Figure 53*) is one that counts to 10 (the 10 states from 0 through 9 in the decimal system) then resets itself and starts over again. Even though the circuit is similar to those already studied, it is particularly important because it is very common in computers, calculators, and in other circuits where a decimal count is needed.

The BCD counter can be either asynchronous or synchronous and it usually counts in the 8421 binary code. Therefore, it is basically the same as any other four-bit counter, except that it has some special gating that limits its count to 10 states (a modulus of 10).

The BCD counter functions like any other 8421 counter from the count of 0 through the count of 9 (1001). Whereas an 8421 counter with more than 10 states advances to 1010 on the next clock pulse, the BCD counter instead resets itself back to 0000.

To accomplish the special counting process of a BCD counter, two special connections must be made. First, to keep FF2 in the 0 state on the next clock pulse after the 1001 state, $\overline{Q}$ of FF4 is connected to gate U1. The $\overline{Q}$ output of FF4 is logic 0, and thus the J and K inputs to FF2 are logic 0, and FF2 does not change to logic 1 at the next clock pulse edge. Second, to reset FF4 to 0, Q of FF1 is connected directly to K of FF4. As a result of this connection, the K input is alternately high and low during the counts of 0 through 6, but the J input of FF4 is continuously low, and thus FF4 remains in the 0 state. At count 7, however, all inputs to U3 become high, and there is a high signal at both J and K of FF4. Thus with the next clock pulse (count 8), FF4 toggles to the 1 state. It remains in the 1 state after one more clock pulse

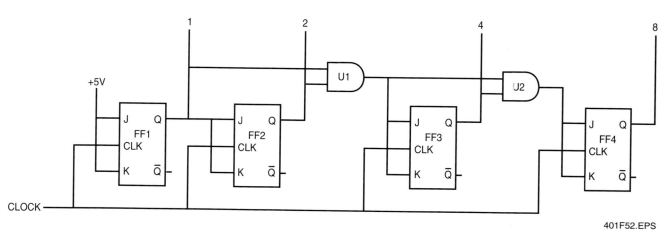

Figure 52 ◆ Ripple carry counter.

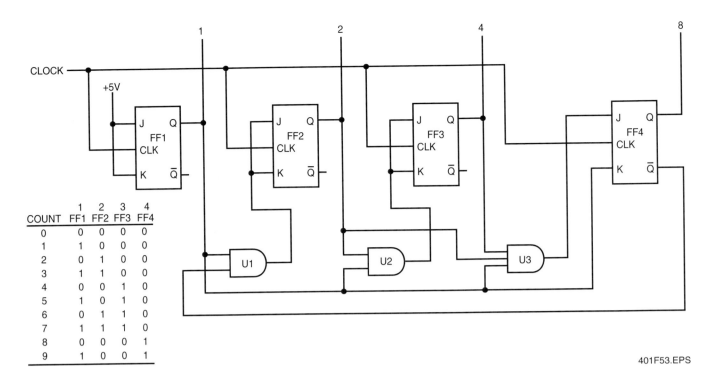

COUNT	1 FF1	2 FF2	3 FF3	4 FF4
0	0	0	0	0
1	1	0	0	0
2	0	1	0	0
3	1	1	0	0
4	0	0	1	0
5	1	0	1	0
6	0	1	1	0
7	1	1	1	0
8	0	0	0	1
9	1	0	0	1

Figure 53 ◆ BCD counter and truth table.

(count 9) because now Q of FF1 is 0 thus removing the logic 1 from both J and K of FF4. At count 9, Q of FF1 goes to 1 again, and thus FF4 has a high K input but still a low J input. Therefore, with the next clock pulse, FF4 resets back to the 0 state.

4.8.0 Ring and Johnson

Shift registers and counters have many characteristics in common. By adding a few logic gates to a shift register to modify the exact path of data through the flip-flops, the shift register becomes a counter. A counter can remember a certain number of different states and can be clocked through these states in a repeatable manner. So-called ring counters and Johnson counters are circuits where the relationship between shift registers and counters is readily apparent. A ring counter, as illustrated in *Figure 54*, is simply a circulating shift register, which is one whose output is connected back to its own input.

A ring counter is commonly preloaded with a logic 1 bit in the first flip-flop and 0s in all other flip-flops, and then it is clocked to circulate the logic 1 bit repeatedly through the register. As a result, the counter outputs a set of unique states that can be used to sequence some equipment or another logic circuit through different operations. For example, it could sequence a machine tool station on an assembly line through steps that are to be repeated over and over again, or it could sequence an electronic calculator through the steps of addition. Ring counters are commonly used as sequencers in computers, decoders, and in a wide variety of other circuits.

A major advantage of ring counters is that, unlike any other counter, they do not require decoding. Any output line can be connected directly to the circuit or the device it is to activate. However, ring counters also have several limitations. First, the circuit does not make very efficient use of flip-flops. A four-bit ring counter can generate only four unique states, whereas a four-bit binary coded counter can generate 16. More generally stated, a ring counter has n states, but an 8421 binary counter has 2n states, where n is the number of flip-flops in the counter. Second, if a flip-flop gets set at the incorrect time due to a noise pulse or a malfunction, an erroneous sequence, or special state, will occur and will continue to circulate without being corrected. What is called a special state is arbitrary because a particular application may well require a set of states other than those shown in *Figure 54*.

Figure 55 shows a ring counter that will correct any erroneous sequence back to the normal count sequence. In this counter, any incorrect bits will simply be shifted through the flip-flops until all have 0 bits in them, except possibly the last. Until this time, the J input of the first flip-flop will be held low and the K input high. After the flip-flops go to 0001 state, the inputs to the first flip-flop

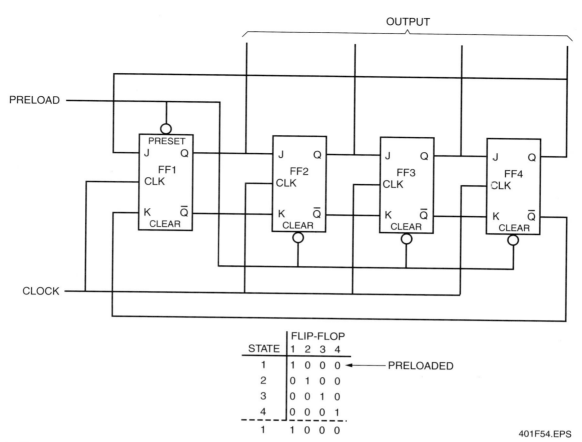

Figure 54 ◆ Ring counter with preload.

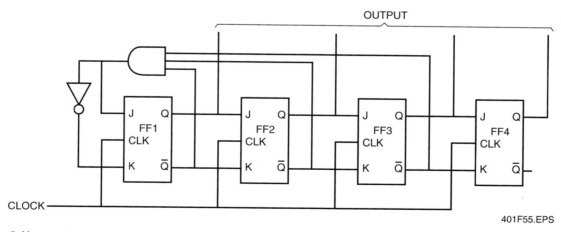

Figure 55 ◆ Self-correcting ring counter.

reverse (J is high, K is low); the next clock pulse sets FF1 and resets FF4, and the normal counting sequence has been restarted. The truth table for this ring counter is the same as the truth table presented for the ring counter with preload. The Johnson counter (also called a switch-tail or twisted ring counter) is a slight modification of the standard ring counter in that the $\bar{Q}$ output of the last flip-flop is connected back to the J input of the first, as shown in *Figure 56*.

As a result of this switching of output-to-input connections, the counter passes through 2n different states, where n is the number of flip-flops in the counter, as shown in the truth table of *Figure 56*. Thus, a four-bit Johnson counter has twice as many states as a four-bit ring counter but only half as many as a straight 8421 binary counter. The Johnson counter also needs a decoder to output a separate signal for each of the 2n states.

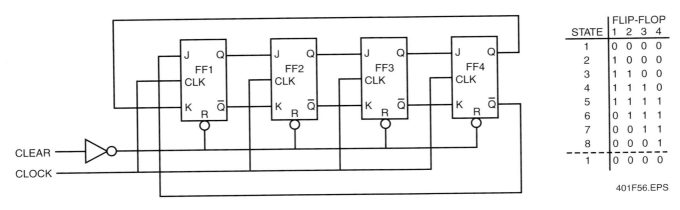

		FLIP-FLOP			
STATE	1	2	3	4
1	0	0	0	0
2	1	0	0	0
3	1	1	0	0
4	1	1	1	0
5	1	1	1	1
6	0	1	1	1
7	0	0	1	1
8	0	0	0	1
1	0	0	0	0

401F56.EPS

Figure 56 ◆ Johnson counter.

Using the truth table shown for the Johnson counter, you can analyze the operation of this counter. Assume all the flip-flops are initially reset and storing 0s. Q of FF4 is connected to K of FF1, and $\overline{Q}$ of FF4 is connected to J of FF1. Therefore, for FF1, J = 1 and K = 0. When the next clock pulse arrives, FF1 will set and store a 1. The output of FF1 is connected to the input of FF2. Therefore, when the second clock pulse arrives, FF1 is still set, but FF2 will also change to the set state and store a logic 1. Because the output of each flip-flop is connected to the inputs of the following flip-flop, this process of counting will continue until the completion of eight clock pulses, at which time the counter is completely reset.

In a manner similar to the straight ring counter, the Johnson counter can be trapped in a self-perpetuating sequence of special states and needs auxiliary logic to force it to change back to a normal state. One way of implementing a correcting circuit is shown in *Figure 57*, with a single AND gate connected to the last two flip-flops in the counter to act as a correcting circuit. This circuit will force any of the eight possible special states back to a normal state within five clock pulses or less. It can be used on a counter with any number of flip-flops.

Another way of forcing a correction is by simply interconnecting the $\overline{Q}$ output of FF1 to the K input of FF4, as shown with the broken line in *Figure 57*, and deleting the connection from $\overline{Q}$ of FF3 to K of FF4. This error correcting scheme can be used within any length counter by changing the connections in the same corresponding positions, but it also constitutes a further modification of the counters in that its count is reduced to $2n - 1$ states. With this modification of the four-bit counter in *Figure 57*, the 0011 state will be followed by 0000, entirely eliminating the 0001 state.

4.9.0 Programmable

A programmable counter is any counter whose modulus or counting pattern can be modified in some way by a control signal rather than by circuit changes. The most common programming signals are those related to presetting the counter to a certain count (thus changing the modulus of the counter) and to controlling whether the counter stops at a certain count or resets itself and starts over again. A counter circuit can be programmed and controlled in a variety of ways, a few of which will demonstrate the basic ideas in counter programming.

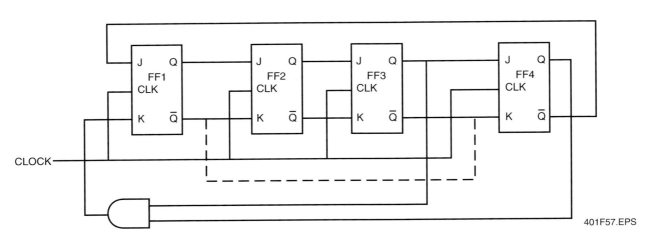

401F57.EPS

Figure 57 ◆ Self-correcting Johnson counter.

INSTRUMENTATION LEVEL FOUR — TRAINEE MODULE 12401-03

In *Figure 58*, an asynchronous four-bit counter can be preset to any desired count between 0 and 15 via gates U1 through U4 by enabling the LOAD signal line (placing at logic 0).

If it is to be used as a modulus 7 counter, it is preset to a count of 8 and allowed to count to 15.

However, notice that at a count of 15, the counter must be preset to 8 again if it is to repeat its counting cycle in the same manner. This requires load control circuitry in addition to that shown.

Figure 59 shows a synchronous four-bit counter that will advance to a preselected count and hold

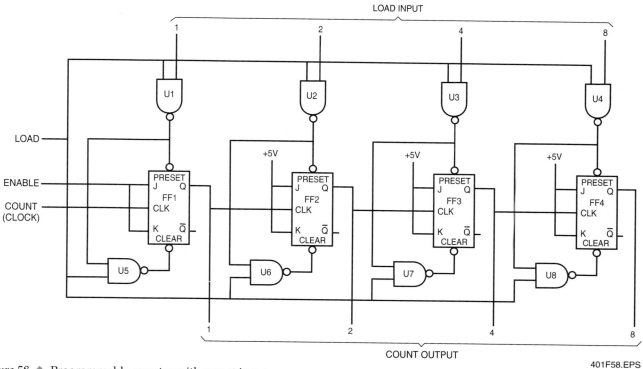

Figure 58 ◆ Programmable counter with preset inputs.

401F58.EPS

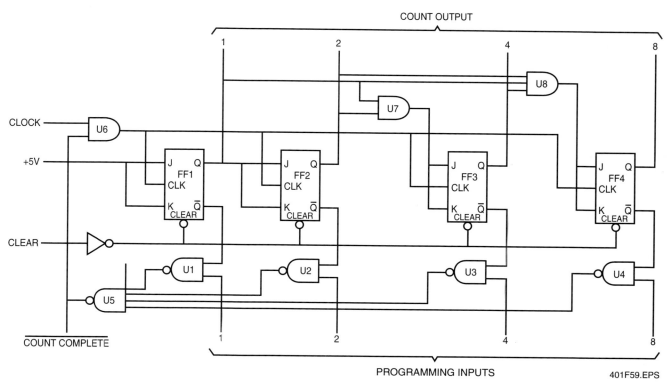

Figure 59 ◆ Programmable counter whose final count can be preselected.

401F59.EPS

there until it is reset. The modulus can be selected with the programming inputs via gates U1 through U4; however, it is done so that the final count (instead of the starting count) is determined. The basic idea is to use the gates U1 through U4 to make a digital comparison: when the output of the counter matches the programming inputs, the clock is disabled and the circuit stops counting. At this point, the count complete signal can be used to reset the counter, or the contents of the counter can be saved.

In addition to these circuits, many other program or control schemes can be used. These are usually designed to fit the needs of a particular circuit by modifying a standard integrated circuit counter with external gating.

5.0.0 ◆ ARITHMETIC ELEMENTS

An arithmetic element is any circuit that can add, subtract, multiply, divide, or perform some other mathematical function with binary numbers. Arithmetic elements lie securely in the domain of computer technology and the related technology of electronic calculators. Even though computers and electronic calculators are not the only digital electronic systems in which arithmetic is performed, it is in those areas that arithmetic elements were developed to their present degree of versatility, speed, and overall sophistication.

5.1.0 Half Adder

One logic circuit that can be used to begin implementing addition is shown in *Figure 60*. The same basic circuit can also be constructed by using a single XOR gate, and both will perform according to the same truth table.

This circuit is the basis of any arithmetic adder. You will recall from the study of binary arithmetic that when two binary 1s are added, the result is 0 with a carry of 1. To generate the carry, the half adder circuit uses an AND gate to sense when A and B are both logic 1.

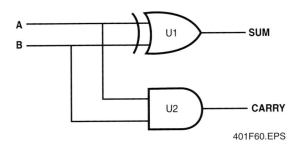

401F60.EPS

Figure 60 ◆ Half adder.

5.2.0 Full Adder

The half adder must be modified still further to become useful in most applications. Although it is entirely adequate for the addition of two binary bits, in general, numbers consist of more than one bit. A separate adder circuit is needed to add each pair of bits, and the addition can be done in parallel. When adding in parallel, however, each bit position affects the addition to the left of that bit position, as shown in the following example:

```
  1 0 1 1   # C (carries)
  1 0 0 1   # A (augend)
+ 1 0 1 1   # B (addend)
1 0 1 0 0
```

In the preceding example, each adder has to be modified to handle three inputs: A, B, and the carry, C. Such a modified circuit is called a full adder (*Figure 61*).

Although the full adder can be constructed in many different ways, it always performs the same function: it adds two bits and a previous carry and generates a sum as well as a carry. This circuit is used in calculators, computers, and many other arithmetic circuits to perform not only addition, but also multiplication, division, and subtraction.

6.0.0 ◆ DECODERS

Decoders available from integrated circuit manufacturers are devices that convert binary, BCD, or some other common code into an uncoded form. Examples of this are the conversion of BCD code received from the arithmetic circuits of a calculator into an output that will turn on one of 10 numbers in a numerical display or the decoding of four binary address bits into an output signal on any one of 16 lines, which access one of 16 words stored in a random access memory (RAM). A category of special integrated circuit decoders, commonly referred to as decoder/drivers, have an output stage in addition to the decoder circuit itself for driving a lamp, a solid-state display, a relay, or another device.

The BCD-to-decimal decoder shown in *Figure 62* has four input lines. Theoretically it could be decoded into 16 unique outputs, but because BCD code uses only the 10 binary counts from 0000 to 1001, the decoder has only 10 output gates.

Another thing to note about the BCD-to-decimal decoder shown is that it does not have a false data rejection capability. If an illegal code (1010 and above) is input into the decoder, the decoder will respond with some output. For example, 1010 will cause the output lines 2 and 8 both to be true. To

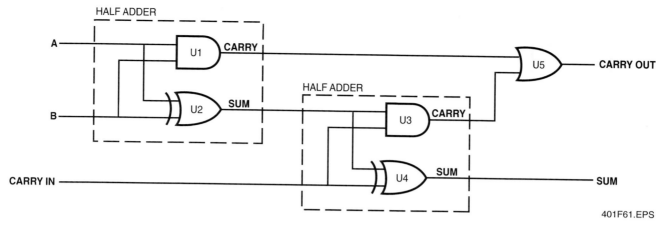

Figure 61 ◆ Full adder.

reject these illegal states, it is necessary to use all five-input AND gates for generating the outputs and connect the gates so that either the true or inverted form of each input signal is supplied to every gate in the decoder. In this way, each AND gate will respond only to one combination of input signals.

The decoder output is often controlled with a strobe line in addition to the inputs to be decoded. The STROBE line in *Figure 62* is connected so that when it is in one logic state, all output lines are disabled, and when it is in the other state, data output can take place. A strobe line does not prevent the input of false data, but it can be used to enable the decoder only when the inputs are known to be valid.

7.0.0 ◆ ANSI/ASQC STANDARDS

The American National Standards Institute (ANSI) provides guidance and conformity in instrumentation drawings and documentation at all stages from design to servicing. The following standards can be used for external quality assurance purposes. These standards are alternative quality assurance models that represent distinct forms of functional or organizational capabilities suitable for two-party contractual purposes.

- *ANSI/ASQC Q91-1987, Quality Systems* – The model for quality assurance in design/development, production, installation, and servicing. For use when conformance to specified requirements is to be assured by the supplier during several stages, which may include design/development, production, installation, and servicing.

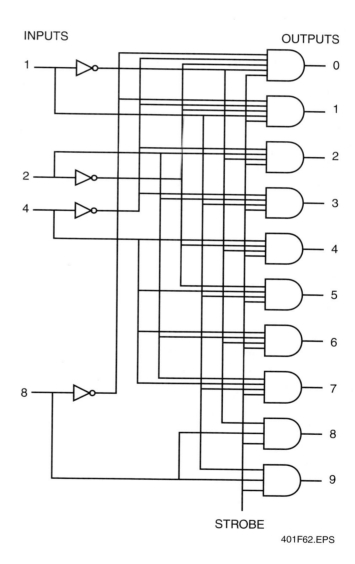

Figure 62 ◆ BCD-to-decimal decoder.

- *ANSI/ASQC Q92-1987, Quality Systems* – The model for quality assurance in production and installation. For use when conformance to specified requirements is to be assured by the supplier during production and installation.
- *ANSI/ASQC Q93-1987, Quality Systems* – The model for quality assurance in final inspection and test. For use when conformance to specified requirements is to be assured by the supplier solely at final inspection and test.

The quality system requirements specified in these standards are complementary (not alternative) to the technical (product/service) specified requirements. These standards are technically equivalent to the International Standards ISO 9001, 9002, and 9003 respectively but incorporate customary American language usage and spelling. It is intended that these standards will be adopted in their present form but can be adapted for specific contractual situations. Q90 provides guidance on adapting these standards as well as selecting the proper quality assurance model.

Summary

There are three basic elements in digital logic: the AND gate, the OR gate, and the inverter. What they do is very simple, but it is essential that you understand them. By interconnecting a number of these gates into circuits, they can perform various increasingly complex functions, such as addition, multiplication, or division of any two numbers, counting, keeping the time of day, and even running a whole computer.

A flip-flop is a device that uses these basic elements for storing information because it has two stable states. It is capable of being in either a high (1) or low (0) state.

Many different types of flip-flops have been presented, and you should now have the knowledge to identify the limitations of each. Each flip-flop presented uses the basic RS latch and modifies it in a different manner so that each has its own particular characteristics and serves a slightly different purpose in logic circuits.

This module has presented the knowledge needed to identify various types of counters and to determine their operation. The relationships and principles should be understood and are listed below:

- *Up counter* – The output of one flip-flop triggers the next at its CLK input.
- *Down counter* – The Q output of each flip-flop causes the following flip-flop to toggle.
- *Up/down counter* – This can count in either direction. The mode of counting is selected by a control signal through logic gates.
- *Synchronous counter* – All flip-flops are clocked by a common clock signal and all change states at the same time.
- *BCD counter* – This counts to 10 then resets itself and starts over again.
- *Ring counter* – This is a shift register used as a counter where the output is connected back to its own input.
- *Johnson counter* – The Q output of the last flip-flop is connected back to the J input of the first.
- *Programmable counter* – The counting pattern can be modified in some way by a control signal rather than by circuit changes.

Shift registers are very important in applications involving the storage and transfer of data in a digital system. A register, in general, is used solely for the purpose of storing and shifting data (1 and 0) entered into it from an external source and possesses no characteristic internal sequence of states.

The flip-flop is the basic element of all shift registers. The storage capability of a register is one of the two basic functional characteristics that make it an important type of memory device. The flip-flops within a register permit the movement of data stored from stage to stage within the register or into or out of the register upon application of clock pulses.

Shift registers can easily convert serial input data to a parallel output or convert parallel input data to a serial output. They can also be used to shift data in either the left or right direction.

1. A(n) _____ gate has an output of a logic 1 only if both of its inputs are a logic 1.
   a. OR
   b. AND
   c. NOR
   d. NAND

2. If several gates are arranged in series, their propagation delays are subtractive.
   a. True
   b. False

3. The logic equation for an OR gate is _____.
   a. $C = A + B$
   b. $(A)(B) = C$
   c. $A = B$
   d. $A = B - C$

4. The Boolean equation for an amplifier is _____.
   a. $C = A + B$
   b. $(A)(B) = C$
   c. $A = B$
   d. $A = B - C$

5. The simplest element of digital logic is the _____.
   a. AND gate
   b. OR gate
   c. amplifier
   d. inverter

6. An AND gate with an inverter on its output is known as a(n) _____.
   a. amplifier
   b. NOR gate
   c. NAND gate
   d. flip-flop

7. A(n) _____ gate has a 0 output when any of its inputs are a logic 1.
   a. OR
   b. NAND
   c. NOR
   d. XOR

8. A(n) _____ gate has a logic high output when either of its inputs are high but not when both are high.
   a. OR
   b. NAND
   c. NOR
   d. XOR

9. A reset-set gate is a type of _____.
   a. amplifier
   b. NOR gate
   c. NAND gate
   d. flip-flop

10. A D latch can be used to prevent _____.
    a. a race condition
    b. synchronous operation
    c. edge triggering
    d. complementary operation

11. A shift register consists of a series of _____.
    a. inverters
    b. flip-flops
    c. NAND gates
    d. NOR gates

12. A(n) _____ counter is a simple device that can be used to count up in binary code.
    a. ripple
    b. parallel
    c. synchronous
    d. ring

13. Calculators typically use _____ counters.
    a. BCD
    b. ripple carry
    c. down
    d. ring

14. An arithmetic element that can perform addition, subtraction, multiplication, and division is known as a(n) _____.
    a. adder
    b. full adder
    c. counter
    d. augend

15. ANSI standards Q91, Q92, and Q93 are technically equivalent to International Standards ISO 9001, 9002, and 9003.
    a. True
    b. False

Trade Terms Introduced in This Module

Amplifier: A device whose output assumes the high state if and only if the input assumes the high state.

AND gate: A device whose output is a logic 1 only when both of its inputs are a logic 1.

Asynchronous: Counters in which the flip-flops change state one at a time.

Combination logic: A circuit with any combination of the different logic gates.

Exclusive OR gate: A device whose output is a logic 1 when either of its inputs is a 1 but not when both are high.

Flip-flop: Any of a number of different types of digital circuits that always has its two outputs in opposite logic levels.

Inverter: An electronic device that changes a digital signal from 1 to 0 or 0 to 1.

MOSFET: Metal oxide semiconductor (MOS) field-effect transistor (FET); a solid-state device often used as a switch.

NAND gate: An AND gate with an inverter on its output.

Nanosecond (ns): One billionth of a second.

Negative-going: When the signal represented by a waveform is traveling to the negative or downward direction.

NOR gate: An OR gate with an inverter on its output.

OR gate: A device whose output is a logic 1 if either of its inputs are a logic 1.

Positive-going: When the signal represented by a waveform is traveling to the positive or upward direction.

Propagation delay: The time interval between the application of an input pulse to a gate and the occurrence of the output pulse.

Race condition: When two digital gates race each other to change states, and the output cannot be predicted.

Timing diagram: A symbolic representation for logic states as they change with time.

Trailing edge triggered: In clock-pulsed circuits, the change of state takes place at the end of the pulse instead of the beginning.

Transistor-transistor logic (TTL): Where low is 0 volts and high is 5 volts.

Truth table: Shows the different input and output combinations of digital gates and circuits.

Additional Resources

This module is intended to be a thorough resource for task training. The following reference works are suggested for further study. These are optional materials for continued education rather than for task training.

Bebop to the Boolean Boogie, 1995. Clive Maxfield. Solana Beach, CA: HighText Publications, Inc.

Digital Fundamentals, 1997. Thomas L. Floyd. Englewood Cliffs, NJ: Prentice Hall, Inc.

NCCER CURRICULA — USER UPDATE

NCCER makes every effort to keep its textbooks up-to-date and free of technical errors. We appreciate your help in this process. If you find an error, a typographical mistake, or an inaccuracy in NCCER's curricula, please fill out this form (or a photocopy), or complete the online form at **www.nccer.org/olf**. Be sure to include the exact module ID number, page number, a detailed description, and your recommended correction. Your input will be brought to the attention of the Authoring Team. Thank you for your assistance.

Instructors – If you have an idea for improving this textbook, or have found that additional materials were necessary to teach this module effectively, please let us know so that we may present your suggestions to the Authoring Team.

NCCER Product Development and Revision

13614 Progress Blvd., Alachua, FL 32615

Email: curriculum@nccer.org
Online: www.nccer.org/olf

❏ Trainee Guide ❏ AIG ❏ Exam ❏ PowerPoints Other _____

Craft / Level: _____ Copyright Date: _____

Module ID Number / Title: _____

Section Number(s): _____

Description: _____

_____

_____

_____

_____

Recommended Correction: _____

_____

_____

_____

_____

Your Name: _____

Address: _____

_____

Email: _____ Phone: _____

Instrument Calibration and Configuration

COURSE MAP

This course map shows all of the modules in the fourth level of the Instrumentation curriculum. The suggested training order begins at the bottom and proceeds up. Skill levels increase as you advance on the course map. The local Training Program Sponsor may adjust the training order.

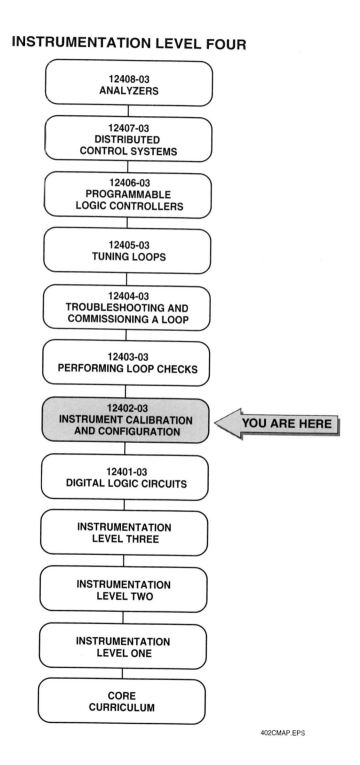

INSTRUMENTATION LEVEL FOUR

12408-03
ANALYZERS

12407-03
DISTRIBUTED
CONTROL SYSTEMS

12406-03
PROGRAMMABLE
LOGIC CONTROLLERS

12405-03
TUNING LOOPS

12404-03
TROUBLESHOOTING AND
COMMISSIONING A LOOP

12403-03
PERFORMING LOOP CHECKS

12402-03
INSTRUMENT CALIBRATION
AND CONFIGURATION   ◁ YOU ARE HERE

12401-03
DIGITAL LOGIC CIRCUITS

INSTRUMENTATION
LEVEL THREE

INSTRUMENTATION
LEVEL TWO

INSTRUMENTATION
LEVEL ONE

CORE
CURRICULUM

402CMAP.EPS

MODULE 12402-03 CONTENTS

1.0.0 **INTRODUCTION** 2.1

2.0.0 **DEFINING CALIBRATION** 2.1

    2.1.0 Measured Variables (Input Energies) 2.2

    2.2.0 Signals (Output Energies) 2.2

    2.3.0 Five-Point Method of Calibration 2.3

3.0.0 **PNEUMATIC CALIBRATION EQUIPMENT AND CALIBRATING PROCEDURES** 2.4

    3.1.0 Calibrating Pneumatic Differential Pressure Transmitters 2.5

    3.1.1 Principles of Operation 2.5

    3.1.2 Using the Wally Box® 2.7

    3.2.0 Electro-Pneumatic and Electro-Mechanical Temperature Transmitters 2.9

4.0.0 **ANALOG CALIBRATION EQUIPMENT AND CALIBRATING PROCEDURES** 2.9

    4.1.0 Analog Differential Pressure Transmitters 2.10

    4.1.1 Hydrostatic Pressure Head 2.11

    4.1.2 Open Vessel Installations (Level) 2.11

    4.1.3 Closed, Pressurized Vessel Installations (Level) 2.12

    4.1.4 Rosemount 1151DP Alphaline® Zero and Span 2.13

    4.2.0 Temperature Transmitters with Analog Output 2.14

5.0.0 **SMART TRANSMITTERS** 2.15

    5.1.0 HART® Communication and Communicator 2.15

    5.2.0 HART® Device Calibration 2.16

    5.2.1 The Input Stage 2.17

    5.2.2 The Output Stage 2.17

    5.2.3 Performance Testing 2.17

    5.2.4 Other Test and Calibration Considerations 2.17

6.0.0 **TRANSDUCERS** 2.18

7.0.0 **CONTROL VALVE POSITIONERS** 2.19

    7.1.0 Pneumatic and Electro-Pneumatic Positioners 2.20

    7.2.0 Smart Positioners (Digital Valve Controllers) 2.22

SUMMARY 2.23

REVIEW QUESTIONS 2.23

GLOSSARY 2.25

REFERENCES & ACKNOWLEDGMENTS 2.26

Figures

Figure 1   Wallace & Tiernan® Model 65-2000 II
            pneumatic calibrator 2.5

Figure 2   Wally Box® calibrator control panel 2.5

Figure 3   Pneumatic DP transmitter 2.6

Figure 4   Pneumatic DP transmitter connected for
            flow measurement 2.6

Figure 5   Siemens® Series 50 pneumatic differential
            pressure transmitter 2.7

Figure 6   Foxboro® 12A Series pneumatic
            temperature transmitter 2.9

Figure 7   Fluke® Model 725 loop calibrator 2.10

Figure 8   Typical DP flow transmitter installations
            (Rosemount Model 1151DP) 2.10

Figure 9   A level transmitter mounted below
            the minimum level 2.11

Figure 10   An open tank installation with a suppressed zero ... 2.11

Figure 11   Compensating for specific gravity of
            liquid in wet leg 2.12

Figure 12   Elevated zero in a wet leg installation 2.13

Figure 13   Location of zero and span adjustment screws
            for a Rosemount Model 1151 transmitter 2.13

Figure 14   Fluke® Model 724 temperature calibrator 2.15

Figure 15   HART® communicator 2.16

Figure 16   Principles of operation for a HART®
            transmitter device 2.17

Figure 17   Cutaway of an I/P transducer 2.19

Figure 18   Mechanical link of a valve actuator
            stem to a positioner 2.19

Figure 19   Operation schematic for a Fisher® 3582 Series
            pneumatic positioner 2.20

Figure 20   Fisher® 3582i Series electro-pneumatic
            positioner 2.21

Figure 21   FIELDVUE® DVC6000 Series digital valve
            controller block diagram 2.22

Instrument Calibration and Configuration

Objectives

When you have completed this module, you will be able to do the following:

1. Define calibration.
2. Discuss the three-point and five-point methods of calibration.
3. Calibrate the following pneumatic instruments using the proper equipment:
   - Differential pressure transmitters
   - Temperature transmitters
4. Calibrate the following 4–20mA instruments using the proper calibration equipment:
   - Differential pressure transmitters
   - Temperature transmitters
5. Define smart instruments.
6. Identify a HART® communicator.
7. Calibrate a smart transmitter using a HART® communicator.
8. Calibrate a transducer.
9. Calibrate the following valve positioners:
   - Pneumatic positioner
   - Electro-pneumatic positioner
   - Smart positioner (digital valve controller)

Prerequisites

Before you begin this module, it is recommended that you successfully complete the following: Core Curriculum; Instrumentation Levels One through Three; Instrumentation Level Four Module 12401-03.

Required Trainee Materials

1. Pencil and paper
2. Appropriate personal protective equipment

1.0.0 ◆ INTRODUCTION

All instruments that provide a proportional output signal based on an input signal or level of energy must be accurately **calibrated** in order to function properly within the loop. Therefore, calibration represents the primary task of any instrumentation technician.

The same methods of calibration for a particular type of instrument can often be shared across the lines of the various process control variables or energies including pressure, flow, level, and temperature. In some cases, however, unique methods must be applied to the same type of instrument because of its function in the loop.

Instruments that require calibration generally fall within three signal groups: pneumatic, analog, and smart instruments. Although digital instrumentation associated with smart instruments and the incorporation of PLCs and fieldbuses have become commonplace in instrument loops, existing pneumatic and analog instrumentation still exists in the industry and must be calibrated. Some examples of control loop instruments that require calibration are transmitters, transducers, and valve positioners.

2.0.0 ◆ DEFINING CALIBRATION

All instruments that require calibration share at least two common characteristics: they have some type of input signal or energy, and they use that input signal or energy to either develop an output signal or convert the energy into some type of other energy (mechanical, electrical, or pneumatic).

Calibration consists of proportionally matching the output signal or level of energy to the input signal or level of energy. This is accomplished by either using proper loop calibration test equipment that provides a variable energy source while reading the output signal using the proper measurement meter or by using **multi-function loop calibration equipment**.

2.1.0 Measured Variables (Input Energies)

Instruments that are in direct contact with the process are designed and manufactured to operate within a specified process input energy range. The input energy may be pressure, temperature, voltage, current, sound decibels, or other forms of energy; however, because the level of this process energy is typically the measured variable from the process, it is not generally a regulated variable at the instrument, and only the simulated input signal should be varied during calibration.

For instance, transmitters are usually designed and manufactured to operate within ranges such as 0–200 **inches of water** (" H_2O) or 0–100 psi. Some operating ranges may include a maximum operating limit and may also specify a minimum operating limit other than zero, such as 60–400" H_2O. The operating range is included on the tag of the instrument, and the instrument must not be installed in any process in which the measured variable does not normally operate within the specified range. If the instrument is installed in a process variable that either exceeds or falls below the specified operating range limits, the instrument will not provide a proportional output.

WARNING!
Exceeding the maximum operating range limit may damage the instrument and can result in personnel injury due to instrument disintegration. Never expose an instrument to a greater maximum operating range limit than that specified by the manufacturer.

2.2.0 Signals (Output Energies)

Output energies from instruments are the usable signals that represent the primary function of the instrument. These output signals also operate within a range, such as 0–60 psig, 3–15 psi, 4–20mA, 1–5VDC, and others; but unlike the input signal or energy, the output signals on most instruments may and should be adjusted (calibrated) so that they proportionally represent the input signal or level at any given point.

NOTE
Some alarming instruments, pressure gauges, or other indicators may not have adjustable output signals or levels; therefore, they cannot be calibrated.

Whenever you adjust the output signal or level to be truly representative of the input signal or level being received into the instrument, you have calibrated the instrument. Two terms that are often misunderstood are range and **span**. The output range of an instrument includes its lower limit and its upper limit, such as 3–15 psi, 0–5 volts, and so on, whereas the span is the difference between the upper limit and the lower limit. In the case of a 3–15 psi range, the span is 12 psi. In calibration, you must apply an input value to an instrument that is representative to a certain percentage of the input span. For instance, if a process operates within a range of 40–250" H_2O, it is said to have a range of 40–250" H_2O and a span of 210" H_2O (250 – 40 = 210). In order to deliver a representative value to the transmitter that is equivalent to 50% of the input span, the calibration equipment must deliver 145" H_2O instead of 105" H_2O because the zero starts at 40" H_2O, so (210 ÷ 2) + 40 = 145. If the transmitter then has a pneumatic output range of 3–15 psi, it has a span of 12 psi because 15 – 3 = 12. But in order to produce an output value that is representative of 50% of the input span, the instrument must deliver 9 psi instead of 6 psi at its output because the output zero starts at 3 psi, and (12 ÷ 2) + 3 = 9.

A process input from a detector or sensor has an operating range, which is the range that suits the process. The minimum value of the process range is considered as a zero input to the instrument. This value may not always be true zero—often it is not—but it still represents zero input to the instrument. When you **zero** the instrument, you are setting its output value at its minimum range limit, which represents a zero input from the process. For example, in a process range of 20–150" H_2O, the zero equivalent is 20" H_2O. If you are using a transmitter that has an input range of 20–150" H_2O and a pneumatic output range of 3–15 psi, a 20" H_2O input to the instrument must produce a transmitter output of 3 psi. When you set the instrument for this value you are zeroing the instrument. You must also set the upper limit of the instrument's output range when the process input is at its maximum limit. In the previous example, the maximum process input of 150" H$_2$O must produce a transmitter output of 15 psi. These are often the only two settings that you must be concerned with in calibrating an instrument; however, adjusting one sometimes affects the other, so instrument calibration requires synchronizing the two settings with each other so that the output span of the instrument is proportional to the process input span. After you have synchronized the lower and upper limits of the transmitter's output range to the upper and lower limits of the process range, you have spanned the instrument.

In some instruments, calibration also requires measuring at other points within the output to make sure the span is synchronized to the input span throughout the entire range. One of two calibration methods is typically used to perform this, depending on the accuracy required and often depending on the choice of the technician: the **three-point method of calibration** in which the instrument output is checked and set against a simulated input signal at three representative percentages of the input span: 0%, 100%, and 50%; and the **five-point method of calibration** in which the instrument output is checked and set against a simulated input signal at five representative percentages of the span: 0%, 100%, 50%, 25%, and 75%.

2.3.0 Five-Point Method of Calibration

When applying the five-point method of calibration to an instrument, the output of the instrument is checked for accuracy and proportionality to the input signal when its value represents 0%, 100%, 50%, 25%, and 75% of the span, in that order. This often requires numerous rechecks and recalibrations of the previous settings and should be performed by zeroing and spanning the instrument's output first (0% and 100% input limits = 0% and 100% output limits) then proceeding to check and set the other three percentages.

The following sequence of steps may be helpful and should be viewed as generic sequential steps for calibrating most instruments:

Step 1   Determine the type of input signal required by the instrument, such as pressure, differential pressure, or temperature, and the equivalent energy, if any, such as millivolts for temperature, pressure for differential pressure, and so on.

Step 2   Check the process range of the instrument to be calibrated (for example, 0–60 psi or 60–200" H$_2$O) to make sure it corresponds to the process input.

Step 3   Acquire the proper calibration equipment that can provide the proper input signal at the required operating range as specified by the instrument.

Step 4   Using the calibration equipment, apply the input signal to the instrument that represents the minimum limit of the operating range.

Step 5   For point one (0%), measure and set the output signal of the instrument being calibrated to correspond to the minimum limit of the output range, as listed on the instrument. For example, use the limit 4mA in a 4–20mA output signal, 3 psi in a 3–15 psi output signal, and 1V in a 1–5VDC output signal.

Step 6   Using the calibration equipment, apply the input signal to the instrument that represents the maximum limit of the operating range.

Step 7   For point two (100%), measure and set the output signal of the instrument being calibrated to correspond to the maximum limit of the output range, as listed on the instrument. For example, use the 20mA limit in a 4–20mA output signal; 15 psi in a 3–15 psi output signal; and 5V in a 1–5VDC output signal.

NOTE
It is common in instrument calibration for the adjustment of one parameter to affect another parameter that had been adjusted previously. It is therefore highly recommended that you repeat Steps 4 through 9 until no readjustment is necessary.

Step 8   Using the calibration equipment, apply the signal to the instrument that represents 50% of the input span.

NOTE
Always make sure you know what value represents 50% of the input span. For instance, in an operating range of 0–200, 50% of the span is obviously represented by 100; but in an operating range of 60–200, 50% of the operating span is represented by 130. The overall span is equal to 140, which is 200 – 60. Because the zero point starts at 60, you must add 60 to 50% of the span (0.5 × 140 = 70, and 70 + 60 = 130), so then 50% of the operating span is represented by an input to the transmitter of 130.

Step 9 For point three (50%), measure and set the output signal of the instrument being calibrated to correspond to 50% of the output span, as listed on the instrument. For example, in a 4–20mA output range, the span is 16mA (20 − 4 = 16), but because the output zero is 4mA, you must add one-half of the span to the zero value (8 + 4 = 12). In a transmitter with 3–15 psi output, 50% of the output span is 9 psi because zero starts at 3 psi.

Step 10 Repeat Steps 4, 5, 6, 7, 8, and 9 until no readjustment is necessary.

NOTE
Steps 1–10 complete the three-point method of calibration.

Step 11 Using the calibration equipment, apply an input to the instrument that represents 25% of the operating span, always remembering where zero starts.

NOTE
If the zero is elevated, meaning the lower limit is more than true zero, you must add that number to one-half of the actual span value. However, if the zero is suppressed, meaning the lower limit is below zero, you must subtract that value from one-half of the actual span value.

Step 12 For point four (25%), measure and set the output signal of the instrument being calibrated to correspond to 25% of the output span, as listed on the instrument. For example, for a 4–20mA output signal, the span is 16, and one-quarter of 16 equals 4, so adding that 4 to the 4mA zero value equals 8mA, so you must use 8mA for the 4–20mA output signal, 6 psi in a 3–15 psi output signal, and 2V in a 1–5VDC output signal.

Step 13 Using the calibration equipment, apply a value that represents 75% of the operating span.

Step 14 For point five (75%), measure and set the output signal of the instrument being calibrated to correspond to 75% of the output span, as listed on the instrument. For example, use 16mA in a 4–20mA output signal, 12 psi in a 3–15 psi output signal, and 4V in a 1–5VDC output signal.

Step 15 Repeat Steps 4 through 14 until no readjustment is necessary.

NOTE
Do not confuse calibrating an instrument with tuning. Tuning has to do with controllers, requires many more parameters to be set, and is covered in another training module in this series.

3.0.0 ◆ PNEUMATIC CALIBRATION EQUIPMENT AND CALIBRATING PROCEDURES

Pneumatic instrumentation is being replaced by newer technology, such as analog and digital instrumentation, including computer-based control systems. However, many industrial facilities are still dotted with pneumatic control loops that have performed for many years and are not high on the priority list for upgrades, either because of their non-critical application or because they have performed well in their function. In order to calibrate these instruments, the proper pneumatic calibration equipment must be available.

Pneumatic calibration equipment must be able to simulate the process input to the instrument being calibrated and measure the pneumatic output signal of the instrument. If one piece of equipment can perform both functions, it is referred to as multi-function pneumatic calibration equipment.

The most commonly used multi-function pneumatic calibration equipment is the pneumatic calibrator manufactured by the Wallace & Tiernan Company, often referred to as the Wally Box®. The Wallace & Tiernan® Model 65-2000 II pneumatic calibrator, one of the latest models of the Wally Box®, is shown in *Figure 1*, and its panel is shown in *Figure 2*.

When this model of the Wally Box® is used as a calibrator in a purely pneumatic loop, it can simulate a variable process pressure input to the instrument being calibrated by connecting an air supply to the air supply port on the calibrator and regulating it with the regulating knob shown in *Figure 2*. This can also be done using the manual air pump built into the equipment.

This model is equipped with an internal power supply that uses a rechargeable 24VDC nickel-cadmium battery pack, or an AC adapter that simultaneously charges the battery pack, that can supply loop power in order to calibrate equipment that uses both electronic and pneumatic signals, such as a transducer. Applying the Wally Box® to specific applications will be covered in the sections covering those instruments.

Older models of the Wally Box® do not include a digital readout (gauge only), hand pump, or 24V power supply. However, companies are available that offer these upgrades to older models of the Wally Box® and are accessible via the Internet by searching for "Wallace & Tiernan Upgrades."

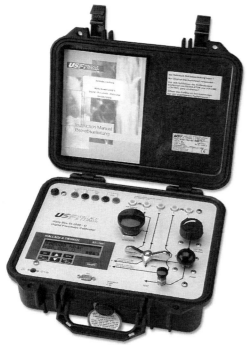

402F01.EPS

Figure 1 ◆ Wallace & Tiernan® Model 65-2000 II pneumatic calibrator.

AIR SUPPLY MUST BE
CONNECTED HERE

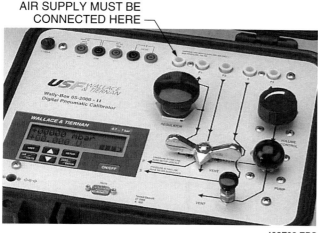

402F02.EPS

Figure 2 ◆ Wally Box® calibrator control panel.

3.1.0 Calibrating Pneumatic Differential Pressure Transmitters

As the word implies, a transmitter transmits or sends a signal. There are several types of pneumatic transmitters. These vary according to the type of process, state of process, variable being monitored, ambient environment, and type of control system. However, the most common type of pneumatic transmitter is the differential pressure (DP) transmitter, as illustrated in *Figure 3*. It can be applied in flow, pressure, and level control applications.

NOTE

Installation and use of DP transmitters in flow, level, and pressure applications were covered in previous levels. Pneumatic differential pressure transmitters can be installed in processes to measure pressure in a line or vessel, flow in a line by measuring the pressure drop across an orifice restriction (differential pressure), or level in tanks or vessels by measuring the difference in pressure between the top and bottom elevations of the tank or vessel.

3.1.1 Principles of Operation

Refer to *Figures 3* and *4* as you follow the principles of operation of a pneumatic DP transmitter installed as a flow transmitter. A pneumatic DP transmitter has a two-sided differential pressure capsule that receives two different levels of pressure and converts the difference between the two pressures into a mechanical movement via a force bar. The movement of the force bar operates a balance beam by pivoting it on a **fulcrum**. A flat plate called a flapper is attached to the balance beam and moves against the opening of a nozzle that is continuously venting air. As the flapper partially closes off the nozzle, backpressure is created in the tubing connected to the nozzle, which is connected on the other end to the top side of a diaphragm in a pneumatic relay. As the backpressure increases and decreases on the diaphragm based on the position of the flapper against the nozzle, the pneumatic relay proportionally regulates a pneumatic output in the relay by means of a spring-loaded valve. The most common pneumatic output signal range is 3–15 psi, which is usually sent to a controller, recorder, gauge, indicator, or pneumatic-operated alarm.

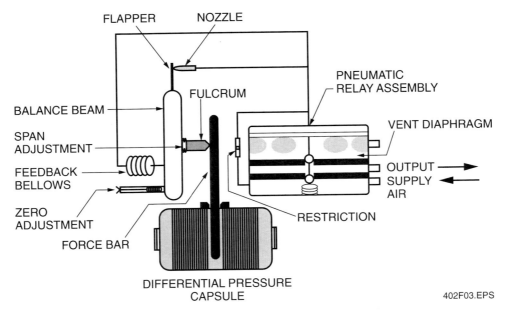

Figure 3 ◆ Pneumatic DP transmitter.

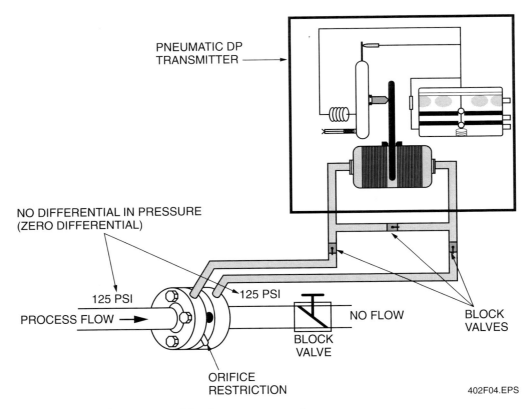

Figure 4 ◆ Pneumatic DP transmitter connected for flow measurement.

NOTE

The pneumatic DP transmitter requires a regulated instrument air supply in order to operate the pneumatic relay and provide air to the nozzle assembly.

NOTE

Calibration should always be performed with the instrument control loop isolated from the process.

Figure 5 ◆ Siemens® Series 50 pneumatic differential pressure transmitter.

Figure 5 shows a pneumatic differential pressure transmitter. This particular series of transmitters, Siemens® Series 50, is available in models having input range and span limits of 0–50" H_2O (range), 5–50" H_2O (span); 0–225" H_2O (range), 20–225" H_2O (span); and 0–850" H_2O (range), 150–850" H_2O (span). It can be applied in pressure, flow, or level applications. Notice that the spans of all three models have minimum span limits that are above the zero minimum range limits. These minimum span limits are referred to as elevated zeros and are necessary in order to make the instrument accurate at the low end of the range. Therefore, the span limits designate both the zero and 100% (span) input levels during calibration.

Differential pressure transmitter capsules are divided into two sides, with one side designated as the high side and the other designated as the low side. The difference in pressures between the two sides is considered the differential pressure and is the process measurement that determines the output signal of the transmitter. When installed as a flow transmitter, the high side is always connected to the upstream side of the orifice restriction, and the low side is always connected to the downstream side of the restriction. If the transmitter is being used to measure pressure in a line or vessel, the low side is vented to the atmosphere. In these applications, the transmitter still measures differential pressure between the high and low sides; it's just that the low side is always zero pressure.

In order to calibrate the transmitter when it is isolated from the process, using either the three-point or five-point method as previously discussed, you must simulate these values of differential pressure within the span limits of the instrument. For instance, in the three examples of the Siemens® Series 50 transmitters, you would set the output pneumatic signal of the transmitter at an equivalent zero output or 3 psi whenever you supply a differential pressure between the two sides of 5" H_2O, 20" H_2O, or 150" H_2O, depending on which one of the models you are calibrating. Of course, for 100% of the span, you would supply a differential pressure of 50" H_2O, 225" H_2O, or 850" H_2O. For the other points of calibration between 0% and 100% of the span, whether three or five points, you would determine the actual range of the span and figure the differential pressures at each of those points. The range of the span for 5–50" H_2O is 45" H_2O (50 – 5 = 45); for the 20–225" H_2O, it is 205 (225 – 20 = 205); and for the 150–850" H_2O, it is 700" H_2O (850 – 150 = 700).

Because you are looking for only the differential pressure between the two sides, it is not necessary to supply pressure to both the high and low sides of the transmitter in order to create a difference between the two sides. Suppose you open (vent) both sides of the capsule to the atmosphere. What is the differential pressure between the two sides at this point? Zero, of course, but in the case of the Siemens® Series 50 DP transmitters, you cannot set the transmitter's output signal at 3 psi (zero equivalent) with an input differential pressure of zero because these instrument spans have elevated zeros. You can leave the low side vented to the atmosphere, however, and supply a pressure to the high side of the capsule that is equal to the minimum pressure of the span shown on the instrument. In the cases of the three Siemens® Series 50 transmitters, it will be 5" H_2O, 20" H_2O, or 150" H_2O, depending on which of the transmitters you are calibrating.

3.1.2 Using the Wally Box®

The calibration process of the pneumatic DP transmitter has been narrowed down to the following two main tasks:

- Provide regulated pressure according to the span of the instrument to only the high side of a pneumatic differential pressure transmitter, while venting the low side in order to provide an accurate differential pressure level.
- Measure the transmitter's pneumatic output signal and adjust it so that it is proportional to and represents the differential pressure input as dictated by the span of the instrument.

Use the following procedure to connect and calibrate an in-service pneumatic differential pressure transmitter with a Wallace & Tiernan® pneumatic calibrator (Wally Box®):

Step 1   Isolate the transmitter from the process by coordinating the isolation process with the proper operations personnel. (This usually involves operating a loop controller in the manual position.)

Step 2   Follow drain and blowdown procedures at the process connections to the transmitter.

Step 3   Verify that the transmitter is isolated from the process.

Step 4   Equalize the static pressure in both sides of the transmitter's capsule by opening the equalizer valve between the high and low sides.

Step 5   Remove the calibration pipe plugs (usually ⅛" or ¼" NPT) from each side of the transmitter.

Step 6   Install the proper tubing connector to accommodate the test tubing supplied with the Wally Box® into the high side's calibration pipe plug hole.

Step 7   Close the equalizer valve between the high and low sides.

Step 8   Again, verify that the two block valves to the process taps are securely closed (high- and low-side manifold valves).

Step 9   Verify that the instrument air supply to the transmitter remains on.

Step 10   Turn the large, black, air regulator knob on the Wally Box® fully counterclockwise in order to set the internal air regulator's output to zero or no output.

Step 11   Connect clean, dry, regulated instrument air to the air supply port on the Wally Box®.

Step 12   Insert one end of the test tubing supplied with the Wally Box® into the P1 port.

Step 13   Connect the other end of this test tubing to the transmitter's high-side calibration tubing connector.

Step 14   Disconnect the transmitter's pneumatic output signal line and cover the open end of the tubing to protect it.

Step 15   Connect one end of the test tubing supplied with the Wally Box® from the S port on the Wally Box® to the transmitter's output tubing connector.

Step 16   Set the multi-port valve handle on the Wally Box® so that the readout on the calibrator reads the pressure that will be applied to the P1 port.

Step 17   Verify the minimum differential pressure of the transmitter's span.

Step 18   Slowly turn the Wally Box® regulator clockwise and watch the readout.

Step 19   Continue to increase the air pressure to P1 using the regulator knob until the minimum differential pressure is shown on the readout.

Step 20   Switch the multi-port valve handle on the Wally Box® so that the readout on the calibrator reads the incoming signal from the transmitter connected to port S.

Step 21   On a 3–15 psi pneumatic signal range, this signal should be 3 psi, which represents the minimum or elevated zero differential pressure being applied to the transmitter's high side.

Step 22   Turn the multi-port valve handle back to the P1 port reading to verify that it has remained at the exact minimum value. Readjust if necessary.

Step 23   Turn the multi-port valve handle back to the S port reading.

Step 24   Locate and turn the zero adjustment screw on the transformer until exactly 3 psi is showing on the readout.

Step 25   Repeat Steps 22 through 24 until no readjustments are necessary.

Step 26   Now apply either the three-point or five-point method of calibration as previously discussed, making sure to switch back and forth between the P1 and S ports to verify that nothing has changed. The transmitter is now calibrated.

3.2.0 Temperature Transmitters

Most older types of temperature control loops that require a pneumatic signal to operate use a temperature transmitter that operates on the principle of differential pressure, with the low side of the capsule open to the atmosphere. *Figure 6* shows one such model, a Foxboro® 12A Series pneumatic temperature transmitter. However, the high-side pressure is generated by a gas-filled tube-and-bulb system (capillary) in which the tube is connected and sealed to a special differential pressure transmitter capsule. The tube and bulb are filled with a gas, such as helium, that expands and contracts with changes in temperature. The bulb acts as the primary detecting element and is inserted directly into the process. As the temperature of the process changes, the gas within the bulb expands or contracts in proportion to the temperature change. The changing pressure of the gas is applied to the high side of the capsule, causing the force-balance mechanism in the transmitter to respond accordingly.

The only field calibration that can be performed on an electro-mechanical temperature transmitter is a range change by elevating or suppressing the zero, using a screwdriver to make the adjustments. In the Foxboro® 12A Series transmitter, the instrument zero can be elevated or suppressed for 100% of the span. However, additional elevation or suppression beyond 100% of the span may require a change in the ambient compensating bellows. To change the span in these situations, the thermal system, range spring, and/or compensating bellows must be changed. This type of change should be performed in the clean environment of a lab or shop and is considered a bench calibration.

Another bench calibration check can be performed on a bulb-type instrument by providing a temperature-controlled bath in which the bulb is inserted to simulate the process, while monitoring the pneumatic output of the signal using a Wally Box® or equivalent pneumatic measuring equipment. As with all pneumatic instruments, the Foxboro® 12A Series temperature transmitter requires its own dry, regulated instrument air supply.

4.0.0 ◆ ANALOG CALIBRATION EQUIPMENT AND CALIBRATING PROCEDURES

An analog signal is a voltage, current, or other form of signal that is continuously variable with time over a range from a minimum value to a maximum value. Some systems, such as PLC, DCS, and Fieldbus technology systems, communicate using digital signals. Digital signals are signals in binary logic form that represent the equivalent analog form. For example, an 8 bit signal will be used to represent a DC voltage that varies from 0 to 255 volts. An 8 bit binary number has 256 possible numbers from 00000000 to 11111111. Therefore the binary number can represent the DC voltage from 0 to 255 in 1 volt increments, the minimum resolution possible with only 8 bits. Analog-to-digital converters (A/D) are used to convert the analog signals into digital form, and determine the number of bits and resolution of the digital signal.

In instrumentation, the most common form of analog signal that is applied in control loops is the current signal, specifically the 4–20 milliampere signal, with 4mA representing the minimum or zero end of a range or span and 20mA representing the maximum or 100% of a range or span. For example, if the process span of a transmitter is 20–200" H_2O, 20" H_2O (0%) of the process should provide a 4mA output when the transmitter is properly calibrated, 110" H_2O (50%) should provide 12mA, and 200" H_2O (100%) should provide a 20mA output under the same calibration settings.

NOTE

There are many brands and models of analog calibration equipment available. Specific brands or models represented in this module are only representative of the type of equipment recommended for calibration, and equivalent brands and models may be used in place of these devices.

One relatively new multi-function calibrator that provides all the necessary functions needed to calibrate an analog transmitter, regardless of the form of input signal, is the Fluke® Model 725 loop

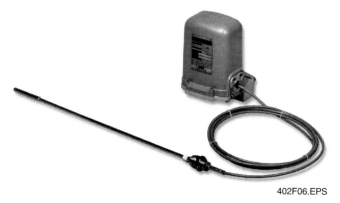

402F06.EPS

Figure 6 ◆ Foxboro® 12A Series pneumatic temperature transmitter.

calibrator shown in *Figure 7*. The Fluke® 725 is capable of measuring an analog output signal while providing a source of signals to transmitters, including volts, mA, millivolts, frequency, and ohms. If coupled with any one of 28 pressure modules designed to operate with the Model 725, it is also capable of measuring pressure and differential pressure, as well as providing a pneumatic source of energy. In any of the calibration modes, and when used with the proper accessory modules, the Model 725 is capable of providing the source and measuring the signal at the same time.

This calibrator can also provide an auto-step or auto-ramp source in which the source function can be automatically set so that it provides a complete range or span in incremental steps, such as 4mA, 9mA, 12mA, 16mA, and 20mA. It may also be programmed to provide automatic steps of source pressure based on any range or span. This feature allows the technician to set the calibrator for auto-step or auto-ramp as an input at one location and check or test other parameters in a different location as the calibrator steps through the various levels of its source.

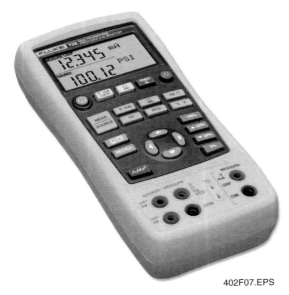

Figure 7 ◆ Fluke® Model 725 loop calibrator.

4.1.0 Analog Differential Pressure Transmitters

Like the pneumatic differential pressure transmitter, the analog DP transmitter can be used to measure flow, pressure, and level, and functions in the exact same manner except for its output, which is analog in nature and not pneumatic.

Many analog signal differential pressure transmitters can be calibrated for suppressed or elevated zero, with some offering up to 500% above suppression and 600% above elevation, depending on the brand and model. The Rosemount Model 1151DP Alphaline® differential pressure transmitter

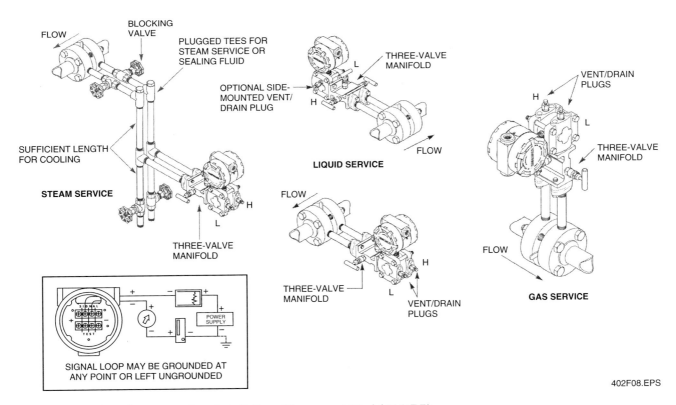

Figure 8 ◆ Typical DP flow transmitter installations (Rosemount Model 1151DP).

can be adjusted for up to 500% suppression or 600% elevation. *Figure 8* shows some typical differential pressure installations of the Model 1151DP, used to measure flows of various processes, including steam, liquid, and gas. Notice the difference between the piping or tubing installations and the location of the taps. These differences in installation are designed to eliminate error caused by condensation building up in the piping or tubing in the processes of gas or steam, and air or gas bubbles in the processes of liquid flow.

4.1.1 Hydrostatic Pressure Head

Differential pressure transmitters used for liquid level applications measure **hydrostatic pressure head**, commonly called head pressure. Factors that must be determined when calibrating DP transmitters in liquid level applications are the liquid level and the specific gravity of the liquid. Hydrostatic pressure head is equal to the liquid's height above the tap multiplied by the specific gravity of the liquid. Pressure head is independent of volume or vessel shape.

4.1.2 Open Vessel Installations (Level)

A differential pressure transmitter can be used to measure level in an open tank. The high pressure side of the transmitter is tubed to the tank and the low pressure side of the transmitter is vented to atmosphere. The transmitter will measure the head pressure above the tank connection point, thereby indirectly measuring the tank level. If the transmitter is mounted below the physical minimum point of the desired level range of the tank, a zero suppression calibration must be performed.

Refer to *Figure 9*, which shows a differential pressure level transmitter (LT) mounted on the side of an open tank or vessel. If the transmitter was mounted at the desired minimum level point, the range of the transmitter would be 0–300". However, in this installation, the transmitter is mounted 75" below the desired minimum level point. This additional hydrostatic pressure must be suppressed by calculating a new range. You can do this by using the formula for head pressure.

Let X = the vertical distance between the minimum and maximum measurable levels (300")

Let Y = the vertical distance between the transmitter's high-side inlet and the minimum measurable level (75")

Let sg = the specific gravity of the liquid in the tank (0.9)

Let h = the maximum head pressure to be measured in inches of water

Let e = the head pressure produced by Y and expressed in inches of water

Let the range = e to e + h

Then h = (X)(sg)

     = 300×0.9

     = 270" H_2O

Then e = (Y)(sg)

     = (75)(0.9)

     = 67.5" H_2O

Range = 67.5" H_2O to 337.5" H_2O

This is shown in graphical form in *Figure 10*.

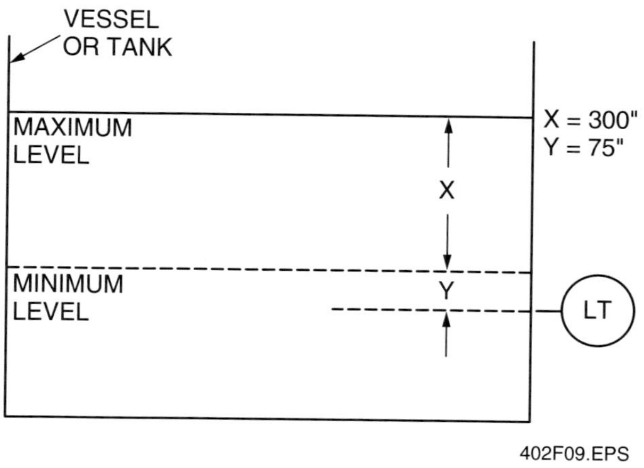

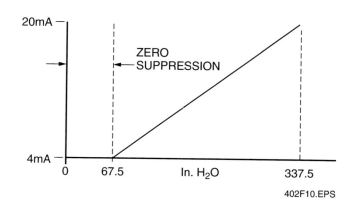

402F09.EPS

Figure 9 ◆ A level transmitter mounted below the minimum level.

402F10.EPS

Figure 10 ◆ An open tank installation with a suppressed zero.

4.1.3 Closed, Pressurized Vessel Installations (Level)

The system pressure above the upper level of a liquid in a closed tank or vessel affects the pressure measured at the bottom of the vessel. The pressure at the bottom of the vessel is equal to the system pressure added to the product of the liquid's specific gravity and height.

Because the differential pressure transmitter is not designed to operate with such overwhelming pressure loaded on one side, the vessel pressure must be subtracted from the pressure at the bottom of the vessel. To accomplish this, both sides of the differential pressure transmitter capsule must be connected to the vessel. The low side of the transmitter is connected to a tap at the top of the vessel, and the high side of the transmitter is connected to a tap at the bottom tap of the vessel. In this installation, the vessel's pressure is applied equally to both the high and low sides of the transmitter, with the resulting differential pressure proportional to the liquid height multiplied by the liquid's specific gravity. The effects of the system pressure are nulled and the transmitter operates as before.

Low-side transmitter piping in a closed, pressurized vessel installation will remain empty as long as the gas above the liquid does not condense. This is referred to as a **dry leg** installation. Procedures applied in determining the range calculations for this type of application are the same as those applied for bottom-mounted transmitters in open vessels or tanks.

If the gas in a closed pressurized vessel condenses, the condensation will eventually fill the low-side piping or tubing. To form a **wet leg** installation, the low-side tubing or piping is purposefully filled with a convenient reference fluid and figured into the calibration in order to eliminate the sure error caused by condensation. This reference fluid will exert a head pressure on the low side of the transmitter capsule, so the zero elevation of the range must be made in order to compensate for its added head pressure. Refer to *Figure 11* as you work through the calculations necessary in zero elevation for a closed pressurized vessel in a wet leg condition.

Figure 11 represents a wet leg installation with the following data:

Let X = the vertical distance between the minimum and maximum measurable levels (325")

Let Y = the vertical distance between the transmitter's high-side inlet and the minimum measurable level (100")

Let Z = the vertical distance between the top of the liquid in the wet leg and the transmitter's bottom tap (575")

Let sg_1 = the specific gravity of the liquid in the tank (1.0)

Let sg_2 = the specific gravity of the liquid in the wet leg (1.1)

Let h = the maximum head pressure to be measured in inches of water

Let e = the head pressure produced by Y and expressed in inches of water

Let s = the head pressure produced by Z and expressed in inches of water

Let the range = $(e - s)$ to $(e + h - s)$

Then h = $(X)(sg_1)$

= 325×1.0

= 325" H_2O

Then e = $(Y)(sg_1)$

= $(100)(1.0)$

= 100" H_2O

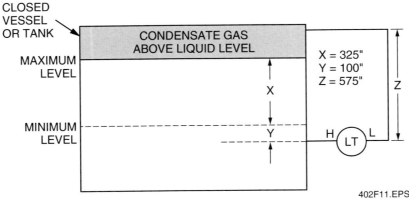

Figure 11 ◆ Compensating for specific gravity of liquid in wet leg.

$$\text{Then } s = (Z)(sg2)$$
$$= 575 \times 1.1$$
$$= 632.5" \; H_2O$$
$$\text{Range} = (e - s) \text{ to } (e + h - s)$$
$$= 100 - 632.5 \text{ to } 100 + 325 - 632.5$$
$$= -532.5 \text{ to } -207.5" \; H_2O$$

This is shown in graphical form in *Figure 12*.

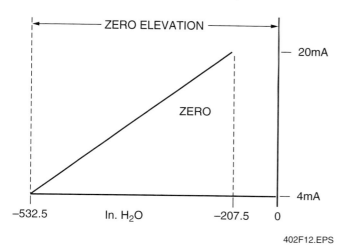

Figure 12 ◆ Elevated zero in a wet leg installation.

4.1.4 Rosemount 1151DP Alphaline® Zero and Span

The span on all Rosemount Model 1151 transmitters is continuously adjustable to allow calibration anywhere between maximum span and one-sixth of the maximum span. For example, the span on a Rosemount Model 1151 Range 4 transmitter can be adjusted between 25" H_2O and 150" H_2O.

The zero on a Rosemount Model 1151 with the E or G output options can be adjusted for up to 500% suppression or 600% elevation. The zero may be elevated or suppressed to these extremes as long as no applied pressure within the calibrated range exceeds the full-range pressure limit. For example, a Range 4 transmitter cannot be calibrated for 100–200" H_2O because 200" H_2O exceeds the 150" H_2O full-range limit of a Range 4.

The zero and span calibration screws are externally accessible behind the nameplate on the terminal side of the electronics housing, as shown in *Figure 13*. To increase the output of transmitter, turn the screws in a clockwise direction. Adjusting the zero adjustment screw does not affect the span setting; however, changing the span adjustment does affect the zero setting. This effect is minimized with those spans that have 0 as their base or

minimum setting. When calibrating for elevated or suppressed zeros, it is much easier on the Model 1151 to make a zero-based calibration first, then compensate for the elevated or suppressed zero by adjusting the zero adjustment screw, because changing it does not affect the span.

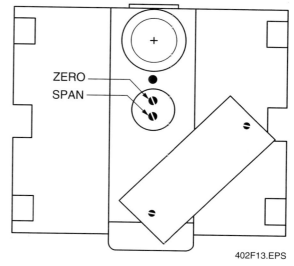

Figure 13 ◆ Location of zero and span adjustment screws for a Rosemount Model 1151 transmitter.

The following calibration procedures are for a Rosemount 1151DP Range 4, but many other types of analog DP transmitters may be calibrated in the same manner. Even though all of the steps are important and must be performed in order, pay close attention to Step 5 to lessen the number of readjustments required due to the span affecting the zero. In this example, the desired calibration is 0–100" H_2O.

Step 1 With zero differential pressure applied to the transmitter, adjust the zero adjustment screw until the output of the transmitter reads 4mA on the Fluke® 725 loop calibrator or equivalent calibration equipment.

Step 2 Apply a differential air pressure of 100" H_2O to the high side of the transmitter, with the low side vented to atmosphere.

Step 3 Adjust the span adjustment screw until the transmitter's output reads approximately 20mA.

Step 4 Release the input differential pressure to zero and readjust the zero output of the transmitter to again read 4mA (±0.032mA).

Step 5 Reapply 100" H_2O to the high side of the transmitter capsule. If the output of the transformer reads greater than 20mA, multiply the difference between the reading and 20mA by 0.25, and subtract the result from 20mA. Adjust the 100% output to this value.

NOTE

If the output of the transformer reads less than 20mA, multiply the difference by 0.25, and add the result to 20mA. Adjust the 100% output to this value.

For example, at 100" H_2O, the transmitter output reads 20.100mA on the calibration equipment; multiplying 0.100 by 0.25 results in 0.025mA, and subtracting 0.025 from 20mA results in 19.975mA. Adjust the transmitter output to 19.975mA with 100" H_2O input differential pressure.

Step 6 Release the pressure to the high-side capsule and readjust the zero setting.

Step 7 Reapply 100" H_2O to the high-side capsule, and repeat Steps 4 through 6 until the full-scale output reads 20mA (±0.032mA).

4.2.0 Temperature Transmitters with Analog Output

Measuring or monitoring temperature plays a key role in many industrial and commercial processes. Examples include measuring cooking temperature in food processing, monitoring the temperature of molten steel in a mill, verifying the temperature in a cold storage warehouse or refrigeration system, or regulating temperatures in the drying rooms of a paper manufacturer.

In an analog temperature control loop, a temperature transmitter uses some type of measuring device to sense the temperature, and then regulates a 4–20mA feedback loop to a temperature control element. The final control element might consist of a temperature control valve (TCV) that opens or closes to allow more steam into a heating process or more fuel to a burner.

The two most common types of temperature sensing devices that provide input to a temperature transmitter are the thermocouple (TC) and resistance temperature detector (RTD). There is a broad range of temperature calibration tools to help you quickly and reliably calibrate temperature instrumentation. Some of the older technology of temperature calibration equipment requires that you use temperature-to-millivolt conversion charts to determine the millivolt source setting for the temperature range of the transmitter. How-

ever, newer calibration equipment, such as Fluke's® line of temperature calibrators, including the Model 724 and 725, allow you to enter the actual temperature limits and type of thermocouple directly into the calibrator. The calibrator then converts the data into the correct millivolt levels.

The Fluke® Model 724 or 725 can be used in calibrating both types of temperature sensing devices—the TC or the RTD. They can perform the following functions pertaining to temperature loops:

- Measure temperature from an RTD probe
- Measure temperature from a TC probe
- Simulate an RTD output
- Simulate a TC output
- Simultaneously simulate an RTD output and read a mA input
- Simultaneously simulate a TC output and read a mA input
- Auto-ramp a temperature signal
- Supply 24VDC loop power to the transmitter

The following example shows how to use the Fluke® 724 temperature calibrator, shown in *Figure 14*, to calibrate a Type K TC transmitter with a range of 0–150°C, generating an output current range of 4–20mA:

Step 1 Connect the 724 test leads to the TC transmitter. The output from the thermocouple jacks on the 724 will simulate a temperature input to the transmitter. The red and black test leads provide loop power to the transmitter and measure the current resulting from temperature changes into the transmitter.

Step 2 Press the calibrator's power button ON.

Step 3 Select the mA button and the LOOP button in order to provide 24VDC to the instrument and measure the transmitter's milliampere output simultaneously.

Step 4 Press the MEAS/SOURCE button until the lower display on the Model 724 indicates the source mode.

Step 5 Depress the TC button until a TC type of K is displayed.

Step 6 Select the °C button for centigrade temperature scale.

Step 7 Set the zero point for this particular application into the calibrator. To do this, set the display initially to 0.0°C. Use the up and down arrow keys to change the output value. Use the left and right

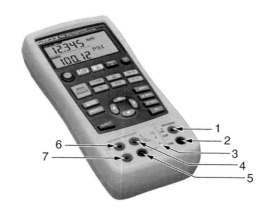

NO.	NAME	DESCRIPTION
1, 2	Measure, mA Terminals	Input terminals for measuring voltage, current, and supplying loop power.
3	TC input/output	Terminals for measuring or simulating thermocouples. This terminal accepts a miniature polarized thermocouple plug with flat, in-line blades spaced 79 mm (0.312 in.) center to center.
4, 5	Source/Measure V, RTD, Ω terminals	Terminals for sourcing or measuring voltage, resistance, and RTDs.
6, 7	Measure 3W, 4W	Terminals for performing 3W and 4W RTD measurements.

402F14.EPS

Figure 14 ◆ Fluke® Model 724 temperature calibrator.

arrows to control the decimal point of the value of the display. When the display reads 0.0, hold down the 0% key on the 724 and observe that 0% is displayed in the lower right corner of the screen. This establishes the zero point for calibration.

Step 8 Set the span point in the calibrator. Set the display to the desired span value for calibration. In this example, the display should read 150°C. Press the 100% key and observe that 100% is displayed in the lower right corner of the screen. This establishes the span point for calibration.

Step 9 To calibrate the temperature transmitter, press the 0% key to source the proper temperature for a 4mA output. Adjust the zero potentiometer until the current reading is 4.00mA.

Step 10 Press the 100% key to source the proper temperature for a 20mA output. Adjust the span potentiometer until the current reading is 20.00mA.

Step 11 Press the 0% key again and adjust the zero potentiometer again, if necessary, to get a 4.00mA output.

5.0.0 ◆ SMART TRANSMITTERS

Smart transmitters operate differently from analog transmitters. A smart transmitter incorporates a microprocessor that stores data about the sensor's specific characteristics in response to pressure and temperature inputs and compensates for these sensor variations. The factory process of generating the sensor performance profile is called **factory characterization**. Today, smart instrumentation transmitters widely use the highway addressable remote transducer (HART®) digital communication protocol to communicate device configuration, status, measurement, and diagnostic information.

5.1.0 HART® Communication and Communicator

The HART® protocol is a globally used open protocol for automation communication among smart instruments. It was developed by Rosemount Inc. in the mid 1980s, but all rights now belong to the independent HART® Communication Foundation, an international organization that supports the protocol and oversees its further development. The HART® protocol can be used with both two- and four-wire HART®-capable devices. Such devices typically include valve positioners and actuators, magnetic flow meters, level devices, liquid and gas analyzers, and control devices.

HART® smart devices are designed to provide a wide range of data accessible using standard commands. Typically, there are about 35 accessible data items standard in every HART® device. These include device status, diagnostic alerts, process variables and units, loop current and percent range, basic configuration parameters, and manufacturer and device information.

The HART® protocol provides two-way digital communications compatible with industry standard 4–20mA analog signaling used by traditional instrumentation equipment. It allows for simultaneous communication of the continuous 4–20mA analog signal with a second digital communications signal superimposed at a low level on top of the analog signal. The digital information, consisting of command and response data, is communicated via American Bell 202 standard frequency shift key (FSK) technology. Data for up to four measurements can be transmitted in a single message. HART® protocol multi-output devices have been developed to take advantage of this. In addition, if only digital communication is used, several devices can be connected in parallel on a single pair of wires. Each device has its own address, so a host device can communicate with each one in turn.

Communication with HART® devices is commonly done using a PC with a compatible HART® modem. It can also be done through HART®-capable multiplexers and I/O systems by using appropriate device drivers. A handheld HART® communicator (*Figure 15*) is commonly used for communication with HART® devices when performing device setup, initial commissioning, and periodic maintenance tasks.

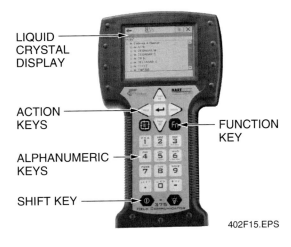

LIQUID CRYSTAL DISPLAY

ACTION KEYS

FUNCTION KEY

ALPHANUMERIC KEYS

SHIFT KEY

402F15.EPS

Figure 15 ◆ HART® communicator.

5.2.0 HART® Device Calibration

As with all devices, HART® devices require periodic calibration. However, unlike many conventional devices, the specific calibration requirements for any particular HART® device depends on its application. Calibration of a HART® device involves the following three functions performed by a microprocessor within the device:

- Calibrating the process variable
- Scaling the process variable
- Producing the process signal

Figure 16 shows the three functional stages of a HART® device microprocessor that are associated with the calibration process. As shown, the input to the first stage is a raw digital count that represents the measured value of the analog process signal being measured. Depending on the device, this measured value can be in millivolts, frequency, capacitance, or some other property. Conversion of the analog process signal to a corresponding digital signal prior to being input to the first stage is done via an analog-to-digital (A/D) converter in the processor.

In the input stage, a built-in equation or lookup table is used to correlate the process signal raw count to the actual property (PV), such as temperature, pressure, or flow. The output of this stage is a digital representation of the measured process signal. This PV value can be accessed and read using a HART® communicator. The form of the lookup table used in this stage is normally provided by the device manufacturer, but most devices also have the capability for making field adjustments. This is called sensor trim.

The second stage converts the digital PV input signal into a corresponding digital milliamp (mA) value representation, with a 4mA value for the lower range and a 20mA value for the upper range. Percent range is also determined in this stage. It does the conversion mathematically, using the range values of the device as they relate to the device zero and span values, in conjunction with the device transfer function. The transfer function is typically linear, but some devices, like pressure transmitters, have a square root function. The output of this stage is a value that can be read with a HART® communicator when reading loop current. It should be pointed out that many HART® devices have the capability to be commanded to a fixed output test mode. This mode causes a specific mA output value to be substituted for the normal output from the second stage.

The output (third) stage converts the calculated mA output value from the second stage into a digital count for subsequent application to digital to

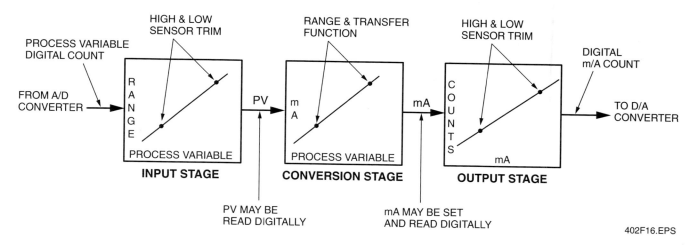

Figure 16 ◆ Principles of operation for a HART® transmitter device.

analog (D/A) converter. There, the actual 4–20mA analog signal is produced such that 0% equals exactly 4mA and 100% equals 20mA. In this stage, internal calibration factors are again used to obtain the correct output signal values. Adjustment of these factors is commonly called current loop trim.

5.2.1 The Input Stage

Calibration of the input stage is required for all device applications. It should be pointed out that if the device is used as a digital device only (current loop output is not used), calibration of the input stage is all that is required. When testing the input stage, a calibrator is used to measure a known input signal, and a HART® communicator is used to measured the resultant output (PV) signal. Because there always is a linear relationship between the input and output signals, and both are recorded in the same measurement units, error calculations are made easier. The test results should meet the accuracy specified by the device manufacturer.

Should the test fail, the stage should be adjusted following the manufacturer's procedure for trimming the input stage. The procedure usually involves one or two trim points. Some devices like pressure transmitters may also have a zero trim, where the stage is adjusted to read exactly zero, instead of low range.

NOTE

Do not confuse a trim with any form of re-ranging or any procedure that involves using zero and span buttons.)

5.2.2 The Output Stage

Calibrating the output stage involves using a HART® communicator to put the device into a fixed current output mode. The test value for the stage input signal is the output mA value that you command the device to produce. A loop calibrator is then used to measure the resulting output current. Because there is a linear relationship between the input and output signals, and both are recorded in milliamps, error calculations are made easier. The test result should meet the accuracy specified by the device manufacturer.

Should the test fail, the stage should be adjusted following the manufacturer's procedure for trimming the output stage. The manufacturer may call this procedure current loop trim, D/A trim, or 4–20mA trim in their service literature. It involves two trim points close to or just outside of 4 and 20mA.

5.2.3 Performance Testing

With both the input and output stages of a HART® device calibrated, it should operate correctly. No calibration is required for the second stage. This is because this stage only performs the calculations necessary to convert the input PV value to a corresponding mA output value. This feature allows the range, units, and transfer functions to be changed without affecting the calibration of the input or output stages.

5.2.4 Other Test and Calibration Considerations

Some other factors you must take into consideration when calibrating HART® devices are described as follows:

- *Damping* – Some HART® devices support a damping parameter that causes a delay between a change in the device input and the detection of that change for output from the device. This delay can exceed the settling time used for the test and calibration of the device. Settling time is the time the test or calibration process waits between setting the input to the device and reading the resulting output. For this reason, the device damping value should always be set to zero before performing any test or adjustment of the device. After the test or calibration is done, the damping value should be reset to its required value.

- *Digital range change* – Many technicians think that using a HART® communicator to change the range of a HART® device calibrates the device. This is a common misconception. Changing the range causes a configuration change only. This is because a change in range affects only the conversion process performed by the second stage (*Figure 16*) of the HART® calibration process. It has no effect on the digital process value as read by a communicator.

- *Zero and span adjustment* – Using only the zero and span procedures used with conventional devices to calibrate a HART® device can yield a 4mA output indicating correct adjustment. However, the internal digital readings are often corrupted. This can be determined by reading the internal digital values using a HART® communicator. Corruption of the internal digital values occurs because the zero and span buttons change the range. This affects only the second stage conversion process of the HART® calibration process. A zero trim procedure should be performed to make the appropriate internal adjustments to correct for a zero shift condition in a HART® device. This adjusts the input stage so that the digital PV value corresponds to a calibration standard.

   To illustrate this kind of a problem, an example is given here. Assume that a technician installs and tests a differential pressure device that was set at the factory for a range of 0–100" H_2O. When the device is tested, it is found that it now has a zero shift of 1" H_2O. Therefore, with both sides of the capsule vented (zero), its output is 4.16mA instead of 4.00mA. When 100 inches of water is applied, its output is 20.16mA instead of 20.00mA. To remedy this, the technician vents both sides and presses the zero button on the device, causing the output to go to 4.00mA. Based on getting the desired 4mA output, the technician assumes that the adjustment was successful. However, when she checks the device using a HART® communicator, the

range is 1–101" H_2O at the test points, and the PV reads 1" H_2O instead of 0.

- *Loop current adjustment* – Technicians often use a handheld HART® communicator to adjust a device's loop current so that an accurate input to the device agrees with some display device on the loop. On some communicators, this procedure is called a current loop trim. Unfortunately, this procedure can also result in corrupted internal digital readings because it masks a calibration problem in the input stage by compensating for it in the output stage.

   To illustrate this kind of a problem, again use the zero shift example described earlier. Assume there is a digital indicator in the loop that displays 0.0 at 4mA and 100.0 at 20mA. During testing, it reads 1.0 with both sides of the capsule vented (zero), and 101.0 with 100" H_2O applied. Using the communicator, the technician performs a loop current trim so that the display device reads correctly at 0 and 100. While this appears to be successful, there is a fundamental calibration problem. The HART® communicator will show that the PV still reads 1 and 101" H_2O at the test points, and the digital reading of the mA output still reads 4.16mA and 20.16mA, even though the actual output from the device is 4mA and 20mA.

6.0.0 ◆ TRANSDUCERS

Another term for a transducer is a signal conditioner. The function of the transducer is to receive a linear signal in one form of energy and convert it into a linear signal of another form of energy. The incoming signal may be pneumatic or in the form of voltage, current, vibration, or sound, while the outgoing signal may be also be any one of these forms of energy but usually not the same as the incoming energy.

Because many control valves operate on pneumatic energy, and a common control loop signal form is the 4–20mA signal, I/P transducers are used to convert or condition the current signal into a linear and proportional signal that the control valve can use. Less familiar transducers that convert a voltage signal to a pneumatic signal are also available and are referred to as E/P transducers.

Calibrating any transducer is fairly simple, as long as the correct calibration equipment is used. Calibrating a transducer amounts to nothing more than supplying the correct form of input signal within the operating range of the transducer and then measuring and adjusting the conditioned output signal so that it is linear and proportional to the incoming signal. Most transducers can be bench-calibrated and taken to the field for installation.

Figure 17 is a cutaway view of a modern I/P transducer. It is an electro-pneumatic device that accepts a 4–20mA signal and converts it to a pneumatic signal. This is accomplished by applying the current signal to a coil that creates a magnetic force or field that causes a flexible arm to move. Like the pneumatic transmitter, the pneumatic section operates on a force-balance principle. In this model, a sapphire ball floats inside a nozzle assembly that is continuously exhausting air out through an orifice in the nozzle. The movement of the sapphire ball within the nozzle is controlled by the flexible arm, which is in turn controlled by the strength of the magnetic field (the level of current).

7.0.0 ◆ CONTROL VALVE POSITIONERS

The function of a control valve positioner is to compare the valve's actuator or stem position to a desired input signal and manipulate the output to the valve's actuator so that its position corresponds with the desired input signal. The positioner is usually linked to the actuator's stem by a mechanical link, as shown in *Figure 18*. The position of the mechanical link inside the positioner determines the stem position, which is compared to the incoming signal and corrected by the positioner by either increasing or decreasing the output signal to the control valve diaphragm.

NOTE

Positioners can be as sensitive as any other instrument. Use care when adjusting them, and use the proper tools.

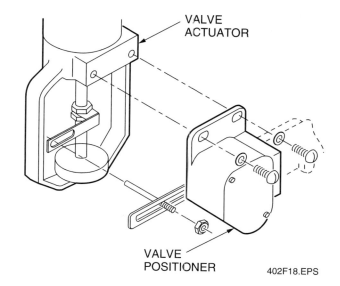

Figure 18 ◆ Mechanical link of a valve actuator stem to a positioner.

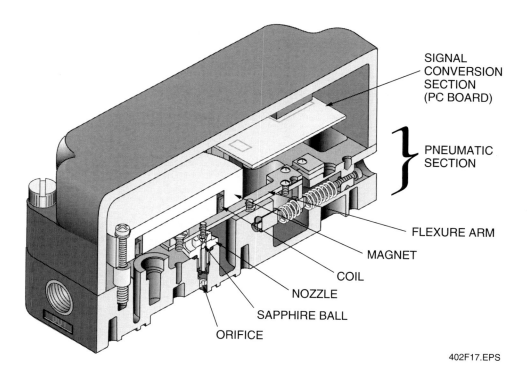

Figure 17 ◆ Cutaway of an I/P transducer.

7.1.0 Pneumatic and Electro-Pneumatic Positioners

A pneumatic valve positioner is generally used with a diaphragm-actuated, sliding-stem control valve assembly. The pneumatic valve positioner receives a pneumatic input signal from a control device and modulates the supply pressure to the control valve actuator, providing an accurate valve stem position that is proportional to the pneumatic input signal.

Figure 19 is an operational schematic of a direct-acting Fisher® 3582 Series pneumatic positioner. The pneumatic positioner receives a pneumatic input signal from a control device. A dry, regulated instrument air supply is connected to the pneumatic relay. A fixed restriction in the relay limits flow to the nozzle so that when the flapper is not restricting the nozzle, air can bleed out faster than it is being supplied.

The input signal from the control device is connected to the bellows. When the input signal increases, the bellows expands and moves the balance beam. The beam pivots about the input axis, moving the flapper closer to the nozzle. As the nozzle backpressure increases, it increases the pressure on the valve actuator diaphragm. This increased output pressure to the actuator causes the actuator stem to move downward. Stem movement is fed back to the beam through the mechanical linkage connected to a cam in the positioner. As the cam rotates, the beam pivots about the feedback axis to move the flapper slightly away from the nozzle. The nozzle pressure decreases and reduces the output pressure to the actuator. Stem movement continues, which causes the flapper to back away from the nozzle until equilibrium is reached.

When the input signal decreases, the bellows contracts (aided by an internal range spring), and the beam pivots about the input axis to move the flapper away from the nozzle. Nozzle backpressure decreases, and the pneumatic relay vents the diaphragm casing pressure to the atmosphere. The actuator stem moves upward due to spring pressure. Through the mechanical linkage and

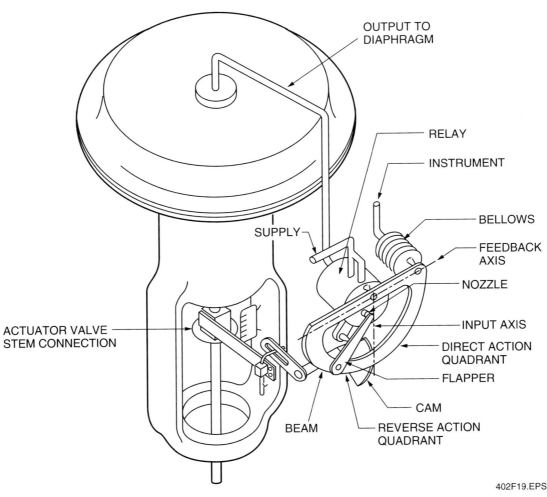

402F19.EPS

Figure 19 ◆ Operation schematic for a Fisher® 3582 Series pneumatic positioner.

cam movement, the stem's movement is fed back to the beam, which repositions the flapper closer to the nozzle. When equilibrium conditions are obtained, stem movement stops, and the flapper is positioned to prevent any further decrease in diaphragm case pressure.

The principle of operation for reverse acting pneumatic valve positioners is similar except that as the input signal increases, the diaphragm casing pressure is decreased. Conversely, a decreasing input signal causes an increase in the pressure to the diaphragm casing. The action of many pneumatic positioners, whether direct or reverse, can be easily changed by a simple field adjustment.

Electro-pneumatic valve positioners are very similar to purely pneumatic positioners except that a DC current signal is received by an electro-pneumatic signal converter (transducer) attached to the positioner. This is shown in *Figure 20*, an illustration of a Fisher® 3582i Series electro-pneumatic positioner.

Calibrating positioners is nearly as simple as calibrating transducers in that the output of the pneumatic relay must be linear and proportional to the input signal. However, in the process of calibrating the positioner's output to the input, the valve's actuator stem position must also be set so that it responds correctly with the positioner output signal.

In the case of direct-acting positioners, the valve's actuator stem should just reach its fully downward position as the positioner's output signal reaches 15 psi. Likewise, at 3 psi output, the actuator stem position should just reach its fully upward position. One of the most important points here is that even the slightest pressure change in the output signal of the positioner should begin a movement in the actuator stem. Just because a valve stem is all the way in the downward position at 15 psi does not mean that the positioner and valve stem movement are calibrated to one another, unless at 14.9 psi, for example, the valve's stem begins a slight movement up. The same goes for the upper end. The valve stem should show movement anywhere within the range of 3 to 15 psi.

With the cover removed from the positioner and a 3 psi or 4mA signal applied to the input of the positioner, depending on type of positioner, check the 0 mark on the cam and the valve seating. At the 0 mark, it is typically desired to have the valve plug just on the seat for direct-acting valves, and the cam at the 0 mark with the positioner still dynamic (not dead-ended). Adjust the zero adjustment so that these parameters are met.

The span setting is accomplished by applying a 15 psi or 20mA signal to the input of the positioner and adjusting the span adjustment screw so that

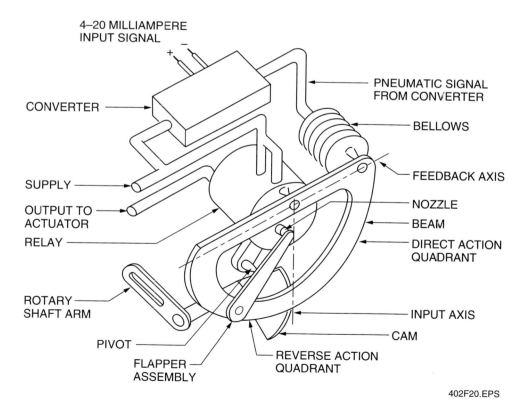

Figure 20 ◆ Fisher® 3582i Series electro-pneumatic positioner.

the upward travel of the stem just reaches its upper limits while the positioner remains dynamic (not dead-ended).

7.2.0 Smart Positioners (Digital Valve Controllers)

As with most instrumentation, control valves and positioners are also being replaced with smart technology instruments. In the area of positioners, such a change to digital technology is the Fisher FIELDVUE® DVC6000 Series digital valve controller, as illustrated in the block diagram in *Figure 21*. It is a communicating, microprocessor-based current-to-pneumatic instrument.

In addition to the traditional electro-pneumatic function of converting a current signal to a valve-position pressure signal, DVC6000 Series digital valve controllers, using HART® communications protocol, give easy access to information critical to

process operation. This can be done using a HART® communicator at the valve or at a field junction box or by using a personal computer or a system console within the control room. Using HART® communication protocol, information can be integrated into a control system or received on a single loop basis.

The digital valve controller uses two-wire, 4–20mA loop power and receives feedback of the valve travel position plus supply and actuator pneumatic pressure. This allows the instrument to diagnose not only itself but also the valve and actuator to which it is mounted. Because they do operate from a traditional 4–20mA control signal, these digital valve controllers can directly replace older analog instruments. Calibration can be performed in the field using the HART® handheld communicator, or it may be performed from the control room if the system is integrated with DCS technology.

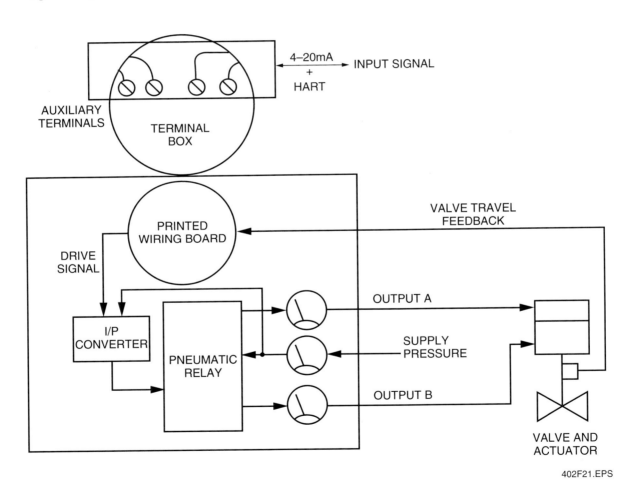

Figure 21 ◆ FIELDVUE® DVC6000 Series digital valve controller block diagram.

Summary

Only those instruments that incorporate both an input and output signal require calibration. The primary instrument that requires calibration is the field transmitter because it is the instrument that receives the first form of energy that represents what is being monitored and controlled in the process control loop in which it is installed.

This module covered the basic transmitters that are common in many installations throughout the industry, including pneumatic, analog, and smart transmitters. Manufacturers often specify calibration procedures for their instruments, and these specific procedures should be followed when calibrating the instrument in order to achieve accuracy and repeatability from the instrument. This module presented some generic procedures that can be applied to most instruments during the calibration process.

Smart instruments must always be calibrated by following the specific steps found in the user's manual for the instrument, as these instruments are generally matched to loop parameters and cannot be calibrated using basic analog calibration procedures. Always use the proper test and calibration equipment when calibrating or testing these instruments, such as the HART® communicator.

This module introduced you to the basic principles of the HART® protocol and provided an introduction into calibrating HART® transmitters and described digital valve controllers, which are rapidly replacing the older technology of analog positioners.

Review Questions

1. All instruments that provide a proportional output signal based on an input signal must be _____.
   a. balanced
   b. linear
   c. calibrated
   d. converted

2. The minimum value of all operating ranges is never a value greater than zero.
   a. True
   b. False

3. Setting the minimum output signal of a transmitter to proportionally represent the minimum input value is referred to as _____.
   a. setting the span
   b. nulling the instrument
   c. zeroing the instrument
   d. adjusting the deadband

4. _____ represents 50% of the span in a range of 50–250" H_2O.
   a. 125" H_2O
   b. 150" H_2O
   c. 175" H_2O
   d. 200" H_2O

5. The Wally Box® is a(n) _____ type of calibration equipment.
   a. smart
   b. HART®
   c. discrete
   d. pneumatic

6. Calibration should always be performed with the instrument control loop isolated from the process.
   a. True
   b. False

7. Differential pressure is the sum of the low-side pressure and the high-side pressure on a DP transmitter.
   a. True
   b. False

8. Wally Boxes® generate dry, regulated instrument air.
   a. True
   b. False

9. If you apply and hold a steady 100" H_2O to the high side of a DP transmitter and open the equalizer valve, the low-side pressure will read _____.
   a. 0" H_2O
   b. 25" H_2O
   c. 50" H_2O
   d. 100" H_2O

10. The bulb serves as a _____ in a temperature transmitter installation that uses a gas-filled tube-and-bulb system.
    a. final element
    b. transmitter low-side input
    c. primary detecting element
    d. transducer

11. In a Foxboro® 12A Series transmitter, the zero can be elevated but not suppressed.

    a. True
    b. False

12. A(n) _____ signal is continuously variable with time.

    a. discrete
    b. digital
    c. pneumatic
    d. analog

13. DP transmitters measure _____ when used for liquid level applications.

    a. hydrostatic pressure head
    b. pneumatic differential pressure
    c. specific gravity
    d. density reduction

14. Pressure above the upper level of a liquid in a closed tank or vessel affects the pressure measured at the bottom of the vessel.

    a. True
    b. False

15. A smart transmitter incorporates a _____ that stores data.

    a. capsule
    b. PLC
    c. microprocessor
    d. sensor

Trade Terms Introduced in This Module

Analog signal: A voltage, current, or other form of signal that is continuously variable with time.

Calibrated: A condition that occurs whenever the instrument's output signal or level of energy is adjusted so that it is representative of the level of the non-adjustable input energy at any given point.

Digital signal: A voltage, current, or other form of signal that is either off or on.

Dry leg: A type of differential pressure transmitter installation used in level measurement in which the low-side transmitter piping or tubing remains empty because the gas in the piping or tubing does not condense and fill the piping or tubing.

Factory characterization: A factory process by which each sensor module in a smart transmitter is subjected to pressures and temperatures covering its full operating range. The sensor module memory stores data generated from this process for use by the microprocessor in correcting the transmitter output during operation.

Five-point method of calibration: A method of calibrating an instrument in which the output level is checked and set against a simulated input level at five points: 0%, 100%, 50%, 25%, and 75%, in that order.

Fulcrum: A mechanical pivot point on which a beam pivots or balances.

Hydrostatic pressure head: A measurement of pressure that is equal to a liquid's height at a given point multiplied by the specific gravity of the liquid.

Multi-function loop calibration equipment: Test equipment that provides an energy source and measuring capabilities in the same equipment housing.

Inches of water: A measurement of pressure that is equal to 0.03612628 psi.

Span: Setting the maximum limit of the output range so that it is at its maximum value when the input process value is at the maximum value of the specified operating range.

Three-point method of calibration: A method of calibrating an instrument in which the output level is checked and set against a simulated input level at three points: 0%, 50%, and 100%.

Zeroing: Setting the minimum limit of the output range so that it is at its minimum value when the input process value is at the minimum value of the specified operating range.

Additional Resources

This module is intended to be a thorough resource for task training. The following reference works are suggested for further study. These are optional materials for continued education rather than for task training.

Applied Instrumentation in the Process Industries, Volume 1, Second Edition. W.G. Andrew, H.B. Williams. Houston, TX: Gulf Publishing Company.

Measurement and Control Basics, Third Edition. T.A. Hughes. Research Triangle Park, NC: Instrumentation, Systems, and Automation Society (ISA).

Maintenance and Calibration of HART® Field Instrumentation. R. Pirret, P.E., Marketing Manager-Process Tools. Everett, WA: Fluke® Corporation.

Figure Credits

Wallace & Tiernan GmbH	402F01 and 402F02
Siemens Energy and Automation, Inc.	402F05
Invensys Foxboro	402F06
Fluke Corporation	402F07 and 402F14
Emerson Process Technologies	402F08, 402F15, 402F19, 402F20, and 402F21
Fairchild Industrial Products Company	402F17

NCCER CURRICULA — USER UPDATE

NCCER makes every effort to keep its textbooks up-to-date and free of technical errors. We appreciate your help in this process. If you find an error, a typographical mistake, or an inaccuracy in NCCER's curricula, please fill out this form (or a photocopy), or complete the online form at **www.nccer.org/olf**. Be sure to include the exact module ID number, page number, a detailed description, and your recommended correction. Your input will be brought to the attention of the Authoring Team. Thank you for your assistance.

Instructors – If you have an idea for improving this textbook, or have found that additional materials were necessary to teach this module effectively, please let us know so that we may present your suggestions to the Authoring Team.

NCCER Product Development and Revision

13614 Progress Blvd., Alachua, FL 32615

Email: curriculum@nccer.org
Online: www.nccer.org/olf

❑ Trainee Guide ❑ AIG ❑ Exam ❑ PowerPoints Other _____

Craft / Level: _____ Copyright Date: _____

Module ID Number / Title: _____

Section Number(s): _____

Description: _____

_____

_____

_____

_____

Recommended Correction: _____

_____

_____

_____

_____

Your Name: _____

Address: _____

_____

Email: _____ Phone: _____

Performing Loop Checks

COURSE MAP

This course map shows all of the modules in the fourth level of the Instrumentation curriculum. The suggested training order begins at the bottom and proceeds up. Skill levels increase as you advance on the course map. The local Training Program Sponsor may adjust the training order.

INSTRUMENTATION LEVEL FOUR

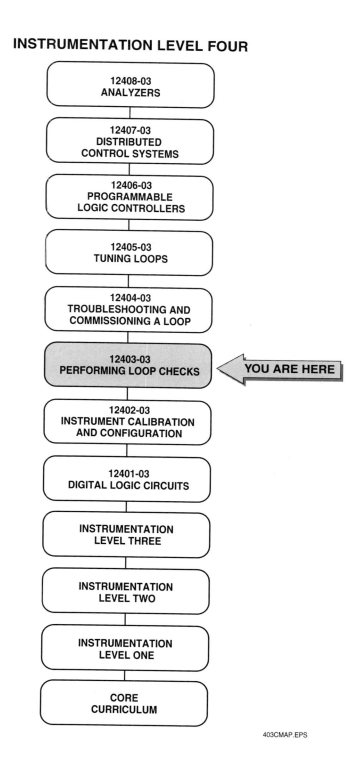

12408-03
ANALYZERS

12407-03
DISTRIBUTED
CONTROL SYSTEMS

12406-03
PROGRAMMABLE
LOGIC CONTROLLERS

12405-03
TUNING LOOPS

12404-03
TROUBLESHOOTING AND
COMMISSIONING A LOOP

12403-03
PERFORMING LOOP CHECKS ⟵ YOU ARE HERE

12402-03
INSTRUMENT CALIBRATION
AND CONFIGURATION

12401-03
DIGITAL LOGIC CIRCUITS

INSTRUMENTATION
LEVEL THREE

INSTRUMENTATION
LEVEL TWO

INSTRUMENTATION
LEVEL ONE

CORE
CURRICULUM

403CMAP.EPS

1.0.0　INTRODUCTION 3.1

2.0.0　VERIFYING MECHANICAL INSTALLATION
　　　 THROUGH VISUAL INSPECTION 3.1

　　2.1.0　Primary Element 3.2

　　2.2.0　Field Transmitter 3.2

　　2.3.0　Field Wiring, Conduit, Fiber-Optic Cable, and Tubing 3.3

　　2.4.0　Control Room Components 3.3

3.0.0　TAG NUMBERS AND LOOP SHEETS 3.4

4.0.0　LOOP CONTINUITY TESTS 3.4

　　4.1.0　Electrical 3.5

　　4.1.1　Fluke 789 ProcessMeter™ 3.5

　　4.1.2　Simplified Methods 3.6

　　4.2.0　Pneumatic 3.6

　　4.2.1　Wallace & Tiernan® (Wally Box®) 3.6

　　4.2.2　Bubbler 3.7

　　4.3.0　Fiber-Optic 3.7

　　4.3.1　Tracer 3.8

　　4.3.2　Optical Time Domain Reflectometer (OTDR) 3.8

5.0.0　PROVING A LOOP 3.8

　　5.1.0　Simulation 3.8

　　5.2.0　Required Test Equipment 3.9

6.0.0　CALIBRATING A LOOP 3.10

　　6.1.0　Conventional 4–20mA Instrument Loops 3.10

　　6.2.0　HART® Instruments 3.10

　　6.2.1　Improper Transmitter Calibration 3.12

　　6.2.2　Communicator Application 3.12

SUMMARY 3.13

REVIEW QUESTIONS 3.13

GLOSSARY 3.15

REFERENCES & ACKNOWLEDGMENTS 3.17

Figures

Figure 1   P&ID 3.2

Figure 2   Proper method of installing tubing to a DP
transmitter on a steam process line 3.3

Figure 3   Loop sheet 3.4

Figure 4   Fluke Model 789 ProcessMeter™ 3.5

Figure 5   Connecting a loop power source meter
to a field transmitter 3.5

Figure 6   Wallace & Tiernan® Model 65-2000
pneumatic tester 3.6

Figure 7   Bubbler 3.7

Figure 8   Fiber-optic cable 3.7

Figure 9   Fiber-optic tracer 3.8

Figure 10   Fluke® Model 725 multifunction process calibrator ... 3.9

Figure 11   Pressure modules for the Fluke® Model 725 .... 3.9

Figure 12   Differential pressure connection using a Fluke®
differential pressure module with a Fluke®
Model 725 multifunction process calibrator 3.10

Figure 13   HART® communicator 3.10

Figure 14   Fisher-Rosemount Model 3051
Smart Transmitter 3.11

Figure 15   Connecting a HART® communicator
to transmitter terminals 3.13

Performing Loop Checks

Objectives

When you have completed this module, you will be able to do the following:

1. Verify mechanical installation.
2. Verify correct tag numbers according to loop sheets.
3. Perform continuity checks on both electrical and pneumatic loops.
4. Prove a loop.

Prerequisites

Before you begin this module, it is recommended that you successfully complete the following: Core Curriculum; Instrumentation Levels One through Three; Instrumentation Level Four Modules 12401-03 and 12402-03.

Required Trainee Materials

1. Pencil and paper
2. Appropriate personal protective equipment

1.0.0 ◆ INTRODUCTION

Once an instrumentation loop has been installed, it must be thoroughly inspected for proper mechanical installation and for verification that the components installed belong in the loop. This can be accomplished by matching the tag numbers on the individual components with the tag numbers found on the corresponding loop sheet.

The next step is to check the loop for continuity from the primary element to the final element. This includes continuity through the transmitter, any installed transducer, through the controller or PLC, and finally to the control valve in the field.

Regardless of type—electrical, pneumatic, or fiber-optic—all loops must be intrinsically sound before activation.

Next, the loop must function properly, with each component interacting with the others as intended. Checking this interaction is referred to as **proving a loop**. In order to prove a loop, the operation of the loop must be simulated, as closely as possible, using the actual parameters and values that it will be subjected to during normal operation.

Finally, the loop must be calibrated so that the designed relationship is set between the input signal from the process and the output signal to the final element. All these stages are part of an overall process known as **commissioning a loop**.

2.0.0 ◆ VERIFYING MECHANICAL INSTALLATION THROUGH VISUAL INSPECTION

Construction and installation of instrumentation loops, like most phases of construction, are usually subjected to time lines that require rapid and sometimes less than perfect installation methods and techniques. Also, the materials and components installed may have been damaged during installation or received from the supplier in an inoperable or damaged state. For these reasons and many more, it is necessary to verify the mechanical installation of all instrumentation loops before assuming they are ready for calibration or normal operation.

The first stage of verifying mechanical installation is a visual inspection of the loop, starting with the primary element and ending with the final element, which in most cases is the control valve. In order to accurately perform a visual inspection you must be familiar with the loop and have reference to both the P&ID (*Figure 1*) for the process and the loop sheet for each particular loop you are going to inspect.

2.1.0 Primary Element

As you learned in previous training modules, the primary element is the first component in a control loop that is in contact with the process being controlled. Depending on the variable, the primary element may be an orifice plate, thermocouple, RTD, float, displacer, bourdon tube, pitot tube, or any one of many types of primary elements whose function is to make either direct or indirect contact with the process being controlled.

The mechanical inspection of these elements goes beyond that of the installation of the element itself. It must also include the installation orientation of the device as well as any tubing, wiring, conduit, fittings, connections, terminals, and methods used in conjunction with its installation. Common faults to look for include the following:

- Orifice plates that are installed backwards, as indicated by the flow indicator on the handle of the plate
- Broken conduit or conduit fittings that may have been damaged after installation
- Damaged thermocouple or RTD heads
- Damaged and/or leaking pneumatic tubing

- Loose or shorted wiring terminals
- Any other obvious primary element installation flaw that might interrupt the proper operation of the loop

2.2.0 Field Transmitter

The field transmitter receives the signal from the primary element and normally sends that signal to either a controller or a PLC. Like the primary element, the transmitter is usually located in the field and is therefore subjected to a greater degree of potential damage than components located in the more secure environment of a control room.

When inspecting a field transmitter, verify that the transmitter is mounted in a location that is most convenient to the primary element. This avoids excessively long runs of conduit or tubing to interconnect the two devices, while making sure that the location provides some physical protection for the transmitter. Transmitters must be rated for extreme ambient temperatures and/or corrosive environments if installed in those types of areas. The newer smart transmitters must be accessible to allow portable handheld communicators to be connected to them in order to program and calibrate the instrument.

Differential pressure (DP) transmitters should be checked for proper tubing and piping. Proper slope, cooling lengths in hot processes, and drain (blow-out) capabilities on both the high- and low-

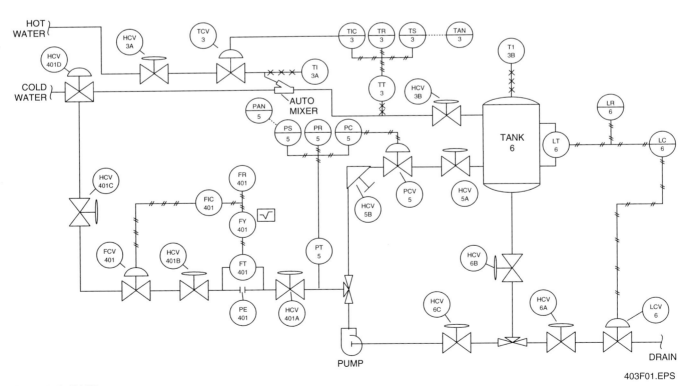

Figure 1 ◆ P&ID.

403F01.EPS

side process connections are needed in order to keep the process lines free of contaminants and condensates. This type of installation is shown in *Figure 2*.

All electrical connections should be checked for tightness of terminals and sound conduit and fittings. Instrument tags should be readily accessible and easy to read. Calibration terminals should be located to allow the technician ease of access without having to expose the technician to potential injury caused by climbing or harsh environments. Visual field readouts on smart transmitters should be oriented to permit ease of reading with minimal glare and obstructions. Instrument stands should be made of material compatible with the ambient environment and fabricated to sufficiently support the instrument while remaining aesthetically pleasing.

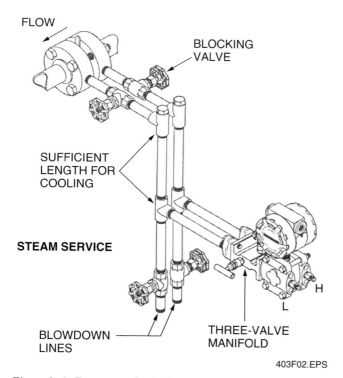

Figure 2 ◆ Proper method of installing tubing to a DP transmitter on a steam process line.

2.3.0 Field Wiring, Conduit, Fiber-Optic Cable, and Tubing

It is common to find damaged wiring, conduit, fiber-optic cable, and tubing in the field environment due to their exposure during construction as well as to daily maintenance activities within an industrial facility. Slightly damaged signal wiring or broken conduit can cause intermittent problems in a control system. Signal distortion is caused by AC noise and/or changes in resistance in the

wiring or cable. Often, it is these small degrees of damage that cause a system or loop to be difficult to control due to the inconsistency of their effects on the instrument signal. While in the field, be on constant lookout for these minor damages, and repair them immediately if possible. At minimum, note them so that once enough of them are listed, repairs can be made to all of them during the same time period. No physical damage, regardless of the degree of damage, should be left unrepaired, as it will only grow in size, as will its effects on the accuracy and response of the system.

Fiber-optic cable cannot be spliced in the way that copper wire can, and aligning fiber cables is difficult and expensive. Fiber is also very fragile and can break if bent. For this reason, if damaged fiber-optic cables are located in the field, it is usually best to replace the entire cable length instead of attempting to repair a section of the cable, unless you are well qualified in the splicing of fiber optics.

2.4.0 Control Room Components

Most components of a loop that are installed in the control room, including single-loop controllers and PLCs, are protected from physical damage. However, these components must always be inspected after installation to ensure that no physical damage was done to them prior to or during installation. Wiring and tubing connections should be checked for tightness and damage. Make sure to check all terminations under, behind, and overhead in cabinets that may not be readily accessible. It is in these areas that many cables, connections, and terminals may not be adequately connected or terminated because of their inaccessibility. Check for proper routing of conductors, cable, and tubing, looking for sharp bends. Restrictions can cause poor signal transmission and can cause a loop to be slow in response or inaccurate. Repair any and all of these types of discrepancies immediately in order to save yourself grief later on when the conditions may worsen.

Always be aware of signal noise and how it develops, both in field and control room wiring. Make sure signal cable shields or drain wires are properly grounded, usually on one end only, and try to provide adequate spacing between alternating current conductors and signal conductors. Never ground a signal cable shield in the control room unless you have physically located the same cable on the other end, verifying that the shield is not grounded on that end. Double grounding can cause **ground loops**, which are current-carrying circuits, to develop in the double-grounded shielded conductor, causing signal noise and distortion.

3.0.0 ◆ TAG NUMBERS AND LOOP SHEETS

Tag numbers are unique numbers assigned to each component in a loop, all sharing the common loop number as part of their tag number. A loop sheet, like the example shown in *Figure 3*, contains all the components that make up a single loop, including their tag numbers and specifications. Every component installed in every loop should be inspected and verified to concur with the information in its loop sheet, including tag number and calibration range. If any component is found to have a tag number on it other than the one found on the loop sheet for that loop, an investigation should be immediately begun to determine the discrepancy. The loop must not be proven until the correct component is installed in the loop.

4.0.0 ◆ LOOP CONTINUITY TESTS

Loop **continuity tests** involve the application of a signal generated by testing equipment to the hard wiring or tubing in a loop to determine if the signal passes through the complete loop without leakage or excessive resistance.

WARNING!

Before performing any continuity test on any loop, regardless of the energy source of the loop, such as electricity or pneumatics, always properly lock out and tag out the source of energy according to the guidelines and policies of your company. Never assume that the circuitry or tubing in the loop will only carry a safe level of voltage, current, or air, as the loop at this phase of testing has not been tested for proper installation. Continuity tests must never be performed with the loop in an energized state, whether the energy is electrical or pneumatic.

Most PLCs and DCS equipment in an instrument loop are designed to operate by digital input signals. You must isolate these components from the instrument loops before testing the loop's hard wiring or tubing to avoid damage to the digital equipment. You can simulate the instrument being in the loop in order to create continuity by connecting the output leads of a loop calibrator or tester directly into the loop.

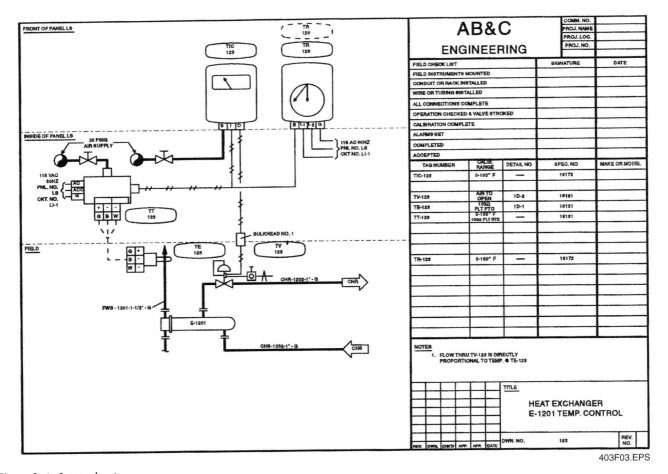

Figure 3 ◆ Loop sheet.

4.1.0 Electrical

Once the circuitry has been locked out and tagged according to your plant's safety policies, and sensitive digital components in the loop such as PLCs and DCS computers have been isolated from the circuitry, you may begin your electrical continuity test on the wiring in the loop. This test will ensure accurate signal transmission once the loop is put into operation.

4.1.1 Fluke 789 ProcessMeter™

There are many test instruments that simulate the output signal of a typical electronic field transmitter. Fluke® offers a complete line of process control test and calibration meters, including the Fluke® Model 705 loop calibrator, 707 loop calibrator, 710 series process calibrator, 712 RTD process calibrator, 714 thermocouple process calibrator, 715 volt/mA calibrator, 725 multifunction process calibrator, and the 787 process meter. Although most of these meters are capable of providing a power source and an input function, it is best to use one that provides a 24VDC source and milliampere input function so that the loop may be tested with the operational loop power locked out during the testing.

NOTE

Fluke® is not the only manufacturer that offers electrical loop testing and calibration meters. Other manufacturers provide a complete line of meters that may also be used for these tests and checks.

An example of a Fluke® meter that may be used for electrical loop continuity checks is the Fluke Model 789 ProcessMeter™ shown in *Figure 4*. The 789 is a handheld, battery-operated tool for measuring electrical parameters, supplying steady or ramping current to test process instruments, and providing a +24 voltage loop power supply. It has all the functions of a digital multimeter plus current output capability.

Because the loop's operating power source should be locked and tagged out during continuity check, you can use the Model 789 or similar test equipment to simulate the loop's operation by placing the meter in the Loop Power mode and connecting the output leads of the meter to the field transmitter. While in Loop Power mode, the meter acts like a battery and supplies voltage while the process instrument regulates the current in the loop. At the same time, the metering section on the Model 789 measures the current running through the loop. The meter supplies loop power at a nominal 24VDC.

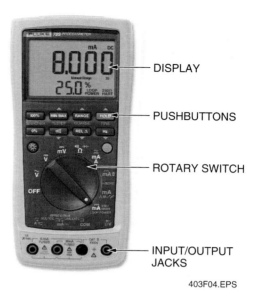

DISPLAY
PUSHBUTTONS
ROTARY SWITCH
INPUT/OUTPUT JACKS

403F04.EPS

Figure 4 ◆ Fluke Model 789 ProcessMeter™.

With the relatively new **smart instruments** that now make up many of the loops controlled by PLCs and DCS, it is necessary to connect a precision (±1%) resistor of 250 ohms into their circuitry in order to simulate the load in the circuit. The Model 789 contains an internal series resistance of 250 ohms that can be switched in for communication with HART® and other smart devices. When the loop power is enabled on the meter, the meter supplies 24VDC and measures current drawn by the loop. This type of test will provide a verifiable electrical continuity check of the loop and at the same time will indicate the proper operation of the loop. The meter's source leads must be connected in series with the instrument current loop, as shown in *Figure 5*.

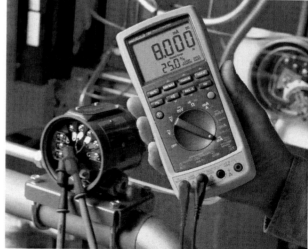

403F05.EPS

Figure 5 ◆ Connecting a loop power source meter to a field transmitter.

4.1.2 Simplified Methods

Because continuity testing an electrical loop can be defined as verifying that the hardwiring and terminations in a loop are intrinsically connected, there are many more simple and less technical methods that may be used to test the wiring from one end to the other. Two people, each equipped with two-way radios and a simple battery-powered continuity tester, can check wiring from one end to the other by intentionally grounding one end of a wire while the person on the other end checks the continuity between the wire and a grounded source.

Another simple method for testing continuity and identifying wires at both ends is the use of battery-powered telephone receivers. These test phones were used extensively in the earlier days of loop testing and were usually made by powering standard telephone receivers with a 9V battery and creating a circuit path through the loop wiring so that two people on opposite ends of the loop wiring could communicate with each other over the phones whenever they were connected to the same pair of loop wires. This method is still used in some facilities but for the most part has been replaced by the more sophisticated and reliable multi-purpose loop testing equipment.

4.2.0 Pneumatic

Even though much of the instrumentation in the field has been replaced by electronic or digital loops, there is still a substantial amount of old, reliable pneumatic instrumentation that requires maintenance and calibration. However, very little, if any, new pneumatic instrumentation is being installed in new facilities or expansions of existing facilities. For that reason, this module will not dedicate much time and explanation to testing and calibrating pneumatic instrumentation other than to familiarize the student with the equipment available to test and calibrate pneumatic control loops.

4.2.1 Wallace & Tiernan® (Wally Box®)

One of the more familiar pneumatic testers or simulators used to test and calibrate pneumatic loops is manufactured by Wallace & Tiernan® (now called Wika Instrument Corporation) and is commonly referred to as the Wally Box®. Its design has been changed slightly from its debut, but the basic technology is still applied. The newer 65-2000 portable pneumatic calibrator (*Figure 6*), designed with a digital readout, is an upgrade of the former Model 65-120 that is equipped with a precision

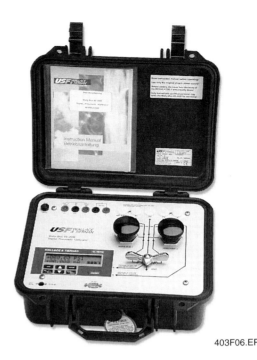

403F06.EPS

Figure 6 ◆ Wallace & Tiernan® Model 65-2000 pneumatic tester.

pressure gauge. Some of the features included in both models are ⅛" pressure connections, two air-regulator controls, and a selector valve. With the selector valve, three different test pressures can be applied individually to the readout. A fourth selector setting vents the gauge to atmosphere.

A filter on the air supply keeps oil and moisture out. Two regulators apply known pressures to the devices under test. The connection block is designed with ⅛" female pipe threaded connections. Small gauges can be threaded directly into the block, or as with larger instruments, connected by flexible tubing.

A built-in pressure-relief valve protects the case against overpressure to 10 times the maximum pressure rating. A separate pressure-relief valve protects the capsule mechanism.

An instrument air supply can be connected to one of the ports on the Wally Box® and routed through the readout or gauge. If the line under test for continuity is plugged or capped on the other end, the line should maintain the applied pressure even when the supply is isolated from it. This condition is monitored by the Wally Box® readout or pressure gauge. Once the air supply is isolated from the charged or pressurized line under test, the gauge or readout (depending on what model Wally Box® you are using) can be monitored to see if the pressure drops. If it does drop, a leak or open end is present in the loop, and it must be repaired before the loop is proven.

However, if the pressure remains the same after a period of time (the length of time that the lines must maintain pressure may be dictated in the specifications), the loop's continuity and integrity are sound and the loop is ready to be proven.

4.2.2 Bubbler

Continuity and integrity of pneumatic tubing and piping can also be checked using a bubbler like the one shown in *Figure 7*. The pneumatic line or tubing must be capped or plugged at one end and the bubbler connected in series into the line at some point. The function of a bubbler placed in a line or system is to provide an indication if air is flowing through it, even at a very small flow rate. If the bubbler is installed in a leak-free line that is capped or plugged and is then pressurized, no air will flow through the system. However, if a tubing fitting or valve is slightly loosened or opened to the atmosphere, air will flow through the bubbler and will be indicated by bubbles appearing in the water-filled glass bowl on the bubbler.

403F07.EPS

Figure 7 ◆ Bubbler.

Continuity of a lengthy run of tubing or piping can be checked by plugging one end, installing the water-filled bubbler on the other end, and then pressurizing the line. No bubbles should appear in the water-filled bowl. If they do, there is a leak somewhere in the system allowing the air to flow. If no bubbles appear after the bubbler is installed, the line is capped or plugged and pressurized.

Then, the plug or cap is slightly loosened on the opposite end of the bubbler to allow a small flow of air to escape into the atmosphere. If bubbles appear in the water-filled glass bowl of the bubbler, the line is both tight and continuous from one end to the other. This is a simple and inexpensive method of testing for both leaks and continuity of a pneumatic system. The fittings may be retightened and the loop proven.

4.3.0 Fiber-Optic

Continuity of fiber-optic cable simply means that the light can travel from one end of the optical cable to the other end. The **core** of the fiber cable transmits the light while the **cladding** contains the light within the core and guides it, even through bends in the cable. The outer covering of the cable is the **buffer**, which is generally a hard plastic coating that protects the glass fiber from moisture and physical damage. These parts of a fiber-optic cable are shown in *Figure 8*.

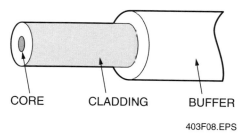

CORE   CLADDING   BUFFER

403F08.EPS

Figure 8 ◆ Fiber-optic cable.

Before testing fiber-optic cable, you should have available the cable layouts for every fiber cable you are going to test. A good practice is to prepare a list of all the cables and fibers before you go in the field and print a copy for recording your test data. Record your test results on the list, and keep it for future maintenance and records. Some optical test equipment is designed with a memory function that will store test results, which can be downloaded later to a computer program.

CAUTION

Fiber-optic sources, including test equipment, are generally too low in power to cause any eye damage. However, it is still a good idea to check cable ends and connectors with a power meter before looking directly into them as some CATV and other systems have very high power that can cause eye damage.

4.3.1 Tracer

Inspect the cable for continuity. Make sure the fibers are not broken by visually tracing the path of a fiber from one end to another, paying special attention to any splices in the cable. A fiber-optic tracer, like the example shown in *Figure 9*, has an end connector that connects to a fiber-optic connector. Continuity is checked by attaching the cable to be tested to the tracer and visually looking at the other end of the cable to see if the light is visible. If it is, the continuity of the cable is good from one end to the other; if you do not see the light, the cable is broken or restricted somewhere along its path. Most tracers, including the example shown, have a range that may exceed two and a half miles. They are inexpensive, so most technicians and fiber installers can own one.

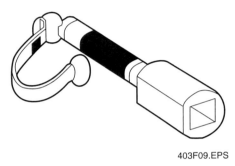

403F09.EPS

Figure 9 ◆ Fiber-optic tracer.

4.3.2 Optical Time Domain Reflectometer (OTDR)

The fiber-optic tracer may not always be effective for testing continuity in fiber-optic cables, especially if the cable is long and contains many splices. In these installations, it may be necessary to use a more sophisticated testing instrument called an optical time domain reflectometer, or OTDR, to test the cable, and especially the splices, for continuity.

Tracers test the cable's continuity directly by applying light to the cable and viewing it from the other end. The OTDR works indirectly on the cable's fibers. It uses the **backscattered light** of the fiber to indicate loss of light. The OTDR works like radar in that it sends a high-power laser light pulse down the fiber and looks for return signals from backscattered light in the fiber itself or reflected light from connector or splice interfaces.

The light the OTDR sees is the light scattered from the pulse passing through a region of the fiber. Only a small amount of light is scattered back toward the OTDR, but with sensitive receivers and signal averaging built into the OTDR, it is possible to test cables over relatively long distances.

An OTDR can detect problems in the cable caused during installation. If there is a broken fiber or a bad splice, the cable will test shorter than its actual length, or a high loss of light will be indicated at that point. If excessive stress is placed on the cable due to kinking or too tight a bend radius, it will look like a splice at the wrong location.

5.0.0 ◆ PROVING A LOOP

In the previous section you learned that performing a continuity check on a loop means that you verify that the loop components and their associated wiring or tubing are soundly connected from one end to the other. You also check that no shorts or loose wires are present in electrical loops and that no leaks are present in pneumatic loops.

The next step in the process of getting a control loop to its operational stage is proving a loop. Proving a loop can be defined as verifying that all instruments in a loop are properly interconnected to one another and are responding to each other as intended and as designed into the loop's process control.

Using a simple single flow control loop as an example, this means that the primary element (orifice plate) creates a differential pressure that is read by the transmitter, which in turn sends its output signal to the next instrument or signal conditioner in the loop and finally on to the controller or PLC. The last step in proving the loop is making sure that the final control element receives the output of the controller or PLC and responds to that output. This is not the calibration stage. In the calibration stage, parameters such as zero adjustments, ranges, spans, and responses are set. The proving stage only verifies that the instruments are interacting with one another as intended.

5.1.0 Simulation

There are only two ways to verify that instruments in a loop are interacting with one another as intended and designed: putting the loop into actual operation or simulating the operation of the loop. You would not want to put a control loop into operation before calibrating it or proving that it works, so you must simulate the loop's operation in order to prove the loop.

As with the other tests previously performed on the loop, it is necessary that you have both the P&ID and the loop sheets available for the loops that you are going to prove, as they list each instrument and their function in the loop. In addition to the proper documentation, you must have the test equipment that has the ability to simulate the loop's operation.

5.2.0 Required Test Equipment

Proving a loop requires test equipment that can supply the loop's power or energy, whether it is air or electrical. The device must also be able to read the output signals of each of the instruments in the loop, regardless of their signal form (for example, volts, milliamperes, and psi). In some unique loops that are equipped with signal conditioners or transducers, it may be necessary to have more than one piece of test equipment on hand in order to read the various signal forms. However, test and calibration equipment that provides all of these functions in one piece of equipment is available.

In order to simulate a loop, your calibration equipment must be able to provide both the form of signal and the level at which the loop would normally operate. For example, most electronic loops are operated by a permanent 24VDC power supply that is capable of supplying many instruments at one time without experiencing a reduction in voltage. Because you are going to simulate the loop, the power source on your test equipment must also be able to handle all of the loads that are interconnected in the loop. Most process control loop calibrators are so equipped.

One multi-channel loop calibrator that provides output in various forms and is capable of receiving multiple input signals in volts, milliamperes, psi, and even differential pressure, is the Fluke® Model 725 multifunction process calibrator (*Figure 10*). It has capabilities to measure and source mA, volts, temperature (RTDs and thermocouples), frequency, ohms, and pressure, using optional pressure modules.

The Model 725 has a split display that lets you view input and output values simultaneously. This allows you to power up a field transmitter from the meter and read its output on the meter at the same time. For valve and current-to-pneumatic transducer tests, you can source mA while measuring pressure. The Model 725 is also equipped with auto-stepping and auto-ramping for remote testing. This function allows you to connect the meter into the loop and set its output to step through the output range so that you may take measurements or readings at various points in the loop to determine if all instruments in the loop are functioning properly. This is a very handy method to use when proving a loop.

The Model 725 is designed to accept a list of accessories, including differential pressure modules of 1 psi, 5 psi, and 15 psi, as well as 1" H_2O and 10" H_2O (*Figure 11*). This means that as long as the differential pressure delivered by the primary element (orifice plate) does not exceed the maximum differential pressure of the module, the process lines can be connected directly to the high and low sides of the differential pressure module, which then can be connected to the Model 725 to provide an actual input to the meter. This type of connection is more applicable during calibration stages, but it can be valuable during the proving phase also, as long as the process is a safe and compatible process for the test module, and the maximum differential pressure of the module is not exceeded. (In addition, the Fluke® Model 725 is also designed to accept a gauge pressure module of either 30 psi or 100 psi maximum gauge pressure.)

There are other loop simulators available from Fluke® and other manufacturers of instrumentation or process control test equipment and meters that provide similar capabilities and functions. Likewise, calibration equipment that is not multi-functional can be used in conjunction with other instruments, where one provides the power source while the other reads the loop input. The choice of test and calibration equipment depends on factors such as cost, familiarity, and preference of equipment available.

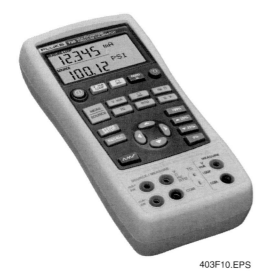

403F10.EPS

Figure 10 ◆ Fluke® Model 725 multifunction process calibrator.

403F11.EPS

Figure 11 ◆ Pressure modules for the Fluke® Model 725.

Figure 12 shows a typical differential flow calibration setup using a differential pressure module in conjunction with the Fluke® Model 725.

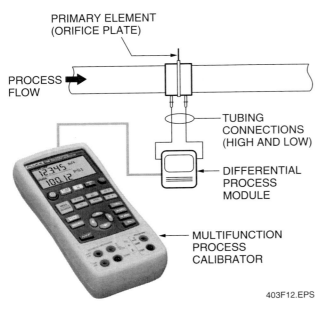

403F12.EPS

Figure 12 ◆ Differential pressure connection using a Fluke® differential pressure module with a Fluke® Model 725 multifunction process calibrator.

6.0.0 ◆ CALIBRATING A LOOP

At this point, it is very important to understand the difference between tuning a loop and calibrating a loop. A loop must be calibrated before it can be tuned. Calibrating a loop is the process of adjusting an instrument to set the correct relationship between the input signal of the instrument and its output signal. Instrument components (such as recorders and indicators) and final control elements (such as control valves) receive an input but have no output. In these cases, calibration involves setting the response or indication of the instrument to an accepted or desired relationship to the input signal.

6.1.0 Conventional 4–20mA Instrument Loops

For a conventional 4–20mA instrument, a multiple-point test using a multifunction calibrator that simulates the input and measures the output is sufficient to verify the overall accuracy of the transmitter. The normal calibration adjustment involves setting only the zero value and the span value because there is effectively only one adjustable operation between the input and output. This procedure is often referred to as a zero and span calibration. If the relationship between the input and output range of the instrument is not linear, the transfer function must be known before expected outputs can be calculated for each input value. Without knowing the expected output values, you cannot calculate the performance errors.

6.2.0 HART® Instruments

Technology in process control has advanced far into digital signaling. Instruments that incorporate digital technology are designed with internal microprocessors so that they may communicate with other digital instruments in the loop, and with the PLC or DCS. In order for a technician to communicate with these instruments to verify or calibrate them, digital communicators must be connected into the loop. One such communicator is the **HART® communicator** (*Figure 13*). HART® is an acronym for highway addressable remote transducer, which provides digital communication to HART®-compatible microprocessor-based (smart) analog process control instruments.

403F13.EPS

Figure 13 ◆ HART® communicator.

For a HART® instrument, like the Rosemount Model 3051 smart transmitter shown in *Figure 14*, a multiple-point test between input and output (as performed on the analog 4–20mA loop) does not provide an accurate representation of the transmitter's operation. Just like a conventional transmitter, the measurement process begins with a technology that converts a physical quantity into an electrical signal. The similarity ends there, however.

Instead of a purely mechanical or electrical path between the input and the resulting 4–20mA output signal, a HART® transmitter has a microprocessor that manipulates the input data. There are typically three calculation sections involved, and each of these sections may be individually tested and adjusted.

The instrument's microprocessor measures some electrical property that is affected by the process variable. The measured value may be in millivolts, capacitance, reluctance, inductance, frequency, or some other property. However, before the value can be used by the microprocessor, it must be transformed to a digital count by an analog to digital (A/D) converter. In the first calculation section, the microprocessor must rely on some form of equation or table to relate the raw count value of the electrical measurement to the actual process variable such as temperature, pressure, or flow. The principal form of this table is usually established by the manufacturer, but most HART® instruments include commands to perform field adjustments. This is often referred to as a sensor trim. The output of the first calculation section is a digital representation of the process variable. When you read the process variable using a communicator, this is the value that you see.

The second calculation section is strictly a mathematical conversion from the process variable to the equivalent milliampere representation. The range values of the instrument (related to the

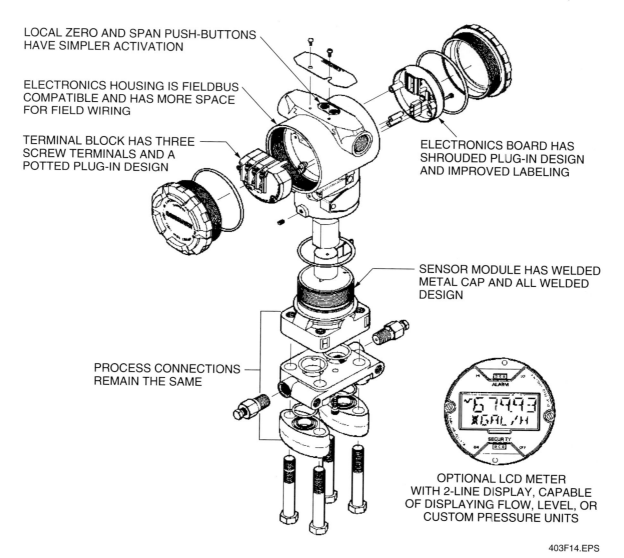

LOCAL ZERO AND SPAN PUSH-BUTTONS HAVE SIMPLER ACTIVATION

ELECTRONICS HOUSING IS FIELDBUS COMPATIBLE AND HAS MORE SPACE FOR FIELD WIRING

TERMINAL BLOCK HAS THREE SCREW TERMINALS AND A POTTED PLUG-IN DESIGN

ELECTRONICS BOARD HAS SHROUDED PLUG-IN DESIGN AND IMPROVED LABELING

SENSOR MODULE HAS WELDED METAL CAP AND ALL WELDED DESIGN

PROCESS CONNECTIONS REMAIN THE SAME

OPTIONAL LCD METER WITH 2-LINE DISPLAY, CAPABLE OF DISPLAYING FLOW, LEVEL, OR CUSTOM PRESSURE UNITS

403F14.EPS

Figure 14 ◆ Fisher-Rosemount Model 3051 Smart Transmitter.

zero and span values) are used in conjunction with the transfer function to calculate this value. Although a linear transfer function is the most common, pressure transmitters often have a square root option. The output of the second calculation section is a digital representation of the desired instrument output. When you read the loop current using a communicator, this is the value that you see. Many HART® instruments support a command that puts the instrument into a fixed output test mode. This command overrides the normal output of the second calibration section and substitutes a specified output value.

The third calibration section is the output section where the calculated output is converted to a value that can be loaded into a digital to analog converter. This produces the actual analog electrical signal. Once again, the microprocessor must rely on some internal calibration factors to get the output correct. Adjusting these factors is often referred to as a current loop trim or 4–20mA trim.

6.2.1 Improper Transmitter Calibration

Using only the zero and span adjustments to calibrate a HART® transmitter—the standard practice with conventional transmitters—often corrupts the internal digital readings. There is more than one output to consider. The digital process variable and milliampere values read by a communicator are also outputs, just like the analog current loop.

Consider what happens when using the external zero and span buttons to adjust a HART® instrument. Suppose a technician installs and tests a differential pressure transmitter that was set at the factory for a range of 0" to 100" H_2O. A test of the transmitter reveals that its zero has shifted from 0" to 1" H_2O. Because of this shift, with both high and low ports vented (zero), its output is 4.16mA instead of 4.00mA; when applying 100" H_2O, the output is 20.16 instead of 20.00mA. In an attempt to remedy this, the technician vents both ports to atmosphere and presses the zero key on the transmitter. The output definitely goes to 4.00mA, so it appears that the adjustment was successful. However, if the technician now checks the transmitter with a communicator, the range will be 1" to 101" H_2O, and the process variable will read 1" H_2O instead of 0".

Zero and span buttons only change the range because the instrument does not know the actual value of the reference input. Only a digital command that conveys the reference value enables the instrument to make appropriate internal adjustments.

The proper way to correct a zero shift condition is to use a zero trim, as previously described. This adjusts the instrument input block so the digital value of the process variable agrees with the calibration standard. When using the digital process values for trending and statistical calculations, disable the external zero and span buttons and avoid using them entirely.

Another observed practice among instrument technicians is to use a handheld communicator to adjust the current loop so that an accurate input to the instrument agrees with some display device on the loop.

Refer again to the zero shift example just described. Suppose there is a digital indicator in the loop that displays 0.0 at 4mA and 100.0 at 20mA. During testing, it read 1.0 with both ports vented to the atmosphere, and 101.0 with 100" H_2O applied. Using the communicator, the technician performs a current loop trim so that the display reads correctly at 0 and 100.

While this appears to be successful, there is a fundamental problem. The communicator will show that the process variable still reads 1" and 101" H_2O at the test points, and the digital reading of the mA output still reads 4.16 and 20.16mA, even though the actual output is 4 and 20mA. The calibration problem in the input section has been hidden by introducing a compensating error in the output section so that neither of the digital readings agrees with the calibration standards.

6.2.2 Communicator Application

For the HART® communicator to function properly, a minimum of 250 ohms resistance must be present in the loop. The HART® communicator does not measure loop current directly. *Figure 15* shows the correct method for connecting a HART® communicator to a compatible device.

If the transmitter is used as a digital-only device, and the current loop output is not used, only input section calibration is required. If the application uses the milliampere output, the output section must be explicitly tested and calibrated. Note that this calibration is independent of the input section and has nothing to do with the zero and span settings.

The calibration approach for a HART® instrument will depend on how the transmitter outputs are used. If only the 4–20mA analog signal is used, it may be treated like an analog transmitter. By using the manual zero and span buttons on the transmitter or by digitally setting the process variable's lower and upper values, the correct relationship between the input sensor and 4–20mA analog output are set.

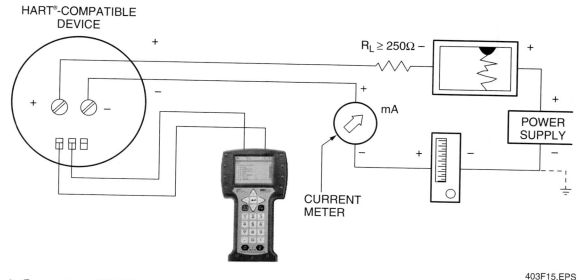

HART®-COMPATIBLE
DEVICE

$R_L \geq 250\Omega$

mA

CURRENT
METER

POWER
SUPPLY

403F15.EPS

Figure 15 ◆ Connecting a HART® communicator to transmitter terminals.

Summary

Once a control loop is installed, a series of inspections and tests must be accomplished before moving on to the final stage of what is referred to as commissioning the loop. This module covered a majority of the steps involved in the inspection phase, including visual inspection. From that point you moved on in the testing process of the loop by verifying that the loop's interconnecting wiring or tubing is intact from one end to the other, which is referred to as continuity. After testing the interconnections, the instruments in the loop must be tested for proper interaction with each other, in order to allow the process control to function properly. This step is referred to as proving a loop. Once the loop is proven, you must calibrate the loop, which involves setting the relationship between the input signal and the output signal.

As technology expands in instrumentation control, so do the tasks involved in the processes of proving and calibrating microprocessor-based smart control loops. You have learned that less complex analog control loops, such as simple 4–20mA loops, are much easier to calibrate in comparison to the smart instrument loops now replacing these analog loops. However, thanks to the development of sophisticated testing and calibrating equipment, tasks associated with testing and maintaining these smart loops are made much easier as long as the proper equipment is properly applied. This module has covered some of that test equipment and how to apply it to the newer smart loops.

Review Questions

1. The first task in the mechanical testing of a loop should be _____ the loop.
   a. proving
   b. commissioning
   c. visually inspecting
   d. continuity testing

2. The first component in a control loop to have contact with the process is the _____.
   a. control valve
   b. final element
   c. transducer
   d. primary element

3. The purpose of drains when installing tubing or piping to a differential pressure transmitter is for _____.
   a. blowing out condensate or contaminants
   b. cooling hot processes
   c. venting surges in the process lines
   d. calibrating and equalizing pressure

4. Fiber-optic cable should be spliced using the same techniques applied to copper signal wiring.
   a. True
   b. False

5. Signal cable shields must be grounded on both ends to prevent ground loops.
   a. True
   b. False

6. A _____ contains all of the components and their tag numbers in a single loop.
    a. P&ID
    b. flow diagram
    c. loop sheet
    d. wiring diagram

7. Before performing any continuity test on any loop, _____ the loop.
    a. lock out and tag out
    b. list the instruments in
    c. prove
    d. calibrate

8. A Fluke 789 ProcessMeter™ has the capability of providing an 24VDC output signal while in the loop power supply mode, as well as reading an input signal.
    a. True
    b. False

9. _____ different test pressures can be applied individually to the Wally Box®.
    a. Two
    b. Three
    c. Four
    d. Five

10. When using a bubbler in testing pneumatic lines, bubbles in the water-filled glass bowl indicate _____.
    a. continuity
    b. acceptable installation
    c. good flow
    d. a leak

11. The _____ component of a fiber-optic cable transmits the light.
    a. cladding
    b. buffer
    c. core
    d. shield

12. An OTDR is used to test fiber-optic cable by sending a light source through the cable's core and visually looking for the light at the other end of the cable.
    a. True
    b. False

13. When proving a loop, you should always simulate the loop's operation and not put the loop into actual operation.
    a. True
    b. False

14. The Fluke® Model 725 multifunction process calibrator can also be used to measure differential pressure as long as an accessory differential pressure module is used along with it to convert the differential pressure to a compatible signal for the calibrator.
    a. True
    b. False

15. A calibration test that simulates the input and measures the output is sufficient to verify the overall accuracy of a smart instrument.
    a. True
    b. False

Trade Terms Introduced in This Module

Backscattered light: In a fiber-optic cable, it is light that is scattered by the cable's core and reflected back to its source.

Buffer: A hard plastic coating on the outside of a fiber-optic cable that protects the glass from moisture or physical damage.

Cladding: The outside optical layer of a fiber-optic cable that traps the light in the core and guides it through bends.

Commissioning a loop: Declaring by supportive documentation that a control loop has been inspected, proven, calibrated, and is ready to be put into normal operation. The documentation usually contains loop information including tag numbers, calibration data, and any other pertinent information associated with the loop.

Continuity test: Verifying that a control loop is interconnected properly from one end to the other.

Core: The innermost glass fiber that carries the light in a fiber-optic cable.

Ground loop: A phenomenon caused by a signal cable shield being grounded at both ends, creating an undesirable circuit that permits current induced by electromagnetic fields to flow in the circuit, resulting in instrument signal noise.

HART® communicator: A handheld interface that provides a common communication link to all HART®-compatible, microprocessor-based instruments.

Proving a loop: A process that verifies the proper interactions of the various components of a control loop.

Smart instrument: Instruments or instrument components that contain a microprocessor that has the ability to communicate with other microprocessors in the system.

Additional Resources

This module is intended to be a thorough resource for task training. The following reference works are suggested for further study. These are optional materials for continued education rather than for task training.

www.hartcomm.org. The website of the HART Communication Foundation, an independent, non-profit organization that provides support for application of the HART Protocol.

www.isa.org. The website of the Instrument Society of America.

www.thehartbook.com. A website providing information on HART Communications products, services, and suppliers.

Instrumentation for Process Measurement and Control, 1997. 3rd ed. Norman A. Anderson. Boca Raton, FL: CRC Press.

Maintenance and Calibration of HART Field Instrumentation. R. Pirret, P.E., Marketing Manager, Process Tools. Everett, WA: Fluke Corporation.

Figure Credits

NCCER CURRICULA — USER UPDATE

NCCER makes every effort to keep its textbooks up-to-date and free of technical errors. We appreciate your help in this process. If you find an error, a typographical mistake, or an inaccuracy in NCCER's curricula, please fill out this form (or a photocopy), or complete the online form at **www.nccer.org/olf**. Be sure to include the exact module ID number, page number, a detailed description, and your recommended correction. Your input will be brought to the attention of the Authoring Team. Thank you for your assistance.

Instructors – If you have an idea for improving this textbook, or have found that additional materials were necessary to teach this module effectively, please let us know so that we may present your suggestions to the Authoring Team.

NCCER Product Development and Revision

13614 Progress Blvd., Alachua, FL 32615

Email: curriculum@nccer.org
Online: www.nccer.org/olf

❑ Trainee Guide   ❑ AIG   ❑ Exam   ❑ PowerPoints   Other _____

Craft / Level: _____   Copyright Date: _____

Module ID Number / Title: _____

Section Number(s): _____

Description: _____

_____

_____

_____

_____

Recommended Correction: _____

_____

_____

_____

_____

Your Name: _____

Address: _____

_____

Email: _____   Phone: _____

Troubleshooting and Commissioning a Loop

COURSE MAP

This course map shows all of the modules in the fourth level of the Instrumentation curriculum. The suggested training order begins at the bottom and proceeds up. Skill levels increase as you advance on the course map. The local Training Program Sponsor may adjust the training order.

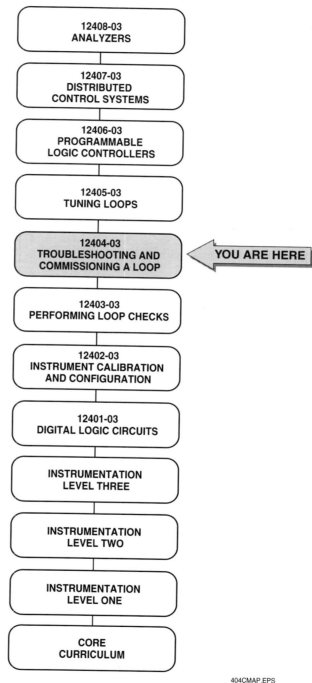

INSTRUMENTATION LEVEL FOUR

12408-03
ANALYZERS

12407-03
DISTRIBUTED
CONTROL SYSTEMS

12406-03
PROGRAMMABLE
LOGIC CONTROLLERS

12405-03
TUNING LOOPS

12404-03
TROUBLESHOOTING AND
COMMISSIONING A LOOP   ← YOU ARE HERE

12403-03
PERFORMING LOOP CHECKS

12402-03
INSTRUMENT CALIBRATION
AND CONFIGURATION

12401-03
DIGITAL LOGIC CIRCUITS

INSTRUMENTATION
LEVEL THREE

INSTRUMENTATION
LEVEL TWO

INSTRUMENTATION
LEVEL ONE

CORE
CURRICULUM

404CMAP.EPS

1.0.0   **INTRODUCTION** 4.1

2.0.0   **FUNDAMENTALS OF TROUBLESHOOTING A LOOP** 4.1

    2.1.0   Analyzing the Loop 4.2

    2.2.0   Identifying the Problem 4.2

    2.3.0   Understanding the Loop and Its Function 4.2

3.0.0   **TROUBLESHOOTING AN OSCILLATING PROCESS** 4.3

    3.1.0   Verifying That a Problem Exists 4.3

    3.2.0   Gathering Information 4.3

    3.3.0   Identifying Possible Causes of the Problem 4.3

    3.3.1   Checking Instruments in the Field 4.4

    3.3.2   Checking the Transmitter 4.5

    3.4.0   Locating the Problem 4.5

    3.5.0   Using a Troubleshooting Flowchart 4.5

4.0.0   **PROVING A LOOP** 4.7

    4.1.0   Interpreting a Loop Sheet 4.7

    4.2.0   Applying Logical Steps in Troubleshooting a New Loop 4.8

5.0.0   **COMMISSIONING A LOOP** 4.10

    5.1.0   Drawings and Documents 4.10

    5.2.0   Commissioning Procedure 4.11

SUMMARY 4.12

REVIEW QUESTIONS 4.12

GLOSSARY 4.15

REFERENCES **4.16**

Figures

Figure 1   Panel graphic 4.3

Figure 2   Loop diagram 4.4

Figure 3   Flowchart 4.6

Figure 4   Loop sheet/wiring diagram 4.7

Troubleshooting and Commissioning a Loop

Objectives

When you have completed this module, you will be able to do the following:

1. Practice universal and methodical troubleshooting techniques in loop tuning.
2. Troubleshoot an oscillating process.
3. Troubleshoot a newly installed control loop.
4. Practice safety procedures when troubleshooting a loop.
5. Commission a loop.

Prerequisites

Before you begin this module, it is recommended that you successfully complete the following: Core Curriculum; Instrumentation Levels One through Three; Instrumentation Level Four Modules 12401-03 through 12403-03.

Required Trainee Materials

1. Pencil and paper
2. Appropriate personal protective equipment

1.0.0 ◆ INTRODUCTION

You learned in previous modules that when a control loop is initially installed, it must go through a series of checks, tests, proving, and calibrations before it is approved and released to operations as complete and calibrated. This loop approval and release to operations is referred to as commissioning a loop. It is usually supported by documentation containing pertinent loop data as well as signatures and dates relating to the checks, tests, proving, and calibration procedures leading up to commissioning.

During the proving process of the loop, problems in the loop may cause it to function improperly. Even after a loop has been commissioned, these types of faults may show up and cause problems. When these situations occur, whether during the proving stage or during operation, the source of the fault in the loop must be located and repaired. This process is referred to as troubleshooting. Troubleshooting may be as simple as recognizing an obviously defective part that can be immediately repaired or replaced, or it may require a systematic approach in locating the problem and then repairing it. This module will examine the universal approaches to troubleshooting, as well as specific loop troubleshooting, and the steps involved in commissioning a loop.

2.0.0 ◆ FUNDAMENTALS OF TROUBLESHOOTING A LOOP

In any loop troubleshooting task, there are basic fundamentals that may be followed in order to successfully remedy the problem. These fundamentals include the following:

- Analyzing the loop
- Identifying the problem
- Understanding the loop and its function in the process
- Identifying the instruments in the loop
- Testing the instruments in the loop
- Repairing or replacing the instruments in the loop

2.1.0 Analyzing the Loop

Troubleshooting a control loop involves more than making the loop work again, although that certainly is the desirable outcome. In order to troubleshoot a loop or process, you must compare its normal condition to its abnormal condition. To accomplish this step, you must have an understanding of the normal and abnormal conditions of the loop, or at least have someone present who can describe the normal condition while you witness the abnormal state. This allows you to make a comparison of the two conditions. The comparison acts as a tool in the troubleshooting process and is your first step in achieving the solution.

Once you have an understanding of the apparent difference between the normal and abnormal states of the loop, you can begin the process of figuring out the possible causes for the abnormality. One of the steps is testing the loop through the use of a loop diagram or other visual aids that graphically represent the sequential operation of the loop. Once the instrument or associated hardwire or wiring that is causing the abnormality is located, the problem can be solved by replacing or repairing the instrument or device, thus returning the loop or process to its normal condition.

2.2.0 Identifying the Problem

Troubleshooting a loop starts with identifying a problem in the loop—a problem recognized by an obvious symptom or through somewhat long-term observation that is then communicated to a technician. Communication through operator feedback is by far the most common form of communicating a problem in an operating loop or process. As a technician responsible for troubleshooting, you must learn to listen to the operator's complaint. Too often, untrained technicians will analyze a loop problem based on their own perception of the loop or process without giving full attention to those who operate or observe the loop or process on a daily basis. Valuable troubleshooting input can be lost due to poor listening skills. Operators may not always use words or terms that technically describe the actual operation of the loop or process, but their descriptive explanations of what's going on as they see or hear it can be instrumental in locating the problem and determining the cause.

Questions you might ask an operator include: Did this loop just begin to malfunction or is it a **chronic** problem? Does the problem go away when the controller is in the manual position? Is the problem worse or better if the weather is warm, cold, dry, or damp? Has anyone tried to fix the problem before? Who was the technician who worked on this loop before?

If you do not ask the operator about the problem, many tasks may have to be repeated in order to get to the point already experienced by the operator. This doesn't mean that you should not see for yourself what transpires when the problem occurs, but asking questions can provide valuable information that can be used as a starting point in the troubleshooting process. Another troubleshooting tool is the documented history or maintenance record of the loop. Many facilities keep maintenance records that list problems and remedies of previous faults with the loop, if any have occurred. Always check the maintenance records, if available, for previous problems with the loop and, if applicable, determine if the current problem matches the previous problem. If so, note what was done to remedy the problem in the past.

Recurring problems in the same loop may indicate an underlying cause that has not yet been located or repaired. It may also suggest an unsatisfactory repair or replacement on a previous maintenance call. Likewise, documented notes from previous technicians can serve as a beneficial tool in locating or determining the cause of the problem. If the records indicate that there have been no previous loop problems, reviewing the history of abnormalities on similar loops can often direct you to a possible cause for the abnormality.

Once the operator's complaint has been heard and the history of the loop has been checked, the next step in identifying the problem should include visually inspecting the loop for obvious abnormalities. If the problem still cannot be identified, performing logical test procedures on the loop may be necessary to specifically pinpoint the abnormality. The operator's feedback, history of the loop (if any), and your own observations can help you decide where to begin testing.

2.3.0 Understanding the Loop and Its Function

If you do not know how something works, you cannot troubleshoot it. This does not mean that you must learn how to operate every loop in every process that you are called upon to troubleshoot, but it does mean that you must have a high level of confidence in your ability to understand what the loop or process is suppose to do. You must know what the power source is to the loop and what types of signals are transmitted through the loop. If the operation of the loop or process is beyond your immediate knowledge, it is up to you to familiarize yourself with its operation by reviewing P&IDs, loop sheets, operation and maintenance instructions, and other drawings and documents; questioning operators and other technicians or supervisors; and observing the process.

WARNING!

Never attempt to troubleshoot a loop in a process without knowing what the loop's function is in the overall process picture. Not only will you possibly have an unsuccessful troubleshooting outcome, you could subject yourself and others in the plant and general area to potential injury or hazards by inadvertently causing undesirable and dangerous changes in the process. You must understand the instruments and their individual functions in the loop so that by your hands-on testing and analysis, you do not cause an instrument to generate a change in the process that would be detrimental to all concerned. If you do not feel confident in your level of knowledge of a loop or process, you should not attempt to troubleshoot it.

3.0.0 ◆ TROUBLESHOOTING AN OSCILLATING PROCESS

Troubleshooting a dynamic or oscillating process takes a good working knowledge of the system as well as patience. Because the process may be constantly changing, it is difficult to compare static readings. The basic procedure for troubleshooting this type of process follows.

3.1.0 Verifying That a Problem Exists

In many cases, trouble in a system is discovered by an operator. The operator may observe symptoms of a problem, or an alarm may indicate that something is wrong. In this example, the operator has discovered an oscillating process that showed up on the recorder controller.

When this type of trouble is discovered, the first step is to verify that there is a problem in the system, and then to learn something about the problem, such as the following:

• How was the problem discovered?
• When did the trouble start?
• Is the trouble constant or intermittent?

Observe the system in operation. This allows you to see the symptoms and to verify how the system works.

3.2.0 Gathering Information

In addition to finding out about the trouble, you must gather information about the system and how it is designed to operate. There are three important sources of information: the operator, the panel graphic, and the loop diagram. The operator can provide information about system operation.

The panel graphic shown in *Figure 1* is another source of information. While its primary function is to serve as a control center for the operator, the panel graphic also provides some information about the process.

The loop diagram provides detailed information about the components and design of the loop. The loop diagram in this example (see *Figure 2*) provides the following information:

• The flow element (FE 112) is an orifice.
• The flow transmitter (FT 112) senses the differential pressure across the orifice and sends a pneumatic signal to the flow recording controller.
• The recording controller (FRC 112) sends a pneumatic signal to the flow control valve.
• The flow control valve (FCV 112) responds to the controller signal to control the flow rate of the process fluid.

3.3.0 Identifying Possible Causes of the Problem

Once the symptoms of the problem have been determined and you are acquainted with how the system should work, you can begin to identify possible causes of the problem.

For example, the symptom is an oscillating process in a flow control loop. There are several possible causes for this type of problem, including the following:

• The transmitter may be malfunctioning and sending improper signals.
• A faulty controller could be causing the oscillations. This condition can occur if the controller is not tuned properly.

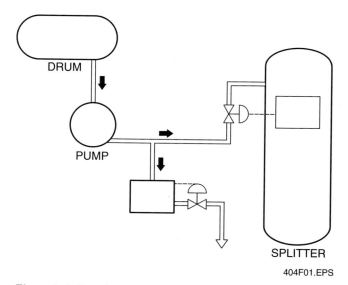

Figure 1 ◆ Panel graphic.

404F01.EPS

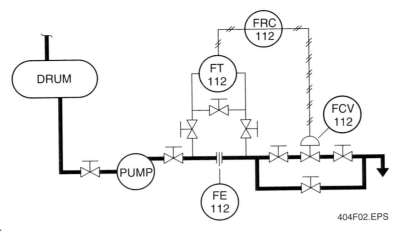

Figure 2 ◆ Loop diagram.

404F02.EPS

- The control valve could be sticking or the actuator may not be operating properly.
- The process itself may be oscillating due to a change in process fluid characteristics.

In this example, the cause of the problem could be in the controller, the transmitter, the valve, or the process. In this case, controller operation is the easiest and most likely initial check to make. A controller can often be ruled out as the cause of the problem simply by shifting it to manual and observing the controller and process response. The manual mode bypasses the controller's automatic components to produce an output that is independent of these components. When in manual, the oscillation should stop if the controller is the problem. If it continues, the controller is probably not the cause of the problem. One way to check a controller is as follows:

- Move the controller transfer switch to the center or seal position.
- Match the output to the setpoint. This is done to reduce upset during the transition from automatic to manual.
- Move the switch to manual.

If the output stabilizes and the input continues to oscillate, the controller is receiving an oscillating signal. This is an indication that the controller is probably not the cause of the problem.

If the controller is not the cause of the problem, the process of identifying the cause is continued by checking out the other possible causes. In this example, the valve, transmitter, and process remain to be checked.

3.3.1 Checking Instruments in the Field

Before going to the field, it is good practice to check the loop again. This information can help determine the most efficient way to approach the problem. Making visual checks and observing gauge readings can frequently identify or rule out possible causes of a problem.

 NOTE
Be sure that receiver-type gauges at transmitters are working correctly.

Use the following steps to check instruments in the field:

Step 1   Visually check the output gauge on the transmitter. In this situation, the gauge is oscillating, indicating that the transmitter output is oscillating. Oscillation in the transmitter output could be due to a problem in the transmitter, the valve, or the process.

Step 2   Check the instrument air supply gauge to the transmitter. If the transmitter is receiving a pulsating supply, it could cause oscillation in the transmitter output. In this case, the gauge reading is steady.

Step 3   Visually check the valve. By looking at the valve stem, you can see if it is moving or not. It should be steady because the output from the controller in the manual mode is steady. If the stem is moving, there may be a problem with the valve actuator or positioner. Look for air supply problems as well as erratic positioner outputs. In this case, the stem is steady, which rules out the valve and actuator as possibly causing the problem.

At this point, the controller and the valve are ruled out as cause of the oscillation.

3.3.2 Checking the Transmitter

Transmitter operation can be checked by first performing a zero check, which is done by equalizing the pressure on both sides of the differential pressure transmitter. With a zero differential input pressure, the transmitter output should have a signal equivalent to zero. If the output drops but continues to oscillate, the transmitter is probably the cause of the problem. One method for performing a zero check on a differential pressure transmitter is as follows:

Step 1 Open the bypass or equalizer valve. This should cause the output reading to go to zero and stop oscillating. This indicates that the transmitter is responding to a change in differential pressure.

Step 2 Close the valve on the high-pressure side of the transmitter.

Step 3 Close the valve on the low-pressure side of the transmitter.

Step 4 With the transmitter on bypass, a stable zero output indicates that the transmitter is at least responding to changes in differential pressure.

Step 5 Once its position has been verified, the transmitter should be put back into service.

Step 6 Open the low-pressure valve.

Step 7 Open the high-pressure valve.

Step 8 Close the bypass valve.

In this example, as differential pressure is restored, the output begins to rise, and the oscillation appears again. At this point, it seems that the cause of the problem is in the process itself. There are some possible instrument problems, such as a loose control valve plug, or a transmitter that does not malfunction until a differential pressure is sensed. However, checking these problems is a time-consuming job. It is most efficient to make a quick check of the process first.

3.4.0 Locating the Problem

When the cause of a problem appears to be in the process, the loop diagram or the P&ID can provide information about the equipment and its function in the process loop. Referring back to *Figure 2*, you can see that there is a drum with a pipe going from the bottom through a suction valve to a pump. From the pump, the pipe goes to a discharge valve and to the orifice plate or primary flow element.

The cause of the problem could be in the process piping, one of the valves, or the pump. Checking these possibilities first may avoid the costly time involved in removing and disassembling instruments.

In this example, the operator found that the suction valve between the tank and the pump was partially closed. The pump was trying to pump more fluid than the valve would allow, causing the pump to cavitate.

A partially blocked valve can cause abnormally low pressures to occur in the pump suction as it attempts to draw liquid in. These low pressures can cause vapor bubbles to form and subsequently collapse at the pump discharge. This process is called cavitation. Cavitation leads to pulsating flow and pump pressure. When the pump cavitated, this caused oscillation in the pump discharge. The oscillation was accurately measured by the instrument system, but the action of the controller could not correct the oscillation.

Because the cause of the problem was in the process and not in the instrumentation, it was fixed by the operator, who opened the valve that was partially closed. With the valve fully open, the process stabilized.

Once the repair or change has been made, you should verify that the problem has been corrected. Verification that the problem has been repaired usually involves making the same check that was performed when verifying the problem existed. In this example, this check involved the following steps:

- Move the controller transfer switch to the seal position.
- Match the output to the setpoint.
- Move the switch to the automatic position.
- Observe the input and the output. Both should be tracking without oscillation. If oscillation occurs, the problem has not been corrected.

3.5.0 Using a Troubleshooting Flowchart

The information gathered from this example problem and the method used to find the failure can be applied to a flowchart, as shown in *Figure 3*.

Some key points about flowcharts are as follows:

- Diamonds represent decision-making points.
- Squares (or rectangles) represent facts or solutions.
- Circles (or ovals) represent a reference to another source.

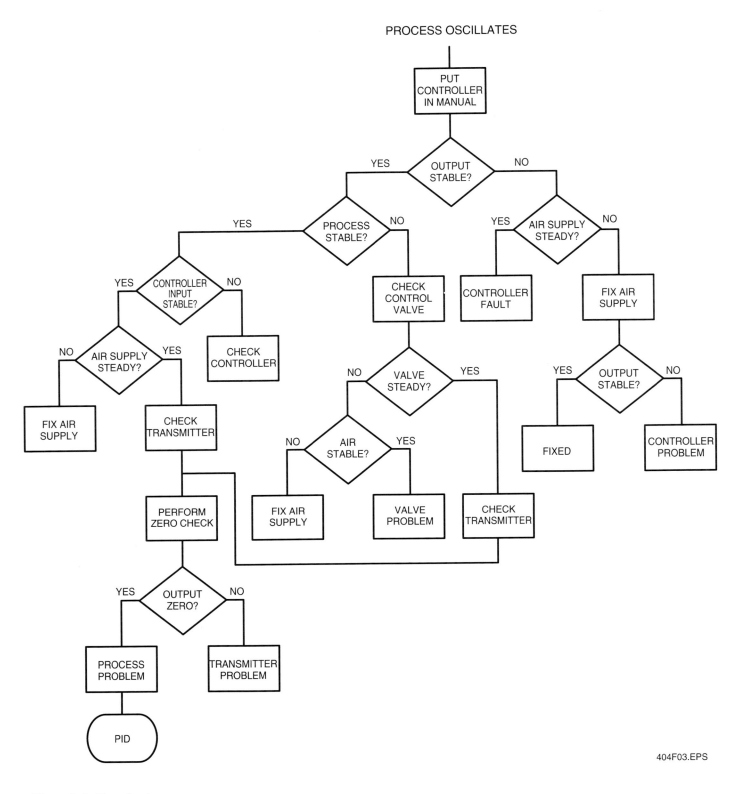

Figure 3 ◆ Flowchart.

404F03.EPS

4.0.0 ◆ PROVING A LOOP

A newly installed loop that is not yet functioning as an operational process control loop and is isolated from the process is referred to as an **uncommissioned loop**. Often times, after a loop has been installed, it fails to prove itself as a functioning loop when put through a simulated test run, which is referred to as **proving a loop**.

In proving or troubleshooting a new loop, there should be no dependency on the actual process in order to verify correct operation of the control loop, either at the primary element or at the final element location. During proving or troubleshooting, use the appropriate test equipment designed to simulate the variable process properties.

At the final element end, which is normally a control valve, the control valve is typically isolated from the process through valve manifold manipulation so that the valve's response (movement) to the loop's signal can be verified. However, during the proving or troubleshooting procedures, all wiring (or tubing in pneumatic loops) should be connected, and the loop should receive its power from the 24VDC power supply or instrument air supply (or both, in some cases).

4.1.0 Interpreting a Loop Sheet

In order to troubleshoot any control loop that passes through multiple zones of interconnected wiring (or tubing, in the case of pneumatic loops) and contains several components in various locations, you must reference the loop sheet for that loop and be able to interpret it before you can take a logical approach to locating a problem in the loop. At this moment, interpret *Figure 4*, which is a drawing of a loop sheet/wiring diagram showing all the components and interconnected wiring in a pressure control loop.

There are three zones depicted on the drawing: the field process area, the panel junction area, and the control room panel. As would be expected, the pressure transmitter (PT1) and the pressure control valve (PCV1) are located in the field process area. Likewise, because the transmitted signal in this loop is a 4–20mA analog signal and PCV1 is a pneumatic control valve, there is also an **I/P transducer** located near the control valve that converts the analog signal received from the controller into a proportional pneumatic signal that can be used by the valve. In this particular loop, there is also a pressure gauge (PI1) located near the location of PT1.

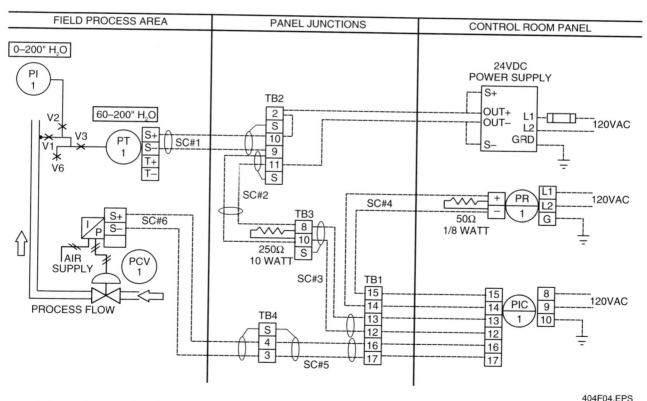

Figure 4 ◆ Loop sheet/wiring diagram.

404F04.EPS

All of the field wiring interconnects with the control room wiring in the panel junction zone. This zone is typically located somewhere in the control room, usually behind the panel or in a junction box enclosure mounted on a wall in the control room. In any event, it should be in an easily accessible location. Finally, the control room panel holds the loop's power supply, a pressure recorder, and a pressure indicating controller. You should be able to see why it may be very difficult to troubleshoot the newly installed loop without this drawing.

The next few sections are pertaining to the loop illustrated in *Figure 4*; however, most of these procedures, suggestions, and recommendations in troubleshooting will pertain to most loops, including analog, digital, and pneumatic loops, with minor differences applied. Naturally, a pneumatic loop will be driven by an adequate air supply and not a 24VDC power supply; however, the power source must still be verified for correct output, regardless of its nature.

4.2.0 Applying Logical Steps in Troubleshooting a New Loop

WARNING!
Although lockout and tagout procedures should be outlined in all facilities, they are not always followed by all personnel. At no time during troubleshooting a problem in any loop should changes be made to the current position of any valve, switch, circuit breaker, interrupting device, or termination without first following established lockout/tagout procedures. An existing problem or additional work may be going on that warrants the current position. That is why it is very important to never troubleshoot a loop without first isolating the loop from the process.

In this example loop, assume that all you know is that during the loop proving, PCV1 did not respond to the simulated primary element signal from the test equipment and acts as if it is not even part of the loop. With a problem like this, any component or wiring connection in the loop could cause that symptom. So where do you begin?

You begin by going back to the first troubleshooting fundamental, which is identifying the problem. Because you are proving the loop for the first time, there's a good chance that you installed the loop, and you should be able to recognize the problem as no response at the control valve.

The second basic troubleshooting fundamental is understanding the loop and how it is supposed

to operate, including all wiring and tubing interconnections. In order to satisfy this fundamental, you must use the loop sheet/wiring diagram.

The next focus should be on the control valve because its non-responsiveness is the only symptom showing up that indicates a problem in the loop. This leads into the next basic fundamental: testing the instruments in the loop. In testing the loop, you could begin at the power supply and verify that you have 24VDC. However, because the power supply in this loop is in the control room, and the symptom is showing up in the field at the control valve, start at the location where the symptom of the problem shows up—the control valve.

NOTE
Troubleshooting must be performed using the proper control loop test equipment. If you do not have loop simulating test equipment available, troubleshooting will take much longer and can only be accomplished by trial-and-error methods, which are not the recommended methods in performing reliable loop troubleshooting. In situations where the proper test equipment is not available, loop troubleshooting should be left up to instrumentation personnel set up for that type of work, or the proper test equipment should be acquired.

The best place to begin is to make sure that an adequate air supply is connected to and providing instrument air to the valve actuator, valve positioner (if installed), and I/P transducer. Without adequate air supply to all three of these components, the valve will not operate properly. This should be checked directly at the air connections to the control valve actuator, positioner, and transducer with either a precision air test gauge (0–60 psi works well) or the proper pneumatic test equipment (Wallace & Tiernan® or equivalent). If the proper air supply is not present, trace the piping or tubing, and locate the interruption.

CAUTION
Do not open a closed instrument air supply valve until you have determined why the valve is in a closed position.

If the proper air supply is present at all three components, but the valve has no response to the loop, the next step is to verify 24VDC loop power to the transducer by measuring between S+ and S− with a DC voltmeter or loop tester. If 24VDC is present, you have eliminated any problems in the power supply circuitry, but you should now check the proper operation of the I/P transducer by disconnecting the loop wiring and simulating a varying 4–20mA signal into the S+ and S– terminals. The valve should **stroke** as you vary the input current to the transducer between 4–20mA. If the valve responds proportionally, the transducer and valve are okay, and the problem is somewhere else in the loop, specifically in the signaling circuit that regulates the current (such as the controller). If the valve does not respond, either the transducer is defective or the valve is stuck or defective.

The transducer can be checked by disconnecting the transducer outlet air line to the valve and connecting a precision pressure gauge to the outlet. The air pressure output of the transducer should vary proportionally between 3–15 psi as you vary the current input to the transducer between 4–20mA. If the transducer functions properly, the problem is in the valve, and the valve should be replaced. If the transducer does not function properly, replace it.

NOTE

Pay close attention to the fact that the 24VDC supply is looped in series (and not paralleled) through the control loop. It originates at the 24VDC S+ terminal of the power supply located in the control room and literally loops its way through the entire loop, returning to the S– of the power supply. Therefore, an absence of 24VDC power at any instrument does not tell where the problem is, it only indicates an open loop. Intentionally varying the current in a 24VDC-supplied loop by changing the overall loop's resistance is what causes the final element to respond. Any unintentional change in resistance, whether more or less, in a control loop will cause problems in the loop because the current flow is no longer totally controlled by the loop.

If 24VDC is not present at the I/P transducer, which is the assumption in this problem, return all wiring and tubing in the field to its normal condition, proceed to the loop's power supply in the control room, and verify that you have 24VDC across OUT+ and OUT– (or S+ and S–). Of course, if 24VDC is not present at the output of the loop's power supply, you must verify that you have 120VAC to the power supply. If not, determine why and restore AC power. If there is AC power at the input and no DC voltage on the output, replace the power supply following all safety procedures. In this case, you can assume that you

have a good power supply but still no 24VDC at the I/P transducer. At this point, you have determined the following:

- *Problem*: The valve does not respond.
- *Test*: You have proper air supply to all pneumatic instruments.
- *Test*: You do not have 24VDC at the I/P transducer.
- *Test*: You have 24VDC at the output of the power supply.
- *Test*: The control valve is okay.
- *Test*: The I/P is okay.
- *Conclusion*: You have an open circuit somewhere in the loop's wiring, and you must begin a systematic test at various terminal points.

Based on these conclusions, you can continue to perform a logical test as follows, using the loop sheet/wiring diagram:

Step 1   Test for 24VDC between terminals 2 and 11 on TB2, testing hardwiring from the 24VDC power supply to TB2.

Step 2   Test for 24VDC between terminals 10 and 11 on TB2, testing to ensure the jumper is installed between terminals 2 and 10 on TB2.

Step 3   Test for 24VDC between terminals 9 and 10 on TB2, testing field shielded cable SC#1 from TB2 to pressure transmitter (PT1) and also testing the internal circuitry of PT1 for an open loop.

Step 4   Test for 24VDC between terminals 8 and 10 on TB3, testing panel junction shielded cable SC#2 from TB2 to TB3.

Step 5   Check for the installation of a 250Ω, 10W resistor between terminals 8 and 10 on TB3. This resistor must be in place in order for the loop to calibrate and operate properly.

Step 6   Test for 24VDC between terminals 12 and 13 on TB1, testing panel junction shielded cable SC#3 from TB3 to TB1.

Based on these tests and checks, repair or replace any defective wiring and terminations. If no problems are located, and 24VDC is present at all of these locations, you have determined the following:

- *Problem*: The valve does not respond.
- *Test*: You have proper air supply to all pneumatic instruments.
- *Test*: You have 24VDC at the output of the power supply.
- *Test*: The control valve is okay.
- *Test*: The I/P is okay.

- Terminations and voltage levels on TB1 (except terminals 14, 15, 16, and 17), TB2, and TB3 are okay.
- SC#1, SC#2, SC#3, and SC#4 are okay.
- A 250Ω, 10W resistor is installed between terminals 8 and 10.
- A jumper is in place between terminals 2 and 10 on TB2.
- *Conclusion*: The problem is confined to one or more of the following components:
  - PIC1 controller
  - Wiring from TB1 to PIC1
  - Wiring from TB1 to TB4
  - Wiring from TB4 to I/P terminals in the field

Based on these conclusions, you can continue to perform a logical test as follows on the remaining terminal points and wiring, referencing the loop sheet/wiring diagram:

Step 1 Test for 24VDC between terminals 12 and 13 on the PIC1 terminal strip, testing the hardwiring between TB1 and PIC1.

Step 2 Test for 120VAC between terminals 8 and 9 on the PIC1 power terminal strip, testing for AC controller operating power.

Step 3 Test for 24VDC between terminals 16 and 17 on the PIC1 terminal strip, testing the output circuitry to the final element of PIC1.

Step 4 Test for 24VDC between terminals 14 and 15 on PIC1, testing the output circuitry to recorder PR1.

Step 5 Test for 24VDC between terminals 16 and 17 on TB1, testing the hardwiring from terminals 16 and 17 on PIC1 to terminals 16 and 17 on TB1.

Step 6 Test for 24VDC between terminals 3 and 4 on TB4, testing the panel junction shielded cable SC#5 from TB1 to TB4.

Conclusion: In this loop, all voltage tests up to and including TB4 indicated 24VDC.

Based on these conclusions, you can identify the problem as a defective wiring in field shielded cable SC#6. The question may be rightly asked, with the problem located near the I/P, why didn't you start from the I/P's wiring and work back through the loop in testing the wiring? Keep in mind that only because of the fact that you have now completed the troubleshooting steps and located the problem close to the valve does it seem logical that you should have started at the final element end. Where a technician chooses to start the troubleshooting procedures is at the discretion of the technician. The problem in this case could have just as easily been defective wiring at the power supply end. It only appears that you started at the wrong end of the loop because of where the problem was eventually found; but it was only found because you stuck to the logical approach and followed through the steps. Start wherever your instinct tells you to start, as long as you follow the logical steps.

5.0.0 ◆ COMMISSIONING A LOOP

Commissioning is the testing that occurs before any process fluids or chemicals are added into the system. Commissioning includes checks of both the physical installation and system functions. During the testing procedure, each component or function is tested and checked off on a form. Any problems are noted on a punch list and must be corrected before system startup.

5.1.0 Drawings and Documents

All project drawings and documents must be assembled before starting the commissioning process. These include the following:

- *Piping and instrumentation diagram (P&ID)* – The P&ID shows the method of control for a given process, along with approximate device locations and tag numbers.
- *Control room plan, section, and detail drawings* – These drawings show the physical location of all equipment in the control room and in any auxiliary equipment and computer rooms.
- *Control room single-line diagram drawings* – These diagrams show the interconnecting cables and pneumatic connections (but not terminations) between major devices in the control room and those between auxiliary equipment, local control panels, and field devices.
- *Instrument location and conduit plan* – This drawing shows the locations and elevations of field instruments, control panels, and junction boxes. It also shows the associated raceway system (electrical conduits and cable trays).
- *Pneumatic location and tubing plan* – This drawing shows the location of the air manifold and pneumatic devices, including the routing of pneumatic tubing.
- *Control panel layout and wiring drawings* – These drawings show the size and location of all panel devices, along with their electrical and pneumatic interconnections. They also depict the control panel termination points for field device connections, which are helpful when troubleshooting open circuit problems in loops.

- *Power distribution panel schedule drawing* – This schematic diagram shows the voltage, amperage, and locations of all circuit-interrupting devices associated with the instruments and systems. It is helpful when attempting to locate problems with equipment supply power.
- *Hardware installation detail drawings* – These drawings show the actual mounting, orientation, and connection of each instrument and control device. These drawings can be helpful when troubleshooting instruments that require special installation procedures.
- *Loop diagrams* – These are the most referenced drawings during the proving, troubleshooting, and commissioning processes. They show the connection details of the control system components arranged into a loop.
- *Instrument index* – This lists the instrument tag numbers for all instruments or control system components, including the associated drawing numbers, materials, specification sheets, and other related information for each component.
- *Specification sheets* – These documents are written in a standard Construction Specification Institute (CSI) format. They contain the technical details for one or more tag-marked components.

5.2.0 Commissioning Procedure

Commissioning normally involves component testing followed by functional loop testing. Single component devices, such as switches, relays, and pressure or temperature gauges not associated with a loop are checked as stand-alone devices during the commissioning procedure. Each system will include detailed commissioning checklists. The actual procedure depends on the system under test. A typical procedure includes the following tasks:

Step 1   Assemble all drawings and documents related to the project. This includes P&IDs, loop diagrams, wiring diagrams, tubing and conduit plans, section and detail drawings, instrument lists, panel schedules, and specifications.

Step 2   Perform pressure tests per the installation design values. Test parameters include duration, test medium, and pressure level.

Step 3   Verify that the instruments have been calibrated. This is normally done by the instrument vendor but may be required as part of the commissioning procedure.

Step 4   Perform mechanical checks. This involves the following tasks:

- Make sure that each device has been securely installed in the correct location, is identified with a loop number tag, and has sufficient clearance for service and maintenance.
- Check that valves are not installed backwards and stroke all control valves (apply a variable input and verify that they open and close all the way and that there are no spots where the valve appears to stick or operate sluggishly).
- Examine instrument air supply lines for leaks, pinch points, and the presence of dirt or moisture.

Step 5   Perform electrical checks. This involves the following tasks:

- Verify the correct loop supply voltage.
- Ensure that the electrical configuration (voltage) of all instruments match the job specifications.
- Verify point-to-point wiring continuity as compared to the loop diagram.
- Check the wiring for unintentional short circuits to ground.
- Verify that specified ground connections are made and that signal cable shields (on the same cable) are not grounded at more than one point.
- Check the transmitter outputs.
- Verify the correct transmitter output display at the controller or computer control system.
- Check the operation of single component devices such as switches, relays, and pressure or temperature gauges.
- Check the controller settings and adjust if necessary.
- Confirm that the output devices go to a fail-safe mode at failure.

Step 6   Test the operation of all safety devices and interlocks.

Step 7   Verify the operation of the functional loop by simulating process variables as closely as possible.

Step 8   Correct any problems using a systematic process of troubleshooting and repair.

Step 9   When all components are functioning properly and have been checked out, assemble the completed test records. These are supplied with the as-built drawings, vendor data, and instruction manuals as part of the complete system documentation package.

Summary

Locating and repairing loop operational problems requires a systematic troubleshooting process. Proper tools include precision test equipment as well as intangible tools such as observation, listening skills, and logical thinking. Randomly searching for a problem in a loop that shows up as a single symptom in one particular instrument or process is not the most efficient and cost-effective way to correct an abnormal condition.

Troubleshooting and repairing operational problems is only one step in the complete process of commissioning a control loop. Other tasks include pressure testing, instrument calibration, mechanical and electrical checks, and verifying the correct operation of all safety devices and interlocks. When these tests have been made, the operation of the process loop is simulated, and any problems are diagnosed and repaired.

The final step in commissioning is the assembly of all related drawings, documentation, and test records. These documents provide a baseline for future maintenance and repair.

Review Questions

1. Identifying the problem is a basic fundamental of proving a loop.
   a. True
   b. False

2. The most common form of communicating a problem in an operating loop or process is _____.
   a. a written work order
   b. operator feedback
   c. observation
   d. experience

3. When you are troubleshooting a loop, you should never rely on notes from previous technicians who worked on the loop.
   a. True
   b. False

4. It is difficult to compare static readings in an oscillating process when troubleshooting that type of problem because _____.
   a. records are not available
   b. static readings are always the same
   c. the process is constantly changing
   d. the zero does not remain constant

5. When troubleshooting, you should always observe the symptoms of a problem before trying to identify possible causes of the problem.
   a. True
   b. False

6. If a controller in a loop is put in manual mode, the control valve in that loop should be _____.
   a. fully closed
   b. fully open
   c. steady—not moving
   d. modulating

7. A _____ check is performed by equalizing the pressure on both sides of a differential pressure transmitter.
   a. span
   b. range
   c. output
   d. zero

8. If a pump cavitates, its discharge pressure will _____.
   a. become very high
   b. drop to zero
   c. oscillate
   d. hold steady

9. Proving a loop is performed under operational conditions, using the process as the primary element signal.
   a. True
   b. False

10. If a control loop must convert an analog current signal to a pneumatic signal to be used by a control valve, a(n) _____ must be used to make this signal conversion.
    a. P/I transducer
    b. valve positioner
    c. I/P transducer
    d. valve actuator

11. It is always a safe practice to change current process valve positions during troubleshooting procedures.
    a. True
    b. False

12. _____ is the testing required just before process fluids or chemicals are introduced into the system.
    a. Troubleshooting
    b. Proving
    c. Commissioning
    d. De-commissioning

13. A _____ shows the connection details for control system components.
    a. P&ID
    b. loop diagram
    c. control room plan
    d. single-line diagram

14. Safety devices and interlocks should be tested before _____.
    a. performing pressure tests
    b. performing electrical tests
    c. proving the loops
    d. performing mechanical checks

15. The final step in commissioning should be _____.
    a. proving the loops
    b. calibrating the loops
    c. troubleshooting the loops
    d. assembling test record documents

Trade Terms Introduced in This Module

Chronic: An ongoing, recurring, or long-term problem.

I/P transducer: A device that converts a current signal (typically 4–20mA) into a proportional pneumatic signal (typically 3–15 psi).

Proving a loop: A functional control loop test in which the control loop is isolated from the process and signals are simulated using the proper test equipment.

Stroke: Causing a control valve to move through its range of motion by supplying a proportional, variable simulated signal directly into the valve's actuator.

Uncommissioned loop: A control loop in which chemicals, process fluids, or other intermediate products have not yet been introduced into the system that will be under control of the loop.

Additional Resources

This module is intended to be a thorough resource for task training. The following reference works are suggested for further study. These are optional materials for continued education rather than for task training.

www.automationtechies.com. An online resource for automation, process control and instrumentation professionals.

www.isa.org. The website of the Instrument Society of America.

Process Control Instrumentation Technology, 2002. Curtis D. Johnson. New York: Prentice Hall.

Process Control Systems, 1996. F. Greg Shinskey. New York, NY: McGraw-Hill Professional Publishing.

Troubleshooting: A Technician's Guide, 2000. William A. Mostia. Research Triangle Park, NC: Instrument Society of America.

NCCER CURRICULA — USER UPDATE

NCCER makes every effort to keep its textbooks up-to-date and free of technical errors. We appreciate your help in this process. If you find an error, a typographical mistake, or an inaccuracy in NCCER's curricula, please fill out this form (or a photocopy), or complete the online form at **www.nccer.org/olf**. Be sure to include the exact module ID number, page number, a detailed description, and your recommended correction. Your input will be brought to the attention of the Authoring Team. Thank you for your assistance.

Instructors – If you have an idea for improving this textbook, or have found that additional materials were necessary to teach this module effectively, please let us know so that we may present your suggestions to the Authoring Team.

NCCER Product Development and Revision

13614 Progress Blvd., Alachua, FL 32615

Email: curriculum@nccer.org
Online: www.nccer.org/olf

❑ Trainee Guide ❑ AIG ❑ Exam ❑ PowerPoints Other _____

Craft / Level: _____ Copyright Date: _____

Module ID Number / Title: _____

Section Number(s): _____

Description: _____

_____

_____

_____

Recommended Correction: _____

_____

_____

_____

_____

Your Name: _____

Address: _____

_____

Email: _____ Phone: _____

Tuning Loops

COURSE MAP

This course map shows all of the modules in the fourth level of the Instrumentation curriculum. The suggested training order begins at the bottom and proceeds up. Skill levels increase as you advance on the course map. The local Training Program Sponsor may adjust the training order.

INSTRUMENTATION LEVEL FOUR

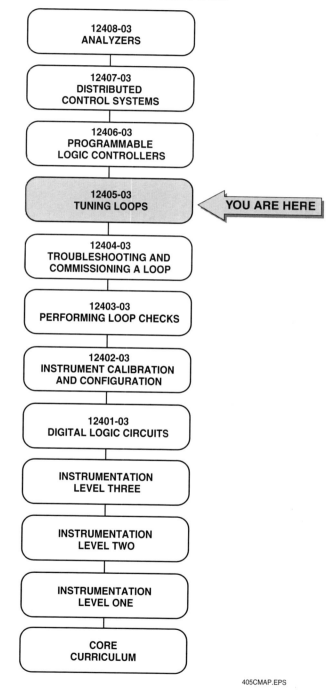

12408-03
ANALYZERS

12407-03
DISTRIBUTED
CONTROL SYSTEMS

12406-03
PROGRAMMABLE
LOGIC CONTROLLERS

12405-03
TUNING LOOPS ◁ YOU ARE HERE

12404-03
TROUBLESHOOTING AND
COMMISSIONING A LOOP

12403-03
PERFORMING LOOP CHECKS

12402-03
INSTRUMENT CALIBRATION
AND CONFIGURATION

12401-03
DIGITAL LOGIC CIRCUITS

INSTRUMENTATION
LEVEL THREE

INSTRUMENTATION
LEVEL TWO

INSTRUMENTATION
LEVEL ONE

CORE
CURRICULUM

405CMAP.EPS

MODULE 12405-03 CONTENTS

1.0.0   **INTRODUCTION** 5.1

2.0.0   **REVIEW OF PROPORTIONAL CONTROL** 5.2

    2.1.0   Proportional/Integral (PI) Control 5.2

    2.2.0   Proportional/Integral/Derivative (PID) Control 5.2

3.0.0   **TERMS AND DEFINITIONS ASSOCIATED WITH LOOP TUNING** 5.3

4.0.0   **BASIC EQUATIONS** 5.3

    4.1.0   Energy Balance 5.3

    4.2.0   Time Constant 5.4

    4.3.0   Complete Response 5.4

    4.4.0   Process Gain 5.5

    4.5.0   Proportional Band 5.5

    4.6.0   Integral Time 5.6

    4.7.0   Derivative Time 5.6

    4.8.0   PID Loop 5.7

5.0.0   **LOOP TUNING METHODS** 5.8

    5.1.0   Ultimate Period/Ziegler-Nichols Closed Loop 5.8

    5.2.0   Dampened-Oscillation 5.10

6.0.0   **OPEN LOOP METHODS** 5.10

    6.1.0   Time Constant 5.11

    6.2.0   Reaction Rate 5.13

7.0.0   **VISUAL LOOP TUNING** 5.14

    7.1.0   Incremental Changes 5.14

    7.2.0   Apparent Instability 5.15

    7.3.0   Sluggish Response 5.15

    7.3.1   Gain 5.15

    7.3.2   Integral Time 5.16

    7.3.3   Derivative Action 5.16

    7.3.4   Cascade Control 5.16

    7.3.5   Sticking Valves (Stiction) 5.17

SUMMARY 5.17

REVIEW QUESTIONS 5.18

GLOSSARY 5.19

REFERENCES 5.20

Figures

Figure 1   First-order lag plus deadtime .5.3

Figure 2   Surge tank 5.5

Figure 3   Proportional and derivative output 5.7

Figure 4   PID output 5.8

Figure 5   Control system at the ultimate gain
or proportional band 5.8

Figure 6   Responses of the measured variable
to a supply disturbance 5.9

Figure 7   Test data for the ultimate tuning method 5.9

Figure 8   Test data for the dampened-oscillation method 5.10

Figure 9   Block diagram of a process being
open loop tuned 5.11

Figure 10   Process reaction curve for an open loop
tested process 5.11

Figure 11   Time constant method for a single capacity
with deadtime approximation 5.11

Figure 12   Test data for open loop time constant method 5.12

Figure 13   Reaction rate method for a single capacity
with deadtime approximation 5.13

Figure 14   Test data for open loop reaction rate method 5.14

Figure 15   Waveform of a sticking control valve (stiction) 5.17

Tuning Loops

Objectives

When you have completed this module, you will be able to do the following:

1. Describe the function and applications of various PID controllers.
2. Apply the appropriate equations and perform closed loop tuning.
3. Perform open loop tuning.
4. Perform visual loop tuning.

Prerequisites

Before you begin this module, it is recommended that you successfully complete the following: Core Curriculum; Instrumentation Levels One through Three; Instrumentation Level Four Modules 12401-03 through 12404-03.

Required Trainee Materials

1. Paper and pencil
2. Appropriate personal protective equipment

1.0.0 ◆ INTRODUCTION

One of the primary points of interest in the overall performance of a control system is the response of the control loop and process acting together. The term *response* is used to describe the dynamic behavior of the entire system after a disturbance, which may occur either as demand or supply. The overall quality of the response is judged by the speed with which the controlled variable returns to the setpoint, by the amount of overshoot that occurs, and by the stability of the system during the transient condition.

The performances that define the quality of response for a control system tend to counteract each other. Generally, an increase in the speed of response decreases the stability of the system. An attempt at achieving too fast a speed of response may actually create a totally unstable system. A stable control system is one that is undergoing **oscillations** that are constant or decreasing in **amplitude**. In an unstable control system, the oscillations are increasing in amplitude. These oscillations normally occur at a certain frequency called the critical frequency. This critical frequency is the one at which there is a 360-degree phase shift around the control loop. The 360-degree phase shift is critical because under this condition, energy or material enters the process in such a way that the oscillations are maintained.

The 360-degree phase shift is not enough to sustain the oscillations. The gain of the control loop must be equal to 1 in order to sustain oscillations at the critical frequency. A control loop gain equal to 1 means that enough energy or material is entering the process to overcome losses. To sustain oscillations, the energy or material must enter the process at the right time, and enough must enter so that the oscillations are constant in amplitude. The control loop gain is adjustable and is normally made at the controller.

Adjusting the controller to match the process is called controller tuning. The goal is to achieve a condition in which the control loop gain is less than 1 at the critical frequency. This condition produces a stable response of the control system during a supply or demand disturbance.

2.0.0 ◆ REVIEW OF PROPORTIONAL CONTROL

The different forms of proportional control were covered in detail earlier in the Level Two module *Process Control Theory*. You may want to review that material before continuing with this module. Some key points about proportional control and variations of proportional control are briefly reviewed here.

As the name implies, proportional control acts to position the final loop control element (a valve, for example) to a position that is proportional to the direction and magnitude of the deviation in the value of the controlled variable (temperature, pressure, flow) from a desired setpoint. It provides a stepless output that can position a control valve or other final control device at intermediate positions as well as fully open and fully shut or on and off. With proportional control, the final control element assumes a definite position for each value of the measured variable; thus it has a linear output-to-input relationship. The setpoint is typically in the middle of a specified proportional range or band, resulting in there usually being an offset between the active loop control point and the setpoint. The proportional band is usually expressed as a percent of full scale or degrees. It may also be referred to as a gain, which is the **reciprocal** of the proportional band.

The main advantage of proportional control is that it immediately produces a **proportional output** as soon as an error signal exists at the controller. This enables the final control element to be repositioned in a relatively short time. The main disadvantage is that a residual offset error exists between the measured variable and the setpoint at all times except at one set of system conditions. This set of conditions is when the final control element is positioned such that the controlled variable exactly satisfies the setpoint.

It should be pointed out that with a proportional controller, it is typically necessary to manually make a small adjustment (manual reset) to bring the controlled variable to setpoint on initial startup or if the process conditions change significantly.

2.1.0 Proportional/Integral (PI) Control

Some control loops use a proportional/integral (PI) mode of control. This mode enhances control by combining the immediate response output characteristics of the proportional mode with an **integral** control function. The integral control function provides for an output that is proportional to the time integral of the input. This means that it continues to change as long as an error exists. It acts only when an error exists between the controlled variable and the reset value. (The reset value is the setpoint for a set time period.) As a result, the integral function acts to eliminate any offset that exists between the measured variable and the setpoint, which can occur with proportional control. The addition of the integral action to proportional control automatically performs a gain resetting. For this reason, PI controllers are sometimes called proportional plus automatic reset, or proportional plus reset controllers.

The main difference between proportional control and PI control is that proportional control is limited to a single final control element position for each value of the controlled variable. With a PI controller, the position of the final control element depends not only on the location of the controlled variable within the proportional band but also on the duration and magnitude of the offset of the controlled variable from the setpoint. That is, under steady state conditions, the control point and setpoint will be the same.

The PI control mode can be adversely affected by sudden large error signals. The large error can be caused by a large demand deviation, a large setpoint change, or during initial system or process startup. Such a large sustained error signal can eventually cause the controller output to drive to its limit. The result is called reset windup. This situation causes loss of control for a period of time.

2.2.0 Proportional/Integral/ Derivative (PID) Control

Proportional/integral/derivative (**PID**) control is the most sophisticated form of loop control. It adds a **derivative** function to the proportional and integral functions used in PI control. The derivative control is also called rate control. It provides an output that is proportional to the derivative (rate of change) of error. This means it acts only when the error is changing with time. As a result, the more quickly the control point changes, the greater the corrective action the derivative function provides. If the control point moves away from the setpoint, the derivative function outputs a corrective action to bring the control point back more quickly than would occur using the integral function alone. If the control point moves toward the setpoint, the derivative function reduces the corrective action to slow down the approach to setpoint, which reduces the possibility of overshoot.

PID control provides the most accurate and stable control when applied in systems that have a relatively small mass and those that react quickly to changes in energy added to the process. PID control is recommended in systems where the load on the process changes often, and the controller is expected to compensate automatically due to frequent changes in setpoint, the amount of energy available, or the mass to be controlled.

3.0.0 ◆ TERMS AND DEFINITIONS ASSOCIATED WITH LOOP TUNING

It is important to be able to recognize and define key terms associated with PID loops and their tuning process. The following terms are important when making changes to a PID loop controller:

- *Deadtime* – The time interval between the start of a change at the process input and when the controlled variable starts to respond. It is a range through which the controlled variable changes without the controller initiating a correction. Deadtime is an inherent property of a process system that can be minimized but not eliminated. It is the most common cause of many closed control loop performance problems, especially if it is relatively long. Deadtime is also commonly called delay.

- *Lag* – The dynamic characteristic of a system where the measured output lags or falls behind the system input. First-order lag is the most common lag in process systems. Any process system above a first-order lag (second-order lag, third-order lag, and so on) is considered a higher order system and is more difficult to tune. As shown in *Figure 1*, many processes can be described as a combination of first-order lag plus deadtime (FOLPDT).

- *Time constant* – The elapsed time or capacity lag starting after the deadtime between the start of a process input change or disturbance and when the resulting process output variable reaches a predefined or steady state value. One time constant is defined as the time required for a process change to reach 63.2% of the total change (*Figure 1*). A full response requires five time constants for completion. It should be pointed out that for a series-lag process system, the overall time constant is the sum of the individual time constants—one for each process lag. Depending on the complexity and nature of the controlled process, time constants can be difficult to measure or estimate.

4.0.0 ◆ BASIC EQUATIONS

The equations discussed here are the most common ones with which you'll need to be familiar. Each equation has a basic function in determining the best settings to obtain a stable process.

4.1.0 Energy Balance

The energy balance equation is used as a simple method to show how a process system receives energy and how it loses energy. In order for any process system to be stable, the equation must be balanced. The equation is written as follows:

$$E_{out} = E_{in} - E_{lost} - E_{stored}$$

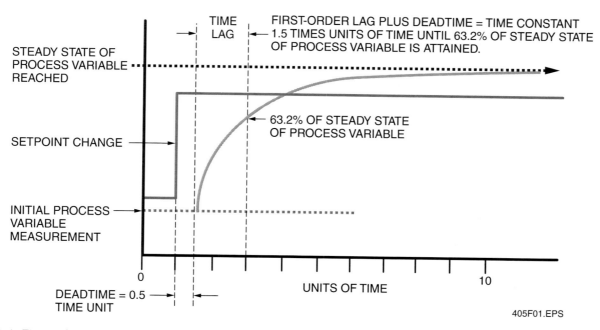

Figure 1 ◆ First-order lag plus deadtime.

Where:

E_{out} = total energy leaving the process system

E_{in} = total energy coming into the process system

E_{lost} = amount of energy the process system loses through heat dissipation and head loss

E_{stored} = amount of energy remaining in the process system

This energy can be in the form of heat, pressure, and flow. As an example, allow hot water entering the shell side of a heat exchanger to increase in temperature. Due to the temperature increase, the energy input to the heat exchanger is increased. This increases the energy transfer to the secondary fluid, raising its temperature. This is the energy output. However, at the same time, more energy is being lost to the surroundings because of the increased temperature differential. Also, the internal components of the heat exchanger are absorbing and storing energy. These three actions continue until equilibrium is reached. At that time, the increased energy input is balanced by an increase in the energy output, the energy lost, and the energy stored. The net result is increased temperature of the secondary fluid out, greater temperature loss to the surroundings, and increased stored energy within the heat exchanger.

This equation is applicable to all systems. It is important to note that only when the equation is balanced will a system be stable.

4.2.0 Time Constant

The time constant of a process is the measurable characteristic representing the capacity lag of the process. A time constant may be defined as the time required for a process to reach 63.2% of the total change. Therefore, a full response requires five time constants for completion. A simple formula for determining the time constant for a pneumatic or fluid system is as follows:

$$TC = V \div F$$

Where:

TC = time constant

V = volume or capacity of the system

F = flow [calculated by the square root of the differential pressure drop across an orifice $(\sqrt{\Delta P})$]

This equation is a basis for all open loop tuning methods, which will be discussed in the next section. An example using this formula follows.

Given that flow through a system is 100 gallons per minute (gpm), the capacity of the system is 1,000 gallons, and the gain equals 1, how long would it take this system to settle into a steady state if a disturbance were introduced? You can find the answer by applying the time constant equation as follows:

$$TC = V \div F$$

V = 1,000 gallons

F = 100 gpm

Therefore:

$$TC = 1,000 \div 100$$

$$TC = 10$$

4.3.0 Complete Response

To find out what percent a process has changed in a given time, you use the complete response equation. It is used when performing open loop tuning and is expressed as follows:

$$CR = A(1/{-}e)(-t/TC)$$

Where:

CR = complete response of the controlled variable at any time (%)

A = amplitude of change applied to the controlled variable (%)

e = base of the natural logarithm, 2.718 (no units)

t = given time

TC = time constant

You can use the complete response method to determine the time constant. Because the total change of the controlled variable is 100%, one time constant should occur when 63.2% of the total change has elapsed. In this example, the time constant is the difference in time between t_1 and t_0. If this time period is one minute, the time constant is one minute. After the first time constant of one minute, the total change of the controlled variable remaining is 36.8%. 63.2% of 36.8% is 23.3%. This 23.3% change is added to the original 63.2% change to give a total change of 86.5%. The time constant now is the difference in time between t_2 and t_1. This time period is again one minute. For this example, it takes one minute for the process to change 63.2% of the total change. By observation, the process completes its full response in five time constants.

If you know the time constant, you can find the value of the controlled variable for any point in time.

Let the time constant be one minute, and the amplitude of total change is 100%. What would be the value of the controlled variable after 2.5 minutes?

Using the equation as follows:

$$CR = A(1/-e)(-t/TC)$$

$$CR = A(1/-2.718)(-2.5/1)$$

$$CR = 100\%(-0.368)(-2.5)$$

$$CR = 92\%$$

The process has completed 92% of its total change in 2.5 minutes.

4.4.0 Process Gain

You can use a different equation to calculate the **gain of process**. This equation is used in most open loop tuning processes. The gain of a system is simply the output change divided by the input change. The equation is written as follows:

K = change in output ÷ change in input

Where:

K = process gain

The changes in input and output are normally expressed as a percentage of the span. This makes process gain a unitless number that describes the percentage change in output that can be expected following a known percentage change in the input. The use of the percent of span notation allows the comparison of gains for processes with different ranges of operation. The process illustrated in *Figure 2A* is a surge tank. The water level remains constant, provided the water flow into the tank is the same as the water flow out of the tank. The span of the inlet flow is 80 gpm, and the span of the tank level is 20'. At some point in time, the inlet flow is suddenly changed from 60 gpm to 80 gpm. The response of the surge tank to the change in inlet flow is shown in *Figure 2B*. The response is one for a single-capacity process. The 20-gpm change has been balanced by an increase of 4' of tank level. The water level remains constant at the new higher level because the input and output flows are again equal. The gain of this process can then be determined as follows:

% change in level = [(14' − 10') ÷ 20']100%
$\qquad$ = 20%

% change in inlet flow = [(80 gpm − 60 gpm) ÷ 80 gpm]100%
$\qquad$ = 25%

$$K = \frac{\%\ change\ of\ output}{\%\ change\ of\ input}$$

$$K = \frac{20\%}{25\%}$$

$$K = 0.8$$

The gain of this process is 0.8. That means that for every 1% change of the span of the inlet flow, the level will change 0.8% of its span. In physical units, an 8-gpm change in inlet flow (10% of span) will cause a 1.6' (8% of span) change in surge tank level.

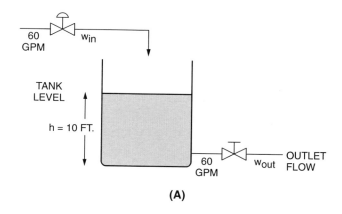

(A)

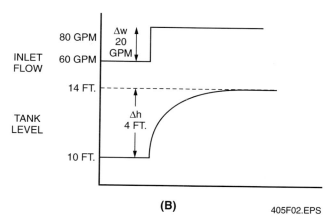

(B)   405F02.EPS

Figure 2 ◆ Surge tank.

4.5.0 Proportional Band

The input band over which the controller provides a proportional output is called the proportional band. It is defined as the change in input required to produce a full range change in output due to proportional control action and is shown by the following equation:

PB = change in input ÷ change in output × 100

For the example given, if a 100% input change causes a 100% output change, the proportional band is shown as follows:

$$PB = \frac{100\% \text{ change in input} \times 100}{100\% \text{ change in output}}$$

Therefore:

PB = 100%

Notice that gain and proportional band are inversely related, shown by the following:

$$\text{Gain} = \frac{100}{PB\%} \quad and \quad PB\% = \frac{100\%}{\text{gain}}$$

This inverse relationship is important to remember. Some proportional controllers have an adjustment that is expressed in units of gain; other proportional controllers have the adjustment expressed in units of percent proportional band.

As an example, if a 4" level change results from a 2" change in valve position, what is the proportional band?

$$PB = \frac{\% \text{ change in input}}{\% \text{ change in output}} \times 100\%$$

or

$$PB = \frac{100\% \text{ change in input}}{50\% \text{ change in output}} \times 100\% = 200\%$$

What is the gain?

$$\text{Gain} = \frac{\text{output change}}{\text{input change}}$$

or

$$\text{Gain} = \frac{2" \text{ valve stroke}}{4" \text{ level change}} = 0.5$$

4.6.0 Integral Time

A device that performs the mathematical function of integration is called an **integrator**, and the mathematical result of integration is called the integral.

The integrator provides a constant that specifies the function of integration. It also provides a linear output with a rate of change that is directly related to the amplitude of the step change input. Mathematically, the equation is written as follows:

$P(t) = (K_i)(\Delta i/t)(T)$

$P(t)$ = output rate of change

K_i = integral time constant

$\Delta i/t$ = percent change in input with respect to time

T = time that change in input lasts

For example, an input ramp changes at a rate of 2% per second and lasts for 10 seconds. The integral time constant is 2 seconds. What is the output of the integrator after 10 seconds? The output rate of change is calculated as follows:

$P(t) = (K_i)(\Delta i/t)(T)$

$P(t) = (2)(2/1)(10)$

$P(t) = 40\%$

Repeats per minute (RPM) is also a measure of how fast a signal is being integrated. Instead of expressing it in terms of time, it is expressed in terms of repeats. The repeats refer to how often a signal is integrated. The higher the RPM, the greater the rate of integration.

4.7.0 Derivative Time

A device that produces a derivative signal is called a **differentiator**. *Figure 3* shows the input-versus-output relationship of a differentiator.

The differentiator provides an output amplitude that is directly related to the rate of change of the input and is a constant that specifies the function of differentiation. Mathematically, the output is expressed by the following:

$P(d) = (T_d)(\Delta i/t)$

$P(d)$ = derivative output

T_d = derivative time

$\Delta i/t$ = percent change in input with respect to time

The derivative constant is expressed in units of seconds, which is simplified from the original unit of %/s. Given a change in input of 40% over a time of 10 seconds and a derivative time of 2 seconds, what is the derivative output?

$P(d) = (T_d)(\Delta i/t)$

$P(d) = (2 \text{ s})(40\%/10 \text{ s})$

$P(d) = 8\%$

The differentiator equation infers that a step change input would theoretically produce an infinite output. In reality, no device can produce an infinite output; therefore, the device rapidly saturates to its maximum output value when a step change is applied to the input. Increasing the derivative constant provides a larger output amplitude for a given input rate of change.

Basically, derivative time is a measure of how much derivative action is added to a process. The larger the number, the more derivative action is added. For example, if a process changes from 0% to 40% in one minute, the addition of a derivative

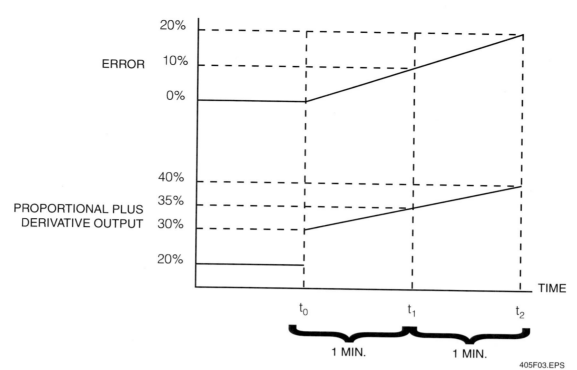

Figure 3 ◆ Proportional and derivative output.

time of one minute would cause the same process to go to 40% immediately, reaching a level of 80% after one minute.

4.8.0 PID Loop

The PID loop equation determines the output from a PID controller. The components of the equation are the proportional output, the integral output, and the derivative output. The equation that describes the PID controller output is as follows:

$$P_{PID} = K_p E_p + K_p K_i E_p \Delta t + k_p T_d (\Delta E_p / \Delta T) + P_0$$

Close examination reveals that this equation is the sum of the individual equations that describe the proportional, integral, and derivative actions.

Because a controller has a derivative section, a linear rate-of-change error signal is used to analyze the open loop characteristics of the controller. Recall that applying a step change to the derivative section results in rapid saturation of the output.

The next equation is therefore rewritten to the proper form for calculating the output at any time during a linear rate of change error input.

$$P_{PID}(t) = K_p(\Delta E_p / \Delta t)(t) + (1/2)K_p K_i(\Delta E_p / \Delta t)(t^2) + K_p T_d(\Delta E_p / \Delta t) + P_0$$

Where:

$P_{PID}(t)$ = proportional plus integral plus derivative controller output at a specified time

$K_p(\Delta E_p / \Delta t)(t)$ = proportional action to a linear rate of change

$(1/2)K_p K_i(\Delta E_p / \Delta t)(t^2)$ = integral action to a linear rate of change

$K_p T_d(\Delta E_p / \Delta t)$ = derivative action to a linear rate of change

P_0 = controller output at the start of the time period of change

Refer to *Figure 4* when reviewing the following problems.

At time t_0, the error signal begins a 5% per minute rate of change. The controller output at time t_0 is as follows:

$$P_{PID}(t_0) = (2)(5\%/min.)(0 \ min.) + (½)(2)(1 \ repeats/min.)(5\%/min.)(0 \ min.)(2) + (2)(1 \ min.)(5\%/min.) + (30\%)$$

$$P_{PID}(t_0) = 40\%$$

This calculation shows that at time t_0 the proportional and integral sections are just beginning to produce output amplitudes; therefore, their outputs at time t_0 are considered to be 0%. The derivative section, however, provides an output at the instant the proportional section output has a rate of change. In this example, a 10% derivative section output is produced and added to the 30% initial controller output for a total controller output of 40% at time t_0.

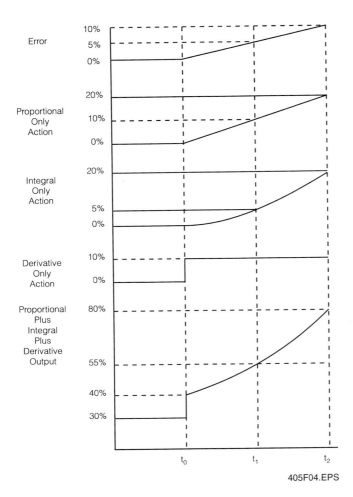

Figure 4 ◆ PID output.

5.0.0 ◆ LOOP TUNING METHODS

There are two basic approaches to tuning loops: closed loop tuning and open loop tuning. Several methods are available.

Open loop tuning methods are based on the same theory that defines open loops, which means there is no feedback. Open loop tuning methods do not even require that the controller be installed, whereas in the closed loop tuning, the controller plays the primary role in the tuning process and is constantly receiving feedback from the process. The following text will look into each approach and some of the equations applied in determining the optimal controller settings.

5.1.0 Ultimate Period/Ziegler-Nichols Closed Loop

The **ultimate period** method for tuning controllers was first proposed by Ziegler and Nichols in the early 1940s. The term *ultimate* is used with this method because it requires the determination of the ultimate gain or proportional band and the

ultimate period. The ultimate gain, Ku, or ultimate proportional band, PBu, is the maximum allowable value of gain or proportional band for which the system is stable. The ultimate period is the period of response with the gain or proportional band set at its ultimate value. During the determination of the ultimate gain or proportional band and the ultimate period, the controller is operated as a purely proportional controller. *Figure 5* shows the response of a control system that is at the ultimate gain or proportional band. As shown in *Figure 5*, the controlled variable is oscillating at a constant frequency. This is the critical frequency.

A 360-degree phase shift exists. The amplitude of the oscillations is constant. The gain of the controller is sufficient to produce a loop gain of one. The controller gain is the ultimate gain or proportional band. *Figure 5* also shows the ultimate period. The ultimate period in this case is the time period from when the controlled variable is at its maximum value to when it is at its maximum value again.

Use these two pieces of information for the ultimate-period method of controller tuning. Use the values of the ultimate gain or proportional band and ultimate period in mathematical formulas to determine the controller settings, but first, you need a method to obtain the necessary information from the control system.

The ultimate-period method is performed with the controller in service. Obtain the required information for tuning the controller by observing the waveform on a recorder connected into the loop.

The first step is to set the controller for proportional control action only. This is needed so that only one of the three possible control actions will cause the system to break into continuous oscillations. The formulas are based on only proportional action causing the oscillations. Achieve proportional-only control by setting the reset adjustment to either minimum repeats per minute or maximum integral time. Set the rate time to minimum, and set the gain of the controller at

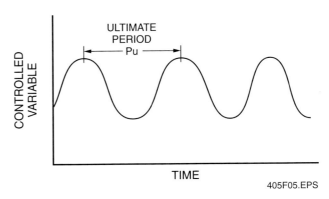

Figure 5 ◆ Control system at the ultimate gain or proportional band.

some low value. This combination of settings is the same as a wide proportional band. Make sure these adjustments are done with the controller in manual mode.

After placing the controller into automatic mode, introduce a disturbance. The easiest way is to introduce a supply disturbance. You can normally do this by causing a change in the setpoint. Observe on recorder the effect of a supply disturbance on the measured variable. Generally, the response is similar to the one shown in *Figure 6*. That is, the oscillations tend to decrease as time continues. This is demonstrated by curve A. However, in some cases, the measured variable responds as indicated by curve B. This curve represents an unstable response. If you obtain this response, the gain has to be decreased or the proportional band widened. A response, such as that shown in curve A, is required as the response after the initial disturbance. The increasing of the gain, or narrowing of the proportional band, and inserting a supply disturbance continue until you achieve continuous constant-amplitude oscillations. Between each adjustment and disturbance, it may be necessary to stabilize the process. The setting of the gain or proportional band adjustment is recorded. This is the ultimate gain or proportional band. You measure the ultimate period from the information contained on the recorder.

Using the ultimate gain or proportional band and the ultimate period, use the following equations to determine the settings that provide a **one-quarter dampened wave** of the measured variable:

- *Proportional (gain) – P*

  $K_p = 0.5Ku$ PB = 2PBu (gain)

- *Proportional (gain) and reset (integral) – PI*

  $K_p = 0.45Ku$ PB = 2.2PBu (gain)

  $T_i = Pu/1.2$ RPM = 1.2/Pu (reset or integral)

- *Proportional (gain) and rate (derivative) – PD*

  $K_p = 0.6Ku$ PB = 1.66PBu (gain)

  $T_d = Pu/8$ (derivative)

- *Proportional (gain) and reset (integral) and rate (derivative) – PID*

  $K_p = 0.6Ku$ PB = 1.66PBu (gain)

  $T_i = Pu/2$ RPM = 2/Pu (reset or integral)

  $T_d = Pu/8$ (derivative)

It should be noted that these equations are **empirical**. A control system tuned using these equations is a very close approximation of a one-quarter dampened wave for the measured variable. A certain degree of trimming is required after you have set the controller by these equations in order to obtain the one-quarter dampened wave.

As an example, consider the test results shown in *Figure 7*. The sustained continuous oscillation occurs for this process when the gain of the controller is 4. This is the same as a proportional band of 25%. A measurement of the time between maximum values of the controlled variable gives an ultimate period of 16 minutes.

Using the appropriate equations, you obtain the following settings:

- *Proportional*

  $K_p = 0.5Ku$ PB = 2PBu

  $K_p = 0.5(4) = 2$ PB = 2(25%) = 50%

- *Proportional and reset*

  $K_p = 0.45Ku$ PB = 2.2PBu

  $K_p = 0.45(4) = 1.8$ PB = 2.2(25%) = 55%

  $T_i = Pu/1.2$

  $T_I = 16 \text{ min.}/1.2 = 13.3 \text{ min.}$

  RPM = 1.2/Pu

  RPM = 1.2/16 min. = 0.075 repeats/min.

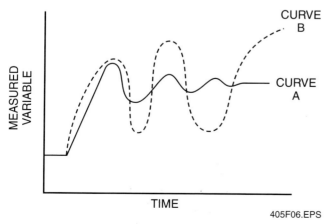

Figure 6 ◆ Responses of the measured variable to a supply disturbance.

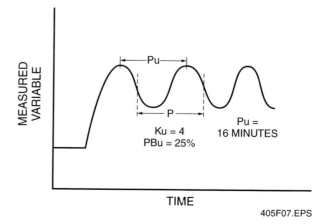

Figure 7 ◆ Test data for the ultimate tuning method.

- *Proportional and rate*

  $K_p = 0.6Ku$ $\qquad$ $PB = 1.66PBu$

  $K_p = 0.6(4) = 2.4$ $\qquad$ $PB = 1.66(25\%) = 41.5\%$

  $T_d = Pu/8$

  $T_d = 16 \text{ min.}/8 = 2 \text{ min.}$

- *Proportional and reset and rate*

  $K_p = 0.6Ku$ $\qquad$ $PB = 1.66PBu$

  $K_p = 0.6(4) = 2.4$ $\qquad$ $PB = 1.66(25\%) = 41.5\%$

  $T_i = Pu/2$

  $T_i = 16 \text{ min.}/2 = 8 \text{ min.}$

  $T_d = Pu/8$

  $T_d = 16 \text{ min.}/8 = 2 \text{ min.}$

  $RPM = 2/Pu$

  $RPM = 2/16 \text{ min.} = 0.125 \text{ repeats/min.}$

5.2.0 Dampened-Oscillation

The **dampened-oscillation method** is a modification of the ultimate period method. This method was proposed by P. Harriott in the 1960s. The method was developed because certain processes cannot tolerate continuous oscillations. In this method, you adjust the gain or proportional band using steps that are similar to those used in the ultimate period method until you obtain a response curve with a one-quarter **decay ratio**, as shown in *Figure 8*. You obtain the response by using only the proportional portion of the controller and adjusting the reset and rate portions to minimum value. The data you obtain by this method is only the period of the oscillation. Knowing this, use the following equations to set the reset and rate adjustments:

- *Reset*

  $T_i = P/1.5$ $\qquad$ $RPM = 1.5/P$

- *Rate*

  $T_d = P/8$

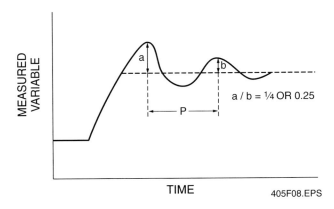

Figure 8 ◆ Test data for the dampened-oscillation method.

Once you make the reset and rate adjustments, readjust the gain or proportional band to provide a one-quarter dampened wave response of the measured variable.

6.0.0 ◆ OPEN LOOP METHODS

In contrast to closed loop methods, open loop techniques require that only one disturbance be imposed upon the process. In actuality, the controller is not in the loop during the testing of the process. These methods seek to determine the characteristics of the process.

From the process characteristics, you obtain the proper values for the controller settings. In general, most processes are multiple capacity. This requires that you use an approximation. The most common approximation is the single capacity with deadtime approximation.

Obtain the process information needed for the open loop tuning methods from the process reaction curve. The generation of this curve requires that you introduce a disturbance into the control system. This again is normally a supply disturbance. *Figure 9* shows the block diagram for a process being tuned by the open loop method.

An inspection of the block diagram in *Figure 9* does not show much difference between this arrangement and the one for closed loop tuning. However, the major difference is that the controller is in manual and not automatic. With this condition, the control loop is effectively broken because the automatic signal does not reach the final control element. The system cannot respond to changes in the controlled variable when in this configuration. The other change is that besides indicating the measured variable, the recorder indicates controller output rather than setpoint. You can use an arrangement of this type to obtain the process reaction curve for a process.

To determine the process reaction curve, bring the process to a steady state level. Ideally, the steady state level should be one where the controlled variable is at the setpoint. You may then place the controller in manual. Introduce a supply disturbance by changing the manual output of the controller. The response of the process to the supply disturbance is recorded by the recorder that receives the measured variable signal. The output of the controller is also recorded to provide a reference for when the disturbance was initiated. After the process reaction curve has been recorded, you should return the controlled variable to the setpoint by manual operation. It is easier to obtain a process reaction curve than it is to determine the ultimate gain.

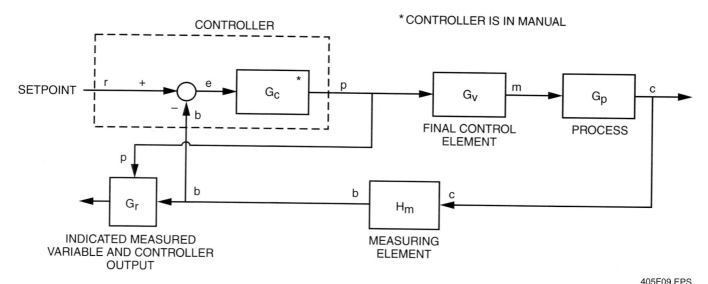

Figure 9 ◆ Block diagram of a process being open loop tuned.

Figure 10 shows a typical process reaction curve for a process that has been open loop tested. The curve is one for a multiple capacity process. This curve is approximated using the single capacity with deadtime approximation.

You employ two methods to extract the information provided by the single capacity with deadtime approximation. The methods only differ in the way the time constant is determined.

6.1.0 Time Constant

The first method is shown in *Figure 11* and is referred to as the **time constant method**. The information required from the process reaction curve is the time constant, process gain, and deadtime. The time constant is the time it takes the process to

reach 63.2% of its total change. This time is shown in the figure. The process gain is the change in the output divided by the change in the input. The change in output is the change in the measured variable. This change must be converted to % of span. This value is used, rather than the units of the measured variable, because a change of 20°F for a 100°F span is greater than a change of 20°F for a 200°F span. Therefore, a 20°F change in the measured variable does not indicate the magnitude of the change as well as the percentage of span does. A 20°F change for a 100°F span is 20% of span, whereas a 20°F change for a 200°F span is 10% of span. The 20% span change is a larger change with respect to the measuring span of the instrument

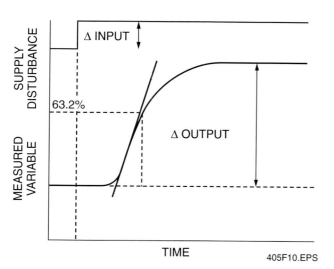

Figure 10 ◆ Process reaction curve for an open loop tested process.

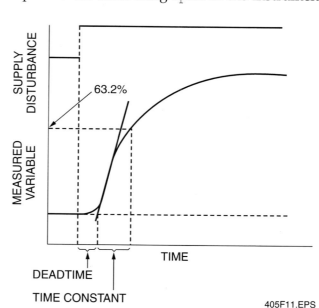

Figure 11 ◆ Time constant method for a single capacity with deadtime approximation.

than the 10% change, even though both changes are 20°F. The input change is the percent of span change in the controller output. Using the % of span requirement for the output change, the process gain may be written as follows:

$$K = \frac{\text{change in measured variable in \% of span}}{\text{change in controller output in \% of span}}$$

Where:

K = process gain

You determine the deadtime the same way as presented earlier. The deadtime is shown in the figure and is abbreviated t_d. Use this information and the following equations to tune the controller for a one-quarter dampened wave response of the measured variable:

- *Proportional*

  $K_p = t/(t_dK)$ PB = 100%/K_p

- *Proportional and reset*

  $K_p = 0.9t/(t_dK)$ PB = 100%/K_p

  $T_i = 3.33t_d$ RPM = 1/3.33t_d

- *Proportional and reset and rate*

  $K_p = 1.2t/(t_dK)$ PB = 100%/K_p

  $T_i = 2.0t_d$ RPM = 1/2t_d

  $T_d = 0.5t_d$

As an example, consider the results of the testing shown in *Figure 12*. The total change in the measured variable is 40°F. The output of the controller has a range of 3 to 15 psi. The change in the output of the controller is 1 psi. To determine the time constant, you need to know the value of the measured variable at 63.2% of the total change. Because the total change is 40°F, 63.2% of this change is 25.3°F. The 63.2% value of the measured

variable occurs when the temperature reaches 65.3°F. This value is plotted on the vertical axis in the figure. The value of the time constant is read from the figure. For this example, the time constant is 1.6 minutes. The deadtime can also be read from the figure. It is 0.2 minutes. The only piece of information remaining is the process gain. Since the span is 100°F, and the total change is 40°F, the change in percent of span is 40%. The controller output has changed 1 psi of its 12 psi span or 8.33% of span change. The process gain can be determined with the following equation:

$$K = \frac{\text{change in measured variable in \% of span}}{\text{change in controller output in \% of span}}$$

$$K = 40\%/8.33\% = 4.80$$

Using the information obtained from the process reaction curve, you can make the following settings (these are Ziegler-Nichols equations):

- *Proportional*

  $K_p = t/(t_dK)$

  $K_p = 1.6 \text{ min.}/(0.2 \text{ min.})(4.80) = 1.67$

  PB = 100%/K_p

  PB = 100%/1.67 = 60%

- *Proportional and reset*

  $K_p = 0.9t/(t_dK)$

  $K_p = 0.9(1.6 \text{ min.})/(0.2 \text{ min.})(4.80) = 1.5$

  PB = 100%/K_p

  PB = 100%/1.5 = 66.67%

  $T_i = 3.33t_d$

  $T_i = 3.33(0.2 \text{ min.}) = 0.666 \text{ min.}$

  RPM = 1/3.33t_d

  RPM = 1/3.33(0.2) = 1.5 repeats/min.

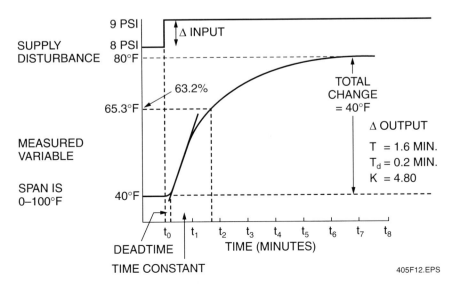

Figure 12 ◆ Test data for open loop time constant method.

- *Proportional and reset and rate*

  $K_p = 1.2t/(t_dK)$

  $K_p = 1.2(1.6 \text{ min.})/(0.2 \text{ min.})(4.80) = 2.0$

  $PB = 100\%/K_p$

  $PB = 100\%/2 = 50\%$

  $T_i = 2.0t_d$

  $T_i = 2.0(0.2) = 0.4 \text{ min.}$

  $RPM = 1/2.0t_d$

  $RPM = 1/2.0(0.2 \text{ min.}) = 2.5 \text{ repeats/min.}$

  $T_i = 0.5t_d$

  $T_d = 0.5(0.2 \text{ min.}) = 0.1 \text{ min.}$

6.2.0 Reaction Rate

The second method of open loop tuning is known as the **reaction rate method** and is shown in *Figure 13*. In this method, the information required from the process reaction curve is the slope of the maximum rate of rise line, which is termed the *reaction rate* and is given the symbol R_r. Deadtime is the other required piece of information. The reaction rate, R_r, is a change in the measured variable in % of span divided by a change in time. However, the reaction rate depends on more than process. It is also affected by the magnitude of the change in the disturbance. A large disturbance produces a steeper reaction rate. To make the reaction rate independent of the magnitude of the disturbance, the magnitude of the change in the measured variable is divided by the disturbance. Normally the disturbance is a change in the controller output in percent of span. In equation form, it is stated as follows:

$$R_r = \frac{\text{change in measured variable in \% of span}}{\text{change in time} \times \text{change in controller output in \% of span}}$$

Where:

R_r = reaction rate

As before, the deadtime is read directly from the figure. With the reaction rate and the deadtime, use the following equations to determine the controller settings that produce a one-quarter dampened wave in the measured variable:

- *Proportional*

  $K_p = 1/(R_rt_d)$ \qquad $PB = 100\%/K_p$

- *Proportional and reset*

  $K_p = 0.9/(R_rt_d)$ \qquad $PB = 100\%/K_p$

  $T_i = 3.33td$ \qquad $RPM = 1/3.33td$

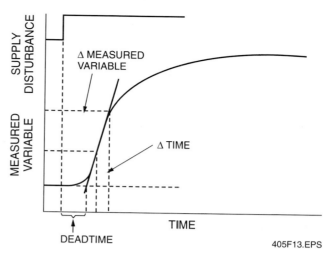

Figure 13 ◆ Reaction rate method for a single capacity with deadtime approximation.

- *Proportional and reset and rate*

  $K_p = 1.2/(CR_rt_d)$ \qquad $PB = 100\%/KP$

  $T_i = 2.0t_d$ \qquad $RPM = 1/2.0t_d$

  $T_d = 0.5t_d$

Figure 14 gives an example of this tuning method. As shown in the figure, the total change in the measured variable is approximately 40°F. Using a 10°F change in the measured variable, with a 100°F span and a 3 to 15 psi controller, you may determine the reaction rate by the following:

$$R_r = \frac{\text{change in measured variable in \% of span}}{\text{change in time} \times \text{change in controller output in \% of span}}$$

$R_r = 10\%/(0.5 \text{ min.})(8.33\%) = 2.4\%/\text{min.}$

The supply disturbance is the change in the controller output of 8.33% of span. The time period for the 10°F change is 0.5 min. This time is determined by using the maximum rate of rise line. The deadtime is read directly from the figure. The deadtime is again 2.0 min.

Use the information just obtained from the process reaction curve to determine the following settings. The equations used here are also Ziegler-Nichols equations:

- *Proportional*

  $K_p = 1/(R_rt_d)$

  $K_p = 1/(2.4\%/\text{min.})(0.2 \text{ min.}) = 2.1\%$

  $PB = 100\%/K_p$

  $PB = 100\%/2.1 = 47.6\%$

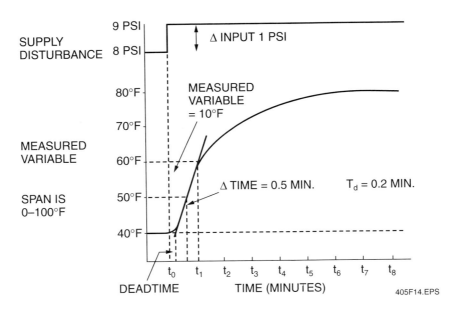

Figure 14 ◆ Test data for open loop reaction rate method.

- *Proportional and reset*

  $K_p = 0.9/(R_r t_d)$
  $K_p = 0.9/(2.4\%/\text{min.})(0.2 \text{ min.}) = 1.875$

  $PB = 100\%/K_p$
  $PB = 100\%/1.875 = 53.3\%$

  $T_i = 3.33 t_d$
  $T_i = 3.33(0.2 \text{ min.}) = 0.666 \text{ min.}$

  $RPM = 1/3.33 t_d$
  $RPM = 1/(3.33)(0.2 \text{ min.}) = 1.5 \text{ repeats/min.}$

- *Proportional and reset and rate*

  $K_p = 1.2/(R_r t_d)$
  $K_p = 1.2/(2.4\%/\text{min.})(0.2 \text{ min.}) = 2.5\%$

  $PB = 100\%/K_p$
  $PB = 100\%/2.5 = 40\%$

  $T_i = 2.0 t_d$
  $T_i = 2.0(0.2 \text{ min.}) = 0.4 \text{ min.}$

  $RPM = 1/2.0 t_d$
  $RPM = 1/(2.0)(0.2 \text{ min.}) = 2.5 \text{ repeats/min.}$

  $T_d = 0.5 t_d$
  $T_d = 0.5(0.2 \text{ min.}) = 0.1 \text{ min.}$

Both methods of open loop control tuning yield a very close approximation of a one-quarter dampened wave. As with the closed loop methods, a certain amount of trimming is required to achieve an exact one-quarter dampened wave. The major difference between the two methods for open loop tuning is the information gathered to set the proportional adjustment of the controller. Either method is reliable.

7.0.0 ◆ VISUAL LOOP TUNING

Visual loop tuning involves making changes while monitoring the effects of the changes on the process. It does not require formulas or complex mathematics. However, it should only be performed by experienced technicians.

 CAUTION

Do not attempt visual tuning unless you are completely knowledgeable and aware of the operational risks involved with variations in the operation caused by your tuning experiments. If you lack this experience, obtain help from an experienced operator in the process that you are about to tune.

7.1.0 Incremental Changes

Changes of approximately 40% in gain, derivative time, and integral time almost always lead to a significant and visible change in closed loop response. A good start is to increase gain in a sequence with values of 1, 1.4, 2, 2.8, 4, 5.6, 8, and 10. Even though the last step (10) does not represent a 40% incremental increase of the previous value, it allows the sequence to be repeated through the next decade. Because the percent change is constant for each adjustment, you can learn to anticipate the magnitude of change in response that will occur, whether the gain step is starting from 0.1 or 100.

These 40% stepped-increment values are usually fine enough to eventually predict the response to the change. In some situations, you may want to finish by splitting a pair of these values. Keep in mind that such precise tuning may sometimes create problems if the process shifts slightly and becomes less stable. Successful tuning is based on clearly understanding when you are getting close to optimal control or when you are nearing the point of driving the process into instability.

7.2.0 Apparent Instability

When attempting to visually tune a loop having an apparent instability, it is first necessary to make sure that the cause of the loop oscillations is inherent to the loop you are tuning and is not being caused by another process. After receiving permission from the unit operator, one way to do this is to place the loop controller for the loop you are tuning in the manual mode. If the process tends to stabilize in the manual mode, then the controller most likely requires tuning.

 CAUTION
Never take any action involving live instrumentation without explicit awareness and permission of the unit operator. Furthermore, when you are communicating information, be certain that the operator is paying full attention.

You can do visual tuning of a loop having an apparent instability by incrementally stepping the output in manual mode then immediately setting the controller back to auto mode while observing the trends of controller output and the feedback from the process variable. Modern instrumentation may include digital control systems (DCS) equipment that makes this easy; however, in some older single-loop controllers, you may need to watch indicating gauges or digital readouts.

If the controller you are visually tuning is equipped with setpoint tracking in which the setpoint varies with the process, you should disable this feature so that the setpoint remains stationary during the visual tuning process.

One safer method to stop oscillations is to increase the integral time, thereby reducing the integral gain. This is especially true when the integral time is less than one-half of the oscillation period. With incremental increases in the integral time, you should obtain a smooth sinusoidal oscillation. After this point occurs, any additional incremental increases should result in an increased oscillation period without any increase in the amplitude. Following this, the integral time should continue to be increased

until the oscillation damps out or stops lengthening in period.

Normally, if the integral time is longer than the oscillation time, the integral action does not play much of a role at the oscillation period and is minimally involved in the oscillation. An exception to this occurs in linear integrating processes like level controls. There, a 90-degree (or more) phase shift is produced over a wide range of period, and the additional phase shift of integral action can push the total phase shift toward 180 degrees, which sustains the oscillation. In these types of processes, the integral effect that causes the process to return to setpoint is performed by the process itself. For this reason, the integrating function of the controller must be disabled by setting it well above the closed loop response period of the process thus preventing the integral function from contributing any phase shift to the process control. Once the integral time is set long enough that it is not involved in the oscillation, it is safe to try lowering the gain until the oscillation is damped out. Following this, you can try stepping the output as mentioned earlier, without expecting that a sustained oscillation will result.

7.3.0 Sluggish Response

Two conditions that usually exist in a loop that is experiencing sluggish response are the lack of derivative action and a long integral time relative to the process response time. The upcoming sections will examine tuning areas and procedures that are intended to improve sluggish response.

7.3.1 Gain

A good starting point in addressing sluggish response is to first adjust the gain by changing the controller output in steps. You can do this by placing the controller in manual mode then stepping the output by whatever is allowable. If you are not sure what is allowable, a good practice is to start with very small steps, and let the operator of the process know what you are doing. Knowing what is allowable allows you step the output in somewhat larger steps. The advantage of making larger steps is that any minor nonlinearities and noise tend to have less impact on the trends. After making a stepped change to the output, put the controller back in auto mode and watch the effects on the process. Also, note the controller's reaction to the change in output. The objective is to apply as much gain as you can without producing a series of echoes of the original response.

If the process reaction to the stepped change in the output causes the process response to swing more than one cycle before approaching an equilibrium value, the gain most likely is set too high. Try lowering the gain by about 40%. On the other hand, if the controller output and process response settle at new values without producing anything resembling a sinusoidal swing, then the gain is probably set too low. Try raising it by 40%.

Repeat this process, alternating the direction of the output step each time until a gain setting is reached that causes approximately one cycle of output swing before the output reaches an equilibrium value. This gain value represents one that addresses the disturbance in the middle range of period that is near the process natural closed loop response time, and it is referred to as the ultimate period when using the Ziegler-Nichols methods of loop tuning.

NOTE

A sign that the gain is too high is a process oscillation that is not centered across the setpoint value.

7.3.2 Integral Time

While still in manual mode, and only after the gain is set at its optimum value, shorten the integral time on the controller in an effort to cause the output value to ramp back to setpoint. The effects of stepping the output and allowing the gain to respond to the changes in output will cause the automatic reset action of the integral to attempt to find the original output value that puts the process exactly back on setpoint. A reliable assumption is that the integral time setting will be on the order of the closed loop time constant of the process with the gain optimized by the previous gain adjustment process.

However, if you want to see the integral effect clearly, start the integral time at approximately double that time then reduce the integral time in 40% steps. This will cause the ramp time back to setpoint to be reduced in 40% steps, but the peak disturbance will reduce or increase slightly, depending on the deadtime relative to the dominant closed loop time constant.

The best range of integral time is reached when the process ramps back to the setpoint about half as fast as it moved away from setpoint from the output step. Too short of an integral time is indicated by an oscillation that is centered across the setpoint with a period that lengthens as the integral time is lengthened.

7.3.3 Derivative Action

Derivative action is often avoided in some controllers because of the controller's adverse response to it. However, in controllers that respond well to derivative action and whose process demands using it, the benefits of derivative action are greater stability and faster response.

In order to determine if the process that is being controlled would benefit from derivative action, you should consider the following conditions:

- Having arrived at reasonable settings for gain and integral time, the process should display a deadtime, acceleration time to full rate of change time, inflection, **over the hump** time, and settling time back to setpoint.
- Observe the inflection point between the accelerating phase and the reverse curvature as the process goes over the hump. The greater the acceleration time relative to the over the hump time, the more benefit the process gets from the derivative action.
- The only adverse ingredient in the loop that prevents the process from getting the full benefit of derivative action is the deadtime. If it is much longer than the accelerating phase, it will cause all the output effects from the derivative action to occur after the point at which they would have done some good. This results in the controller storing echoes of the step response in the process dead time delay.

In summary, derivative action is beneficial whenever the accelerating phase of the process step response is a significant fraction of both the deadtime and the closed loop over the hump time. It should be pointed out that if the control loop exhibits a jerky output or sinusoidal oscillations that make the process measurement look like a flight of stairs as it goes over the hump, this is a sign that the derivative action in a control loop should be reduced.

7.3.4 Cascade Control

The visual tuning methods described up to this point tested the response of a process to output steps (effectively, load disturbances) having a constant setpoint. When a controller has its setpoint delivered as the output of another controller, or cascaded, there is an effective increase in loop gain. That is, as the inner or slave controller is driving the process toward its setpoint, the outer or master controller may be driving the setpoint toward the process value. This can be compared somewhat to shooting at a target that is rapidly approaching the end of a gun, which can result in overshooting the target. Also it is less important that the slave controller reach perfection in the shortest possible time because its setpoint will not be there by the time the output of the controller gets the process there. So, inner cascade loops tend to work best with somewhat less gain and somewhat longer integral time than if the loop were shooting for a fixed setpoint.

7.3.5 Sticking Valves (Stiction)

An example where visual tuning might be more effective than the Ziegler-Nichols methods of tuning might be in a loop where sticking or **stiction** is present in the control valve. Ziegler-Nichols tuning relies on linear response in order to be effective; stiction usually causes a nonlinear effect on the process.

Stiction may also produce an oscillation in the process that has a period proportional to integral time; but when tuning this loop, if the integral time is increased in an effort to stabilize the oscillation, the increased integral time will probably result in a longer oscillation period.

Sinusoidal oscillation, a by-product of stiction, is not representative of either a controller that is improperly tuned or a linear process. In sinusoidal oscillation, the controller output will generally show a triangular wave as the integral action keeps ramping the output up and down. On the other hand, in a control loop with stiction present, the process variable will often indicate a form of square wave with a fairly constant amplitude. This wave is representative of the size of the sudden jump when the valve breaks loose, as shown in *Figure 15*.

> **NOTE**
>
> A loop with a sticking control valve should be visually tuned to limit the oscillation caused by the sudden change in valve position. In these cases, the valve should be repaired or replaced, or a valve positioner should be installed if one is not already installed.

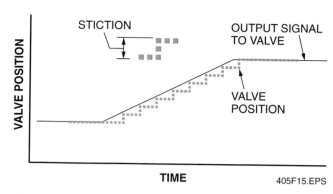

Figure 15 ◆ Waveform of a sticking control valve (stiction).

Summary

In process control loops, disturbances in the normal operation of a process can be caused by many things, including other processes or even faulty loop components. The response or behavior of the process control loop to the disturbance is directly related to parameters of the loop's tuning.

Many of the control loops in process control allow for the manipulation of specific controller functions that, when used individually or in combination with each other, cause the controller's output signal to change in various ways once a disturbance occurs. There is no one common combination of controller parameters that can be applied to all processes because the variables in each process dictate specific parameters in order to function as designed and produce the desired effects or product quality. Setting these parameters for optimal process control based on the demands of the process is referred to as loop tuning.

Loop tuning can be as simple as increasing only one function, such as gain, to either a slightly higher or lower setting and watching the response of the controller as it adjusts its output accordingly. On the other hand, loop tuning can involve complex formulas that use loop characteristics such as deadband, lag, and time constants. These equations must be applied in order to properly set controller functions, such as gain, integral time, and derivative, to arrive at the optimal set of parameters for a particular loop.

This module introduced the most common methods used in both types of loop tuning as well as the definitions of those factors that equate to complex tuning.

Review Questions

1. The dynamic behavior of a system after a disturbance has occurred is called _____.
   a. feedback
   b. response
   c. oscillation
   d. amplitude

2. Proportional control occurs within a specified band around the _____.
   a. setpoint
   b. gain
   c. valve position
   d. error

3. The _____ control function in a PID-controlled loop provides an output that is proportional to the _____ of error.
   a. integral; rate
   b. proportional; reset value
   c. derivative; rate
   d. derivative; reset value

4. The time interval between the initiation of an input change or disturbance to a process and the start of the resulting process variable response is called _____.
   a. integral
   b. deadtime
   c. lag
   d. time constant

5. The first time constant of process response includes the process deadtime.
   a. True
   b. False

6. In order for any system to be stable, its energy equation must be _____.
   a. solved
   b. unequal
   c. balanced
   d. equal to 1

7. The _____ equation is used to find out what percent a process has changed in a given time.
   a. PID loop
   b. time constant
   c. process gain
   d. complete response

8. Changes in input and output are normally expressed as a percentage of the _____.
   a. gain
   b. span
   c. total change
   d. input versus output

9. The inverse relationship of proportional band is called _____.
   a. integral
   b. energy
   c. loss
   d. gain

10. The derivative constant of a controller is expressed in units of _____.
    a. gain
    b. percentage
    c. seconds
    d. minutes

11. Open loop tuning methods rely on constant feedback from the process.
    a. True
    b. False

12. The ultimate gain or ultimate band is the maximum allowable value of gain or proportional band in which the system is _____.
    a. not responding
    b. at its highest amplitude
    c. oscillating
    d. stable

13. The process reaction curve that is used in open loop tuning is generated by introducing _____.
    a. a time lag
    b. feedback
    c. a disturbance
    d. deadtime

14. Visual loop tuning is performed by applying equations and formulas.
    a. True
    b. False

15. If the controller you are visually tuning is equipped with setpoint tracking, you should _____.
    a. increase the tracking to its maximum
    b. ignore its output
    c. disable the tracking feature
    d. use the various setpoints in your calculations

Trade Terms Introduced in This Module

Amplitude: In control systems, it is the amount of change (height) in a signal, whether output, input, or valve position, as represented by a graphical waveform.

Dampened-oscillation method: A decrease in the *amplitude* of an oscillation or wave motion with time.

Decay ratio: The ratio of one oscillation to the previous oscillation.

Derivative: With derivative action, the controller output is proportional to the rate of change of the process variable or error.

Differentiator: The controller device that produces a *derivative*-based signal.

Empirical: Data that is collected from actual experience rather than calculations.

Gain of process: Gain of process is defined as the change in input divided by the change in output.

Integral: With integral action, the controller output is proportional to the amount and duration of the error signal.

Integrator: A device in a controller that produces an *integral*-based signal.

One-quarter dampened wave: A one-quarter reduction in the amplitude of a signal's waveform.

Oscillation: Repetitive changes in signal amplitude.

Over the hump: A signal represented by a waveform that has reached its peak and is headed in the direction of setpoint or stability.

PID: Abbreviation for proportional/integral/derivative control loop or controller.

Proportional output: The more gain a controller has, the faster the loop response and more oscillatory the process. Gain = 100/(proportional band).

Reaction rate method: In tuning a controller, it is a method where the basis of tuning is the rate at which the process changes from one steady state to another.

Reciprocal: The reciprocal of a value is that number over 1. For example, the reciprocal of 3 = 1/3.

Repeats per minute (RPM): A measure of how fast a signal is being integrated.

Stiction: Stiction in a control valve exists when the static (starting) friction exceeds the dynamic (moving) friction inside the valve. Stiction describes the valve's stem (or shaft) sticking when small changes are attempted.

Time constant method: A method of tuning a loop in which the time it takes the process to reach 63.2% of its total change is used as a factor.

Ultimate period: The period of response with the gain or proportional band set at its ultimate value, which is the maximum allowable value of gain or proportional band that permits stability in the system.

Additional Resources

This module is intended to be a thorough resource for task training. The following reference works are suggested for further study. These are optional materials for continued education rather than for task training.

www.isa.org, the Web site of the Instrument Society of America.

Good Tuning: A Pocket Guide, 2000. G.K. McMillan. Research Triangle Park, NC: Instrument Society of America.

Standards and Practices for Instrumentation and Control: Instrument Loop Design, 1989. Research Triangle Park, NC: Instrument Society of America.

NCCER CURRICULA — USER UPDATE

NCCER makes every effort to keep its textbooks up-to-date and free of technical errors. We appreciate your help in this process. If you find an error, a typographical mistake, or an inaccuracy in NCCER's curricula, please fill out this form (or a photocopy), or complete the online form at **www.nccer.org/olf**. Be sure to include the exact module ID number, page number, a detailed description, and your recommended correction. Your input will be brought to the attention of the Authoring Team. Thank you for your assistance.

Instructors – If you have an idea for improving this textbook, or have found that additional materials were necessary to teach this module effectively, please let us know so that we may present your suggestions to the Authoring Team.

NCCER Product Development and Revision

13614 Progress Blvd., Alachua, FL 32615

Email: curriculum@nccer.org
Online: www.nccer.org/olf

❑ Trainee Guide   ❑ AIG   ❑ Exam   ❑ PowerPoints   Other _____

Craft / Level: _____   Copyright Date: _____

Module ID Number / Title: _____

Section Number(s): _____

Description: _____

_____

_____

_____

_____

Recommended Correction: _____

_____

_____

_____

_____

Your Name: _____

Address: _____

_____

Email: _____   Phone: _____

Programmable Logic Controllers

COURSE MAP

This course map shows all of the modules in the fourth level of the Instrumentation curriculum. The suggested training order begins at the bottom and proceeds up. Skill levels increase as you advance on the course map. The local Training Program Sponsor may adjust the training order.

INSTRUMENTATION LEVEL FOUR

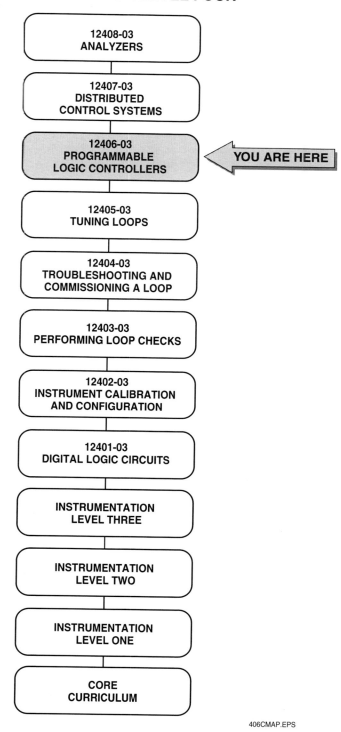

12408-03
ANALYZERS

12407-03
DISTRIBUTED
CONTROL SYSTEMS

12406-03
PROGRAMMABLE
LOGIC CONTROLLERS

YOU ARE HERE

12405-03
TUNING LOOPS

12404-03
TROUBLESHOOTING AND
COMMISSIONING A LOOP

12403-03
PERFORMING LOOP CHECKS

12402-03
INSTRUMENT CALIBRATION
AND CONFIGURATION

12401-03
DIGITAL LOGIC CIRCUITS

INSTRUMENTATION
LEVEL THREE

INSTRUMENTATION
LEVEL TWO

INSTRUMENTATION
LEVEL ONE

CORE
CURRICULUM

406CMAP.EPS

1.0.0 INTRODUCTION 6.1

2.0.0 OVERVIEW 6.2

    2.1.0   Hardwired and PLC Systems 6.2

    2.1.1   Hardwired 6.2

    2.1.2   PLC 6.6

    2.2.0   Hardwired/PLC Systems Comparison 6.6

3.0.0 NUMBER SYSTEMS REVIEW 6.8

    3.1.0   Binary 6.8

    3.1.1   Binary to Decimal Conversion 6.9

    3.1.2   Decimal to Binary Conversion 6.9

    3.2.0   Octal 6.9

    3.2.1   Octal to Decimal Conversion 6.10

    3.2.2   Decimal to Octal Conversion 6.10

    3.2.3   Octal to Binary Conversion 6.10

    3.2.4   Binary to Octal Conversion 6.10

    3.3.0   Hexadecimal 6.11

    3.4.0   Codes (Binary) 6.11

    3.4.1   ASCII 6.11

    3.4.2   BCD 6.11

    3.4.3   Gray Code 6.12

4.0.0 HARDWARE 6.12

    4.1.0   Power Supply 6.12

    4.2.0   Input/Output Modules 6.13

    4.2.1   Discrete 6.13

    4.2.2   Numerical Data 6.15

    4.2.3   Analog 6.15

    4.2.4   Special 6.16

    4.2.5   Remote Adapter 6.17

5.0.0 PROCESSORS 6.17

    5.1.0   Scanning 6.18

    5.2.0   Memory 6.18

6.0.0 SOFTWARE 6.19

    6.1.0   Languages 6.19

    6.1.1   Ladder Logic 6.19

    6.1.2   Boolean 6.21

    6.1.3   English Statement 6.21

    6.1.4   Functional Block 6.21

    6.1.5   Machine Stage 6.21

6.2.0 Ladder Diagram Instructions 6.21

6.2.1 *Relay* 6.21

6.2.2 *Timer and Counter* 6.22

6.2.3 *Arithmetic* 6.24

6.2.4 *Data Comparison* 6.25

6.2.5 *Data Transfer* 6.25

6.2.6 *Program Control* 6.26

7.0.0 **HARDWARE TO PROGRAM CORRELATION** 6.27

8.0.0 **GUIDELINES FOR PROGRAMMING AND INSTALLATION** 6.27

8.1.0 Programming 6.27

8.2.0 Installation 6.29

8.3.0 I/O Wiring 6.29

8.4.0 Dynamic System Checkout 6.30

SUMMARY 6.30

REVIEW QUESTIONS 6.31

GLOSSARY 6.33

REFERENCES & ACKNOWLEDGMENTS 6.34

Figures

Figure 1   Micro PLC 6.3

Figure 2   Standard medium-capacity PLC system 6.4

Figure 3   Large-capacity, multi-tasking PLC system 6.5

Figure 4   Typical human/machine interfaces 6.6

Figure 5   Hardwired lamp circuit 6.6

Figure 6   PLC lamp system 6.7

Figure 7   Process vat control system 6.7

Figure 8   Hardwired vat control 6.7

Figure 9   PLC vat control 6.7

Figure 10   Relay system changes 6.8

Figure 11   PLC system changes 6.8

Figure 12   AC/DC discrete input circuit 6.14

Figure 13   Input module connection diagram 6.14

Figure 14   AC/DC discrete output circuit 6.15

Figure 15   Output module connection diagram 6.15

Figure 16   Analog input module connection diagram 6.16

Figure 17   PLC language comparison 6.20

Figure 18   Input contact symbols 6.22

Figure 19   Output coil symbols 6.22

Figure 20   Logic flow for normally closed and
normally open contacts 6.23

Figure 21   Latching/unlatching coil operation 6.23

Figure 22   Timers 6.24

Figure 23   Counter application 6.24

Figure 24   Arithmetic operations 6.25

Figure 25   Examples of data comparison instructions 6.25

Figure 26   BTD transfer operation 6.26

Figure 27   MOV and MVM transfer operations 6.26

Figure 28   Typical MCR instructions 6.27

Figure 29   Hardware to software correlation 6.28

Tables

Table 1   Decimal-Binary Equivalents 6.8

Table 2   Using the Subtractive Method to Convert
39_{10} to 100111_2 . . . . .6.9

Table 3   Using the Division Method to Convert
351_{10} to 101011111_2 . . . . .6.9

Table 4   Decimal-Octal-Binary Equivalents 6.9

Table 5   Using the Subtractive Method to Convert
149_{10} to 225_8 . . . . .6.10

Table 6   Using the Division Method to Convert
189_{10} to 275_8 . . . . .6.10

Table 7   Example of Binary to Octal Conversion 6.11

Table 8   Decimal-Hexadecimal-Binary Equivalents 6.11

Table 9   Decimal-BCD Equivalents 6.12

Table 10   Comparison of Gray and Binary Codes 6.12

Table 11   Programming Guidelines 6.29

Programmable Logic Controllers

Objectives

When you have completed this module, you will be able to do the following:

1. Describe the function and purpose of a programmable logic controller (PLC).
2. Compare hardwired and PLC systems.
3. Count and convert between number systems.
4. Explain the purpose of binary codes.
5. Describe the purpose of the various power supplies used within a PLC.
6. Explain the general function of an input/output (I/O) module, including the following types:
   - Discrete
   - Numerical data
   - Special
   - Remote
7. Explain the power supply and ground connections to I/O modules.
8. State the function of the PLC processor module.
9. Explain the interrelations between the various microprocessor components.
10. State the characteristics of various types of memory.
11. Describe the characteristics and features of a PLC processor module.
12. Explain the purpose of PLC software and firmware.
13. Describe the features and the differences between PLC programming languages.
14. Describe the features of relay ladder logic instruction categories.
15. Explain the principles used to correlate PLC hardware components to software instructions.
16. Program and install a PLC.

Prerequisites

Before you begin this module, it is recommended that you successfully complete the following: Core Curriculum; Instrumentation Levels One through Three; Instrumentation Level Four Modules 12401-03 through 12405-03.

Required Trainee Materials

1. Pencil and paper
2. Appropriate personal protective equipment

1.0.0 ◆ INTRODUCTION

The programmable logic controller (PLC) has been used in industry in various forms since 1968. Originally, the goal was to use PLCs to replace the relay-based control systems used in manufacturing applications. This would eliminate the high cost of maintaining these inflexible systems. In 1970, with the innovation of the microprocessor, the PLC that was originally used as a relay replacement device started evolving into the advanced PLC of today.

Many of the PLCs currently available are simply called programmable controllers. This is because they perform many more additional functions than the relay logic functions of the original PLCs. Using special input/output (I/O) modules, additional functions include motion control, drive control, and specialized process control that can be programmed using software applications via a standard personal computer. The special I/O modules include high-speed encoder/counter, servo control, stepper control, synchronized axis control, velocity control, and temperature monitoring/control modules, along with common **discrete** digital, AC/DC, analog, and smart transmitter interface modules. Many of the advanced software application programs automatically assign the controller memory locations and input/output terminals as well as adjust the programming sequences. They can also automatically reconfigure these parameters if the controller programming requires changes or corrections. Many of the controller programming software applications are supported by additional emulation software that can be used to test the controller programming before it is actually implemented in the field.

PLCs are currently used in many industrial applications. These include chemical, food, petrochemical, pharmaceutical, and various water/wastewater treatment plants. In these applications, they are used for process control either in an autonomous mode or in conjunction with other computers for overall plant control, data collection, and reporting functions. These functions can also include statistical process control, quality assurance, and process or equipment diagnostics.

2.0.0 ◆ OVERVIEW

A programmable logic controller is a microprocessor-controlled device that can be programmed by the user to perform a specific task. The two basic components of a PLC are the hardware and the software. The hardware includes all the physical components that make up the PLC system and allows connection to external devices. The software is the configuration data that makes up the operating instructions for the PLC hardware.

Examples of PLC hardware for different capacity systems are shown in *Figures 1* through *4*. *Figure 1* shows a micro PLC that can be mounted on a panel or a **DIN** rail. This particular unit can have up to 16 I/O modules with up to 152 I/O connections using various combinations of discrete, AC/DC, and analog modules along with temperature monitoring/control, and high-speed encoder/counter modules. Standard ladder diagram (LAD) language programming is done via a software application through a PC.

Figure 2 is typical of a common medium-capacity PLC system. The particular system is rack mounted and can have up to 125 I/O modules if 7 remote I/O chassis are connected to the main PLC unit via a communications system and appropriate communications modules. Using the maximum number of I/O modules, up to 50,176 I/O connections can be used, depending on the type of I/O modules selected. Besides the usual discrete, AC/DC, analog, and communications I/O modules, a full assortment of I/O modules for motion

control, drive control, and process control are available. This particular system can be programmed via application software and a PC, using LAD, sequential function chart (SFC), or a structured text (ST) language. These languages and others will be explained later in the module.

Figure 3 shows the processing PLC for a large-capacity, multi-tasking PLC system. This particular PLC system is panel mounted and, with a number of remote chassis, can make use of 250 I/O modules of mixed types with up to 125,000 connections, including a maximum limit of 4,000 analog connections. A large number of different I/O modules and special modules, including diagnostic and optical isolation modules, are available for use in motion, drive, and process control applications. In this particular system, motion control is integral with the processor, and servo interface modules are used for each axis of motion. Many of the modules have status indicators on the front of the modules and self-diagnostic capabilities along with software configuration capabilities instead of switches and jumpers. The system can accommodate up to 32 continuous or periodic tasks. Up to 32 programs may be assigned to a task, each with its own data or ladder logic. The programming languages used for this particular system are LAD, SFC, or ST.

Figure 4 shows various human/machine interfaces (HMI) that are used with PLC systems. These are typical of the types of terminals that can be used to monitor and/or program the PLC systems.

2.1.0 Hardwired and PLC Systems

The major hardware components of a PLC are the processor, input module, output module, and power supply. A programming terminal is used to program the processor, but it is not considered a major component because once the processor is programmed, the terminal may be disconnected. The relationships between these major parts and the purpose of a PLC can best be seen by comparing a hardwired circuit to an identical circuit that is PLC controlled.

2.1.1 Hardwired

Figure 5 is a simplified ladder diagram of a hardwired circuit used to control two lamps. Switch 1 and switch 2 are normally open pushbutton switches. Switch 1 will send power to lamp 1, and switch 2 will send power to lamp 2. When switch 1 is closed, lamp 1 will light. When switch 2 is closed, lamp 2 will light.

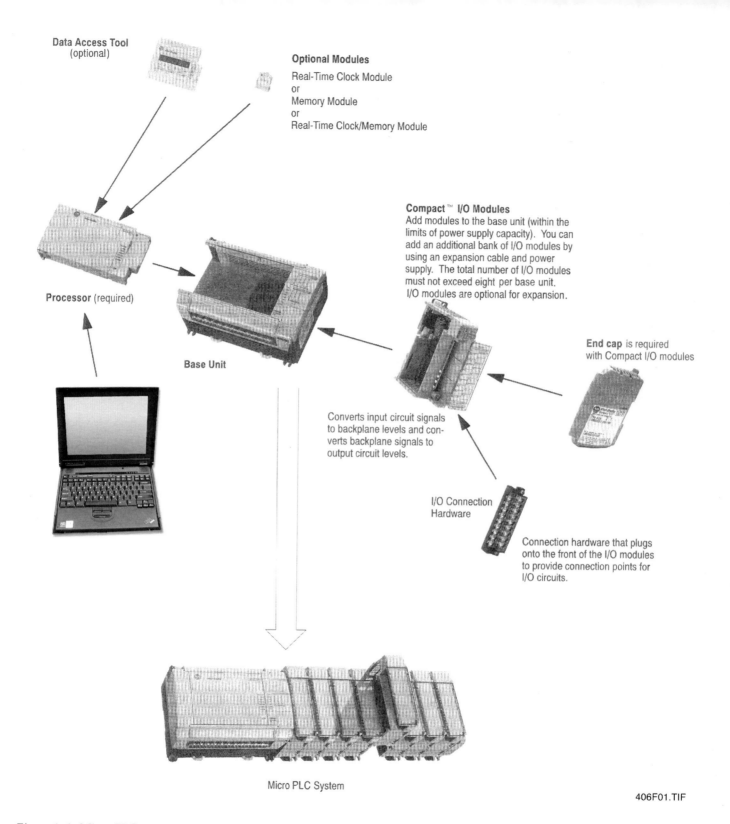

Data Access Tool
(optional)

Optional Modules

Real-Time Clock Module
or
Memory Module
or
Real-Time Clock/Memory Module

Compact ™ I/O Modules
Add modules to the base unit (within the limits of power supply capacity). You can add an additional bank of I/O modules by using an expansion cable and power supply. The total number of I/O modules must not exceed eight per base unit. I/O modules are optional for expansion.

Processor (required)

Base Unit

End cap is required with Compact I/O modules

Converts input circuit signals to backplane levels and converts backplane signals to output circuit levels.

I/O Connection Hardware

Connection hardware that plugs onto the front of the I/O modules to provide connection points for I/O circuits.

Micro PLC System

406F01.TIF

Figure 1 ◆ Micro PLC.

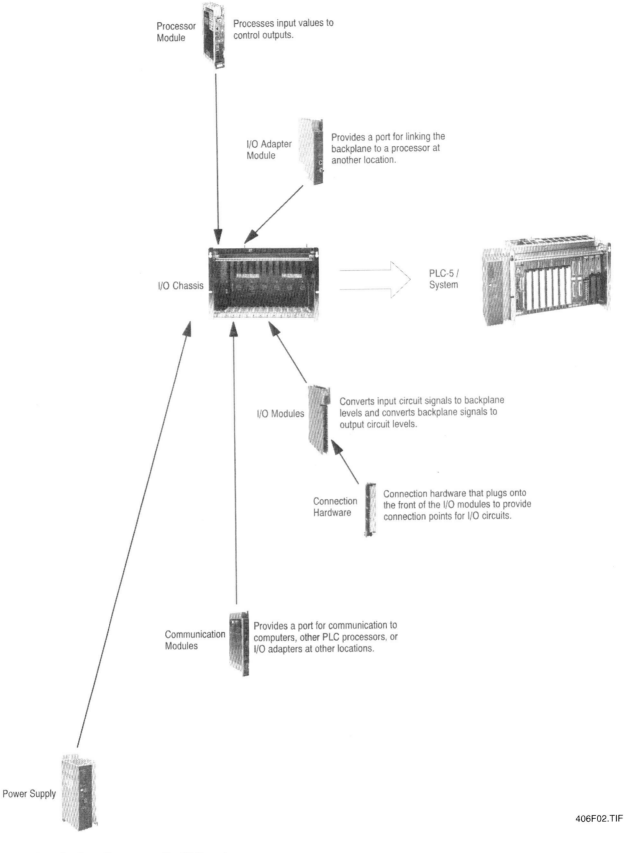

Processor Module — Processes input values to control outputs.

I/O Adapter Module — Provides a port for linking the backplane to a processor at another location.

I/O Chassis

PLC-5 / System

I/O Modules — Converts input circuit signals to backplane levels and converts backplane signals to output circuit levels.

Connection Hardware — Connection hardware that plugs onto the front of the I/O modules to provide connection points for I/O circuits.

Communication Modules — Provides a port for communication to computers, other PLC processors, or I/O adapters at other locations.

Power Supply

406F02.TIF

Figure 2 ◆ Standard medium-capacity PLC system.

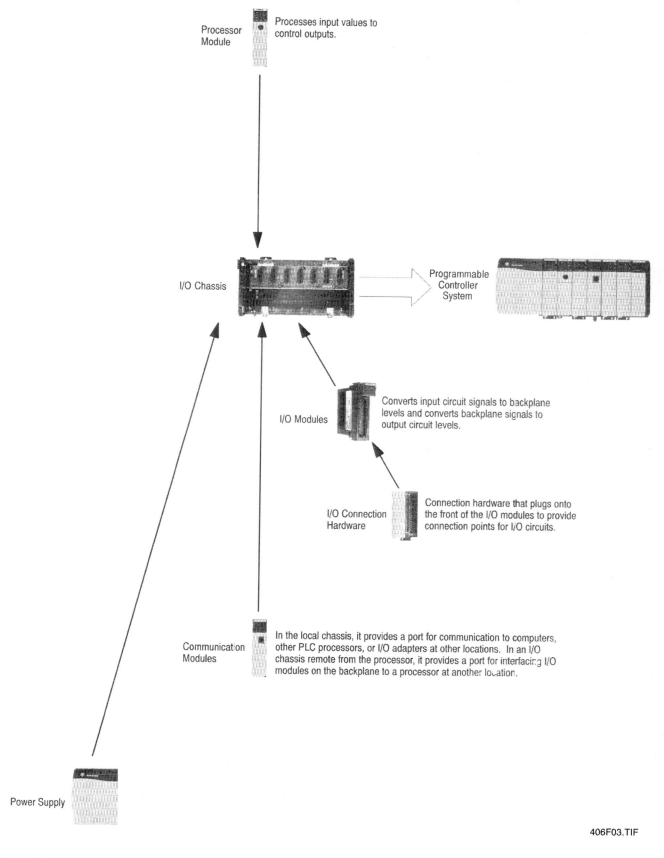

Processor Module — Processes input values to control outputs.

I/O Chassis

Programmable Controller System

I/O Modules — Converts input circuit signals to backplane levels and converts backplane signals to output circuit levels.

I/O Connection Hardware — Connection hardware that plugs onto the front of the I/O modules to provide connection points for I/O circuits.

Communication Modules — In the local chassis, it provides a port for communication to computers, other PLC processors, or I/O adapters at other locations. In an I/O chassis remote from the processor, it provides a port for interfacing I/O modules on the backplane to a processor at another location.

Power Supply

406F03.TIF

Figure 3 ◆ Large-capacity, multi-tasking PLC system.

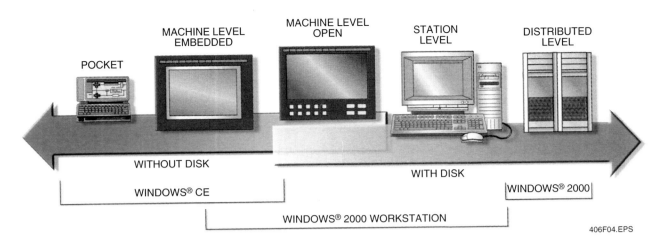

Figure 4 ◆ Typical human/machine interfaces.

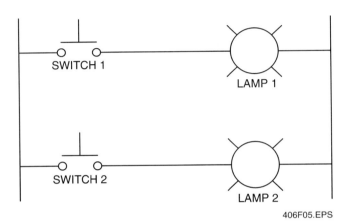

Figure 5 ◆ Hardwired lamp circuit.

2.1.2 PLC

Figure 6 shows the same components connected to a PLC. From this diagram, you can see several differences. First, switches are not connected directly to the lamps; instead, the switches are connected to input modules, and the lamps are connected to output modules. Another difference is that the input modules and output modules are not connected to each other directly: they are connected to the processor.

The processor is programmed to connect switch 1 to lamp 1, and switch 2 to lamp 2. This is accomplished by entering a program into the processor using a terminal. The program looks very similar to a standard electrical ladder diagram. The operation of the hardwired lamp circuit and the PLC-controlled system seem identical. When switch 1 is closed, lamp 1 lights, and when switch 2 closes, lamp 2 lights. The versatility of configuring inputs and outputs and the ease of configuration can be considered a major difference.

2.2.0 Hardwired/PLC Systems Comparison

In the hardwired system, the electrons flow from the voltage source, through the switch, to the correct indicator lamp. Electrical power simply follows the wire conductors to the lamp. When the switch is opened, power is interrupted and the light goes out.

In the PLC-controlled system, electrical power comes from the voltage source, through the switch, into the input module. The input module senses the presence of this voltage and in turn, sends a low-voltage signal into the processor. The voltage from the switch is isolated from the voltage signal that the module sends into the processor. This isolation is necessary because the fragile processor chip operates at very low voltage and current levels. Isolation is generally provided by an electronic component known as an opto (optical) coupler.

The signal received by the processor will cause a low-voltage signal to be sent to the output module as directed by the program. The program directs the processor-transmitted signal to the appropriate output terminal for lamp 1 when it receives a signal from the input module terminal connected to switch 1. When switch 2 is activated, the processor completes a similar action, but this time the signal is sent to the output module terminal for lamp 2. An observer of both hardwired and PLC-controlled systems would not notice any difference in system operation. In both systems, switch 1 controls lamp 1, and switch 2 controls lamp 2.

A big advantage of a programmable logic controller becomes evident when a change is needed in the circuits previously discussed. For example, if you needed to change the circuits of a hardwired system to have switch 1 control lamp 2, and switch 2 control lamp 1, it would take several minutes to rewire them and would involve exchanging the

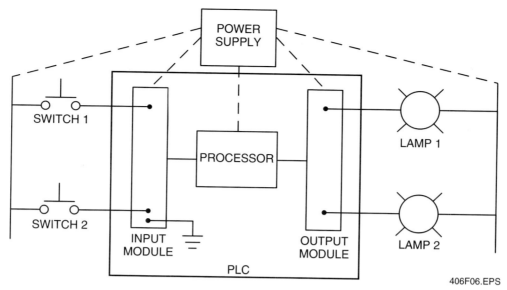

Figure 6 ◆ PLC lamp system.

wires at the switches or the lamps. With a PLC, a simple editing operation can make these changes internal to the program. This eliminates the need for rewiring and takes only a fraction of the time and cost required to change a hardwired system.

A practical application demonstrating the flexibility of a PLC ladder program can be seen in *Figure 7*. In this system, a vat is filled with liquid. When temperature and pressure conditions are met, a motor comes on and operates a stirrer to mix the liquid.

Figure 8 shows the hardwire method used to operate this system. A pressure switch and a temperature switch are hardwired into the system with a manual override installed.

Figure 9 shows the way the circuit would look programmed into a PLC. The only hardwire points with the PLC are the temperature and pressure switch inputs and the mixer motor output. The contacts and coil within the PLC are labeled to identify each point as unique. The method of numbering differs according to the programming logic used by each manufacturer.

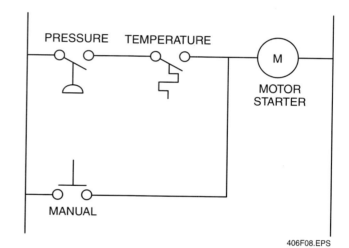

Figure 8 ◆ Hardwired vat control.

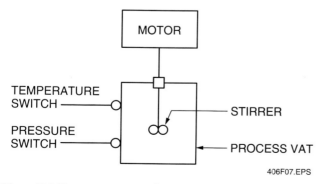

Figure 7 ◆ Process vat control system.

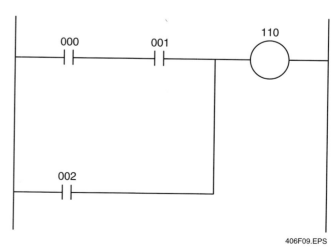

Figure 9 ◆ PLC vat control.

It is very easy to change the circuit setup in the PLC without ever moving a wire connection. *Figure 10* shows how a traditional circuit would have to be re-wired in order to make temperature a critical path for the motor to work. As you can see, the wiring of the switch must be physically changed, which could involve extensive work, depending on its location.

Figure 11 shows how the PLC-controlled circuit can be configured to perform the same function. As you can see, there is no need to physically touch the wiring. All that has to be done is a programming change on the PLC.

3.0.0 ◆ NUMBER SYSTEMS REVIEW

PLCs, being microprocessor-controlled devices, use number systems that easily correspond to the digital operation of the processor. The primary number systems that are used are **binary**, **octal**, and **hexadecimal**. Additionally, various binary codes are used, such as the **American Standard Code for Information Interchange (ASCII)**, to allow the PLC to communicate with other standard external devices.

3.1.0 Binary

The binary numbering system is a base two system. It contains two digits: 0 and 1. Since the binary system only has two symbols, it can be easily used with digital circuits that only have two states, such as off and on.

The binary system starts with 0 representing zero objects. One object is represented by 1. For two objects there is no symbol. As in the **decimal** system, the position concept is used. Two is represented by beginning the column to the right over again with a 0 and putting a 1 in the next column to the left. An equivalent binary system and decimal system are written in *Table 1*.

Note that in going from 3 to 4, both binary columns have been completed and have gone from 11 to 100. In going from 7 to 8, three binary columns have been completed and have gone from 111 to 1000. This is similar to the decimal system when going from 99 to 100.

Each position in the binary number represents the base raised to a power. In base 10, the units position has a power of 10^0, the next position 10^1, and so on. In binary, or base 2, the same reasoning applies. The units position has a power of 2^0, the next position 2^1, and so on. Binary digits are called **bits**. Therefore, the digit in the units position is called the least significant bit (LSB), and so on until the most significant bit (MSB) is reached.

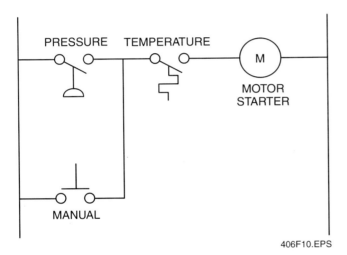

Figure 10 ◆ Relay system changes.

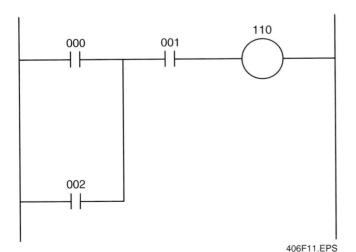

Figure 11 ◆ PLC system changes.

Table 1 Decimal-Binary Equivalents

Decimal	Binary
0	0
1	1
2	10
3	11
4	100
5	101
6	110
7	111
8	1000
9	1001

3.1.1 Binary to Decimal Conversion

Converting from base 2 (binary) to base 10 (decimal) is fairly simple. If a 1 is present in a given position, the weight of that bit is added. If a 0 is present, the weight of that bit is not added. Convert the binary number 110011_2 to its decimal equivalent:

$$110011_2 = (1 \times 2^5) + (1 \times 2^4) + (0 \times 2^3) + (0 \times 2^2) + (1 \times 2^1) + (1 \times 2^0)$$

$$110011_2 = (1 \times 32) + (1 \times 16) + 0 + 0 + (1 \times 2) + (1 \times 1)$$

$$110011_2 = 51_{10}$$

3.1.2 Decimal to Binary Conversion

There are two methods for converting decimal numbers to binary numbers. The two methods are completely different, but regardless of which method you choose to use, the end result will be the same. The two methods are the subtractive method and the division method.

To convert using the subtractive method, start by subtracting the highest possible power of two from the decimal number and place a 1 in position of the power of two used. Repeat this subtraction process on the balance of each subtraction until there is nothing left. Convert 47_{10} to binary using the subtractive method:

$47 - 32 = 15$	$32 = 1 \times 2^5$
	0×2^4
$15 - 8 = 7$	$8 = 1 \times 2^3$
$7 - 4 = 3$	$4 = 1 \times 2^2$
$3 - 2 = 1$	$2 = 1 \times 2^1$
$1 - 1 = 0$	$1 = 1 \times 2^0$

$$47 = (1 \times 2^5) + (0 \times 2^4) + (1 \times 2^3) + (1 \times 2^2) + (1 \times 2^1) + (1 \times 2^0) = 101111_2$$

It may be convenient, when working with the numbers, to set it up as shown in Table 2 when using the subtractive method of binary to decimal conversion. Convert 39_{10} to its binary equivalent.

Thus for the number 39, the 32, 4, 2 and 1 bits must be active (marked with a 1). The decimal numbers are added and the binary coded number is written as 100111_2.

Table 2 Using the Subtractive Method to Convert 39_{10} to 100111_2

Decimal numbers	64	32	16	8	4	2	1
Binary equivalents	2^6	2^5	2^4	2^3	2^2	2^1	2^0
Bit weight (1 or 0)	0	1	0	0	1	1	1

To convert using the division method, start by successively dividing the decimal number by the binary base number (2). The remainder of each of these divisions is marked down and becomes part of the binary number. The remainder is either a 1 or 0. See Table 3 to understand how 351_{10} is converted to its binary equivalent.

Table 3 Using the Division Method to Convert 351_{10} to 101011111_2

	Result	Remainder	
$351 \div 2 =$	175	1	LSB
$175 \div 2 =$	87	1	
$87 \div 2 =$	43	1	
$43 \div 2 =$	21	1	
$21 \div 2 =$	10	1	
$10 \div 2 =$	5	0	
$5 \div 2 =$	2	1	
$2 \div 2 =$	1	0	
$1 \div 2 =$	0	1	MSB

$$351_{10} = 101011111_2$$

3.2.0 Octal

The octal is a base 8 system containing the digits 0 through 7. As with the digital and binary systems, the carry concept is used. A carry of 1 has a weight of 8 because the base is 8. Positional values for octal numbers are 8^0, 8^1, 8^2, and so on for whole numbers. Table 4 gives the first 12 octal numbers and their decimal and binary equivalents.

Table 4 Decimal-Octal-Binary Equivalents

Decimal	Octal	Binary
0	0	0
1	1	1
2	2	10
3	3	11
4	4	100
5	5	101
6	6	110
7	7	111
8	10	1000
9	11	1001
10	12	1010
11	13	1011
12	14	1100

3.2.1 Octal to Decimal Conversion

When converting octal to decimal numbers, use the same principles used to convert binary to decimal numbers except the base number for octal operations is 8 and not 10. Convert 142_8 to its decimal equivalent.

$$142_8 = 1(8^2) + 4(8^1) + 2(8^0)$$

$$142_8 = 1(64) + 4(8) + 2(1)$$

$$142_8 = 64 + 32 + 2$$

$$142_8 = 98_{10}$$

3.2.2 Decimal to Octal Conversion

The two methods of converting decimal to octal numbers are the subtractive method and the division method.

To convert using the subtractive method, start by subtracting the highest possible multiple of eight from the decimal number, and place the multiplier in the position of the power of eight. Repeat this subtraction process on the balance left after each subtraction until there is nothing left. See *Table 5* to understand how 149_{10} is converted to its octal equivalent.

From the powers of eight and positional value numbers shown, 149 is less than 8^3 but greater than 8^2. Therefore, the highest possible multiple of eight to be subtracted from 149 is 8^2 or 64:

$$\begin{array}{r} 149 \\ \underline{-128} \\ 21 \end{array} = 8^2 \times 2 \text{ (MSD)}$$

Repeat this subtraction process on the balance left after each subtraction step until the decimal number has been reduced to 0:

$$\begin{array}{r} 21 \\ \underline{-16} \\ 5 \end{array} = 8^1 \times 2$$

$$\begin{array}{r} 5 \\ \underline{-5} \\ 0 \end{array} = 8^0 \times 5 \text{ (LSD)}$$

$$149_{10} = 225_8$$

To convert using the division method, start by successively dividing the decimal number by the octal base number (8) the same way as for binary.

The remainder of each of the divisions is marked down and becomes a part of the octal number. The remainder of each division step will be a number from 0 through 7.

For example, see *Table 6* to understand how to convert 189_{10} to its octal equivalent.

Table 6 Using the Division Method to Convert 189_{10} to 275_8

	Result	Remainder	
$189 \div 8 =$	23	5	LSB
$23 \div 8 =$	2	7	
$2 \div 8 =$	0	2	MSB
	$189_{10} = 275_8$		

3.2.3 Octal to Binary Conversion

To convert from octal to binary, write the binary equivalent in groups of three digits for each octal digit, beginning with the LSD on the right and succeeding significant groups to the left in increasing order to the MSD. Using *Tables 1* and *4*, convert 43_8 to its binary equivalent.

$$4_8 = 100_2$$

$$3_8 = 011_2$$

$$43_8 = 100_2 + 011_2 = 100011_2$$

Using *Tables 1* and *4*, convert 167432_8 to its binary equivalent.

$$167432_8 = \begin{array}{cccccc} 1 & 6 & 7 & 4 & 3 & 2 \\ 001 & 110 & 111 & 100 & 011 & 010 \end{array}$$

$$167432_8 = 1110111100011010_2$$

3.2.4 Binary to Octal Conversion

When using binary numbers to represent large quantities, many ones and zeros are required. This is tedious and awkward. Other number systems are often used as a shorthand notation for binary numbers. Note that if only three bit positions are considered, eight combinations are possible (000 through 111). Therefore, octal numbers may be substituted directly for 3-bit binary numbers. For example, the binary number 110111100001011001_2

Table 5 Using the Subtractive Method to Convert 149_{10} to 225_8

Decimal numbers	262,144	32,768	4,096	512	64	8	1
Powers of eight	8^6	8^5	8^4	8^3	8^2	8^1	8^0
Octal number	0	0	0	0	2	2	5

Table 7 Example of Binary to Octal Conversion

110	111	100	001	011	001	(binary number divided into groups of three)
6	7	4	1	3	1	(octal substitution for each three bit group)

$$110111100001011001_2 = 674131_8$$

may be converted to octal by the simple procedure of separating the bits into groups of three and substituting octal digits for each group of three bits, as shown in *Table 7*.

3.3.0 Hexadecimal

The hexadecimal system is base 16. In the hexadecimal number system, sixteen different symbols must be used. 0 through 9 are used for the first ten digits, and A through F are used for the remaining six digits, as shown in *Table 8*.

Table 8 Decimal-Hexadecimal-Binary Equivalents

Decimal	Hexadecimal	Binary
0	0	0000
1	1	0001
2	2	0010
3	3	0011
4	4	0100
5	5	0101
6	6	0110
7	7	0111
8	8	1000
9	9	1001
10	A	1010
11	B	1011
12	C	1100
13	D	1101
14	E	1110
15	F	1111

3.4.0 Codes (Binary)

An important requirement of programmable logic controllers is to communicate with various external devices that supply information to the controller or receive information from the controller. This input/output function involves the transmission, manipulation, and storage of binary data that at some point must be interpreted by humans. Although the machine can easily handle this binary data, we require that the data be converted to a readable form.

One way of satisfying this requirement is to assign a unique combination of 1s and 0s to each number, letter, or symbol that must be represented. This technique is called binary coding. In general, there are two categories of codes: those used to represent numbers only and those that represent letters, symbols, and decimal numbers. Several codes for representing numbers, symbols, and letters have been instituted and are standard throughout the industry. Among the most common are ASCII, **binary coded decimal (BCD)**, and Gray codes.

3.4.1 ASCII

Alphanumeric codes (letters, symbols, and decimal numbers) are used when information processing equipment, such as printers or CRTs, must handle the alphabet as well as numbers and special symbols. These characters—26 letters (uppercase), 10 numerals (0 through 9), mathematical and punctuation symbols—can be represented using a 6-bit code, such as $2^6 = 64$. The most common code for alphanumeric representation is the American Standard Code for Information Interchange (ASCII).

The ASCII code can be 6, 7, or 8 bits. Even though the basic alphabet, numbers, and special symbols can be accommodated in a 6-bit code (64 possible characters), standard ASCII character sets use a 7-bit code ($2^7 = 128$ possible characters), which allows lower case and control characters for communication links in addition to the characters already mentioned. This 7-bit code provides all possible combinations of characters used when communicating with peripherals or interfaces. The 8-bit ASCII code is used when **parity** check is added to the standard 7-bit code for error checking.

3.4.2 BCD

Binary coded decimal was introduced as a convenient means for humans to handle numbers that need to be inputted to digital machines and to interpret numbers outputted from the machine. Handling numbers in binary would be extremely tedious because people are much more familiar with the decimal number system. The best solution to this problem is a means of converting a code readily handled by man (decimal) to a code readily handled by the equipment (binary). The result is BCD.

The decimal system uses the numbers 0 through 9, whereas in BCD, each of these numbers is represented by a 4-bit binary number. *Table 9* illustrates the relationship between the BCD code and the decimal numbering system.

Table 9 Decimal-BCD Equivalents

Decimal	BCD
0	0000
1	0001
2	0010
3	0011
4	0100
5	0101
6	0110
7	0111
8	1000
9	1001

To perform this conversion, find the binary equivalent for each decimal digit in the BCD table, and write the binary number beginning with the lowest significant 4-bit group followed by succeeding 4-bit groups in increasing significance to the left. Convert 563_{10} to BCD.

$$563_{10} = \quad 5 \qquad 6 \qquad 3$$
$$\qquad\qquad 0101 \quad 0110 \quad 0011$$

$$563_{10} = 010101100011 \text{ BCD}$$

To perform BCD to decimal conversions, break the BCD number into 4-bit groups beginning with the least significant digit. Find decimal equivalents for each group in the BCD table. The lowest significant decimal digit is equivalent to the lowest significant BCD group. Write the decimal digit symbols in increasing order of significance from right to left. Convert the BCD number 100100110100 to decimal:

$$100100110100 \text{ BCD} = 1001 \quad 0011 \quad 0100 = 934_{10}$$
$$\qquad\qquad\qquad\qquad\qquad 9 \qquad 3 \qquad 4$$

3.4.3 Gray Code

Gray code is used by rotational or linear motion encoders to convert motion positions to a digital value. The encoders are mechanical devices that create the Gray code values as the rotational or linear motion occurs. They are commonly used to

provide angular position information for rotating shafts. Gray code place values do not remain constant like binary code. In binary code, each position value represents the next higher power of 2 from right to left. With binary code, more than one digit can change when going from one number to the next higher or lower number, such as going from 1111 to 10000 or vice versa. Gray code avoids that problem because only one bit changes when going from one number to the next higher or lower number. *Table 10* shows the Gray code compared to binary code for decimal values from 0 to 15. Sequences of Gray code numbers can be easily checked for errors by observing that no more than one digit changes between each number.

Table 10 Comparison of Gray and Binary Codes

Decimal	Gray Code	Binary Code
0	0000	0000
1	0001	0001
2	0011	0010
3	0010	0011
4	0110	0100
5	0111	0101
6	0101	0110
7	0100	0111
8	1100	1000
9	1101	1001
10	1111	1010
11	1110	1011
12	1010	1100
13	1011	1101
14	1001	1110
15	1000	1111

4.0.0 ◆ HARDWARE

As mentioned earlier, the hardware components of a PLC include a power supply, input/output modules, and a processor. Each of these components has a separate function and may perform its function in a variety of ways and under various conditions.

4.1.0 Power Supply

There may be one or more power supplies associated with each PLC system. One power supply provides power to operate the processor and the circuitry internal to the I/O modules. Additional power supplies may be required to provide power to operate the field devices connected to the output modules.

The power supply that provides power to the processor and I/O module circuitry can be either internal to the processor module, mounted inside the rack that contains the processor and I/O modules, or mounted external to the rack. Regardless of location, the function of the power supply is to convert the incoming power, usually 120VAC/60Hz, to a usable level for the modules. Normally, the power supply is protected from the source with a fuse, and an indicator light is used to show power supply operation.

4.2.0 Input/Output Modules

The input/output system is the interface between the field devices and ladder logic programmed in the PLC memory. Each module may either be an input module or an output module. The input module is designed to receive signals from the field input devices and to condition and isolate those signals for use by the processor. An output module conditions and isolates a signal from the processor for use in activating or deactivating an output field device. There are five general classes of I/O modules. They are listed as follows:

- Discrete
- Numerical data
- Analog
- Special
- Remote adapter

4.2.1 Discrete

The most common class of input/output interface is the discrete type. This interface connects to field devices, which will provide an input signal that is non-analog in nature, or field output devices that will require a non-analog signal to control their state. This characteristic limits the discrete I/O interfaces to sensing signals that are on/off or open/closed. To the interface circuit, all discrete inputs are essentially a switch that is open or closed. Likewise, output control is limited to devices that only require switching to one of two states, such as on/off.

Each discrete input and output is powered by a voltage source that may or may not be of the same magnitude or type. For this reason, I/O interface circuits are available at various AC and DC voltage ratings.

If an input switch is closed, a discrete input interface senses the supplied voltage and converts it to a digital signal acceptable to the processor to indicate the status of that device. A logic 1 indicates ON or CLOSED, and a logic 0 indicates OFF or OPEN. In operation, the output interface circuit switches the supplied control voltage that will energize or de-energize the field device on and off. If an output is turned ON through the control program, the supplied control voltage is switched by the interface circuit to activate the addressed output device.

The system design for the discrete I/O modules must match the **sinking** and **sourcing** operations of the field devices connected to the modules. If an inputting device provides current when its output is in a true or ON state, the device is referred to as a sourcing current device. If the inputting device must receive current when its output is in a true or ON state, the device is referred to as a sinking current device. A sinking input module must be used if the field devices connected to it are sourcing current devices, and a sourcing input module must be used if the field devices are sinking current devices. A similar situation occurs with the output modules and the output devices connected to the modules. Sourcing output modules must be connected to sinking output devices, and sinking output modules must be connected to sourcing output devices. Problems can occur if the wrong types of devices are connected to the wrong I/O modules. For instance, if a sinking input module is connected to both sinking and sourcing input devices, the true or ON state of the sinking input devices would never be detected even though a voltage can be measured across the module's corresponding input terminals. Mismatching field devices to the I/O modules could also damage circuits in the devices or modules.

A block diagram of a typical AC/DC discrete input interface circuit is shown in *Figure 12*. Input circuits vary widely among manufacturers, but in general, all AC/DC interfaces operate in a manner similar to that described in this diagram. The input circuit is composed of two primary parts: the power section and the logic section. The power and logic sections of the circuit are normally coupled with an isolator, which electrically separates the two.

The power section of an input interface basically performs the function of converting the incoming voltage from an input sensing device to a DC logic-level signal to be used by the processor. The bridge rectifier circuit converts the incoming signal (AC or DC) to a DC level that is passed through a filter circuit, which will protect against signal debounce and electrical noise on the input power line. The threshold circuit detects whether the incoming signal has reached the proper voltage level for the specified input rating. If the input signal exceeds and remains above the threshold

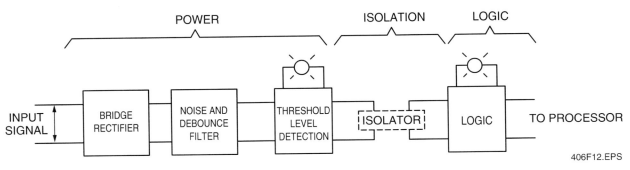

Figure 12 ♦ AC/DC discrete input circuit.

voltage for a duration of at least the filter delay, the signal will be recognized as a valid input.

When a valid signal has been detected, it is passed through the isolation circuit, which completes the electrically isolated transition from AC to logic level. The DC signal from the isolator is used by the logic circuit and made available to the processor for its data bus. Electrical isolation is provided so that there is no electrical connection between the field device power and the controller logic. This electrical separation will help prevent large voltage spikes from damaging the logic side of the interface or the processor. The coupling between the power and logic sections is normally provided by an optical-coupler.

Figure 13 illustrates a sample input device wiring diagram. Most input circuits have an LED indicator to signify that the proper input voltage level is present. An LED indicator may also be available to indicate the presence of a logic 1 when the input voltage is present.

Figure 14 shows a block diagram of a typical AC/DC discrete output circuit. The circuit consists primarily of the logic and power sections coupled by an isolation circuit. The output interface can be thought of as a simple switch through which power can be provided to control the output device.

During normal operation, the processor sends the logic circuit the output status according to the ladder program. If the output is energized, the signal from the processor is fed to the logic section and passed through the isolation circuit, which will switch the power to the field device.

The switching section generally uses a triac or a silicon-controlled rectifier (SCR) to switch the power. A fuse may be provided in the output circuit to prevent excessive current from damaging the AC switch.

As with input circuits, the output interface may provide LED indicators to indicate operating logic and power circuits. If the circuit contains a fuse, a fuse status indicator may also be incorporated. An output connection diagram is illustrated in *Figure 15*.

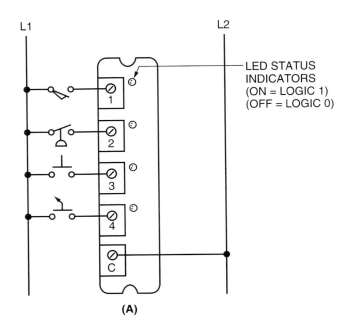

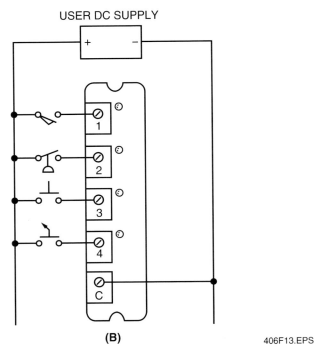

Figure 13 ♦ Input module connection diagram.

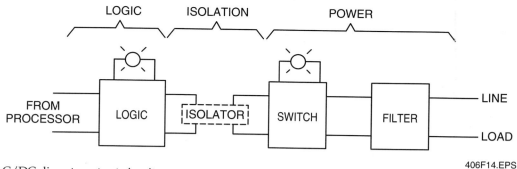

406F14.EPS

Figure 14 ◆ AC/DC discrete output circuit.

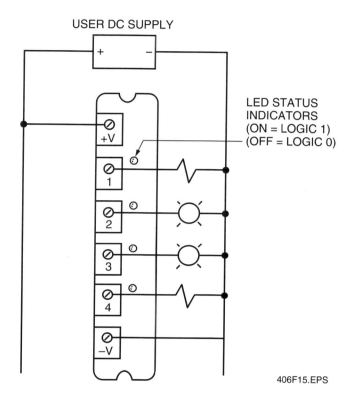

406F15.EPS

Figure 15 ◆ Output module connection diagram.

4.2.2 Numerical Data

With the integration of the microprocessor into PLC architecture in the early 1970s came new capabilities for arithmetic operation and data manipulation. This expanded processing capability led to a new class of input/output interfaces known as numerical data I/O. Numerical input interfaces allowed measured quantities to be inputted from instruments and other devices that provide numerical data, while numerical output interfaces allowed control of devices that require numerical data.

In general, numerical data I/O interfaces can be categorized into two groups: those that provide interface to multi-bit digital devices and those that provide interface to analog devices. The multi-bit interfaces are like the discrete I/O in that the processed signals are discrete. The difference, however, is that with the discrete I/O, only a single bit is required to read an input or control an output. Multi-bit interfaces allow a group of bits to be inputted or outputted as a unit to accommodate devices that require the bits be handled in parallel form or in serial form. The analog I/O will allow monitoring and control of analog voltages and currents that are compatible with many sensors, motor drives, and process instruments. With the use of multi-bit or analog I/O, most process variables can be measured or controlled with appropriate interfacing.

4.2.3 Analog

The analog input module contains the circuitry necessary to accept analog voltage or current signals from field devices. The voltage or current inputs are converted from an analog to a digital value by an analog-to-digital converter (ADC). The conversion value, which is proportional to the analog signal, is passed through to the processor's data bus and stored in a memory location for later use.

Typically, analog input interfaces have a very high input impedance, which allows them to interface to high source-resistant outputs from input devices. The input line from the analog device generally uses shielded conductors. The shielded cable provides a better interface media and helps to keep line impedance imbalances low in order to maintain good common mode rejection of noise levels, such as power line frequencies. The input stage of the interface provides filtering and isolation circuits to protect the module from additional field noise.

The analog value (after conversion) is expressed as a BCD value that ranges from 000 to 9999 or a decimal value from 0 to 32,767 where the low and high counts represent the low and full-scale input signals. A typical analog input connection is illustrated in *Figure 16*.

The analog output interface receives numerical data from the processor, which is translated into a proportional voltage or current to control an analog field device. The digital data is passed through a digital-to-analog converter (DAC) and outputted in analog form. Isolation between the output circuit and the logic circuit is generally provided through optical couplers. These output interfaces normally require an external power supply with certain current and voltage requirements.

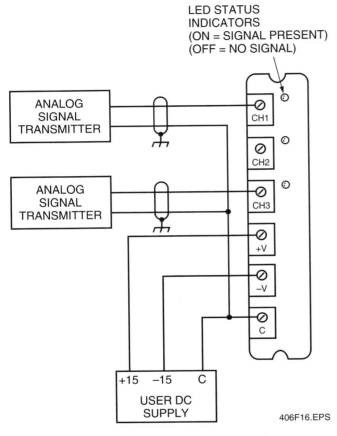

Figure 16 ◆ Analog input module connection diagram.

4.2.4 Special

In the previous sections, the standard discrete, numerical, and analog modules were discussed. They will probably cover the requirements for many of the applications that you will encounter. The remaining will most likely require an advanced special I/O module.

To process certain types of signals or information efficiently, the processor will require special interfaces. These special I/O interfaces, sometimes called preprocessing modules, include those that simply condition input signals or other signals that cannot be interfaced using standard I/O modules. Special I/O modules may also incorporate an on-board microprocessor to add intelligence to the interface. These intelligent modules can perform complete processing tasks independent of the processor.

One example of a special I/O module is the ASCII input/output interface. It is used for sending and receiving alphanumeric data between peripheral equipment and the processor. Typical peripheral devices include printers, video monitors, and displays. This special I/O interface is usually available with communications circuitry that includes an on-board **random access memory (RAM)** buffer and a dedicated microprocessor. The information exchange of either type of interface generally takes place via an RS-232C, RS-422, or a 20mA current loop standard communications link.

If the ASCII interface does not use a microprocessor, the main processor handles all the communications routing, which significantly slows the communication process and the program **scan**. Each character or string of characters to be transmitted to the module or received from the module is handled on a character-by-character basis. The module interrupts the processor each time it receives a character from the peripheral, and then it accesses the module each time it needs to send a message to the peripheral. The communication speed is generally very slow, and for a character to be read, the scan time must be faster then the time required to accept one character.

If a smart ASCII interface is used, the transmission is accomplished between the peripheral and the module also on an interrupt basis but at a faster transmission speed. This is possible since the on-board microprocessor is dedicated to performing the I/O communication. The on-board microprocessor contains its own RAM, which can store blocks of data to be transmitted. When the input data from the peripheral is received at the module, it is transferred to the PLC memory through a data transfer instruction at the I/O bus speed. All the initial communication parameters, such as number of **stop bits**, parity, and **baud rate**, will be hardware selectable or selectable through control software, depending on the interface. This method significantly speeds up the communication process and increases data output.

Other special I/O modules include high-speed encoder/counter, servo control, stepper control, synchronized axis control, velocity control, and temperature monitoring/control modules, along with common discrete digital, analog, and smart transmitter interface modules. Of these, the most common are the encoder/counter, temperature monitoring/control, stepper control, and smart transmitter interface modules, which are briefly described as follows:

- *Encoder/counter modules* – These types of modules provide a high-speed counter external to the PLC processor. Normally they operate independently of the processor and register high speed input pulses while the processor is engaged with other program activities. The registered count value is then periodically sent to the processor during program activities. They are used in process control with turbine flowmeters that produce TTL pulses, or any field device that creates TTL pulses, and with rotational or linear motion encoders. Absolute shaft or linear motion encoders are usually used with modules that receive BCD or Gray code data.

- *Temperature monitoring/control modules* – Temperature monitoring/control modules condition inputs from a number of thermocouples. The thermocouple inputs are filtered, amplified, and digitized by analog-to-digital converters. The digitized signals for each input are processed by an internal microprocessor as an input to a PID loop. The processor performs a PID algorithm for each input, and the output of each loop is sent to the PLC as both a numeric value and as a time-proportioned output (TPO) signal. The PLC program can send either the numeric value to an analog output module or the TPO signal to a digital output module to close the loop.

- *Stepper control modules* – These types of output modules generate a pulse train that is compatible with stepping motor translators. The pulses normally represent distance, speed, and direction commands to a stepping motor. The desired motor position is dictated by the preset count of output pulses, a forward or reverse direction command, and by an acceleration or deceleration command that is determined by the pulse rate. When the output module is initialized by a PLC program instruction, the module will send the output pulses as determined by the PLC program to the translator. The output module will generally not accept any commands from the PLC program until the move is completed as determined by status feedback from the translator. Some output modules may have a provision for an override command that resets the stepper motor to its current position; however, the override must be disabled to continue operation.

- *Smart transmitter interface modules* – These modules allow a PLC to access all of the analog and digital information generated by HART®-compliant field devices. The digital process values can be used to qualify analog device inputs, enabling PLC-based programs to compensate more precisely and control a process more accurately. The HART® (highway addressable remote transducer) communication protocol is the de facto industry standard field communication protocol for instrumentation networks. It was developed by Rosemount.

4.2.5 Remote Adapter

Larger PLC systems allow input/output subsystems to be remotely located from the processor. A remote subsystem is usually a rack-type enclosure in which the I/O modules are installed. The rack generally includes a power supply to drive the logic circuitry of the interfaces and a remote I/O adapter module that allows communication with the processor. Capacity of a single subsystem is normally 32, 64, 128, or 256 I/O points. A large system with a maximum capacity of 1,024 I/O points might have subsystem sizes of either 64 or 128 points, in which case there could be either eight racks with 128 I/O, sixteen racks with 64 I/O, or some combination of both sizes.

Individual racks are normally connected to the processor using a daisy chain or star configuration via one or two twisted-pair conductors or a single coaxial cable. The distance a remote rack can be placed away from the processor varies by manufacturer but can be as much as two miles. Remote I/O offers tremendous savings on wiring materials and labor costs for large systems in which the field devices are in clusters at various locations. With the processor in a central area, only the communication link is brought back to the processor, instead of hundreds of field wires. Distributed I/O also offers the advantage of allowing subsystems to be installed and started up independently, as well as allowing maintenance on individual subsystems while others continue to operate.

5.0.0 ◆ PROCESSORS

The microprocessor that controls and operates the processor module functions in the same basic manner as the microprocessor explained previously. The PLC microprocessor and the memory are both physically contained within the processor module.

Most processor modules have front panel lights or indications to provide the user with status indication of PLC operation. These lights are very useful in troubleshooting. Also provided on most processor modules is a switch used to change the module's mode from RUN to PROGRAM or TEST. Additional connections are usually provided to allow connecting a terminal for programming the PLC and a port for connections to external I/O racks.

5.1.0 Scanning

The processor module controls the PLC by executing the software program. During program execution, the processor reads all the inputs, takes these values and, according to the control logic, energizes or de-energizes the outputs thus solving the ladder network. Once all the logic has been solved, the processor updates all outputs. The process of reading the inputs, executing the program, and updating the outputs is known as a scan. The time required to make a single scan (scan time) can vary from 1 to 100 milliseconds. The use of remote I/O subsystems increases the scan time as a result of having to transmit and receive the I/O update from remote subsystems.

The scan is normally a continuous and sequential process of reading the status of inputs, evaluating the control logic, and updating outputs. The common scan method of monitoring the inputs at the end of each scan is inadequate for reading certain extremely fast inputs. Some PLCs provide software instructions that will allow the interruption of the continuous program scan in order to receive an input or update an output immediately. These immediate instructions are very useful when the PLC must react instantaneously to a critical input or output.

5.2.0 Memory

To execute the system program, the processor must make use of the memory. Within the PLC are several levels of memory. These memory types may be RAM, **read only memory (ROM), programmable read only memory (PROM),** or any other type of memory previously discussed.

The PLC's memory is organized into various areas, depending on how the processor uses the data. This organization of memory is called a memory map. Almost all PLCs have different memory maps. In general, all programmable controllers must have memory allocated for the following six items:

- Executive program
- Processor work area
- Scratch pad
- Diagnostics
- Data table
- User program

The executive program, processor work area, scratch pad, and diagnostics are all system memory devices. These memory locations are not accessible by the user. Their function is to contain the system operations program, program diagnostics, system status, and memory location map and to act as temporary storage for processor calculations.

The data table and user program make up the application memory. The application memory stores programmed instructions and any data that will be used by the processor to perform its control functions. Each processor has a maximum amount of application memory, which varies depending on its size. All data is stored in what is called the data table. Program instructions are stored in the area called the user program.

The data table can be functionally divided into four areas: the input table, output table, internal storage bits, and storage registers. The input table is an array of bits that stores the status of digital inputs, which are connected to input interface circuits. The number of bits in the table is equal to the maximum number of inputs. A controller with a maximum of 64 inputs would require an input table of 64 bits. Each input has a bit in the input table that corresponds exactly to the terminal to which the input is connected. If the input is ON, its corresponding bit in the table is ON (1). If the input is OFF, the corresponding bit is cleared or turned OFF (0). The input table is constantly changed to reflect the current status of connected input devices. This status information is used by the control program.

The output table is an array of bits that controls the status of digital output devices, which are connected to output interface circuits. The number of bits in the output table is equal to the maximum number of outputs. A controller with a maximum of 128 outputs would require an output table of 128 bits. Each output has a bit in the output table that corresponds exactly to the terminal to which the output is connected. Bits in the output table are controlled by the processor as it interprets the control program and are updated accordingly during the I/O scan. If a bit in the table is turned ON (1), then the connected output is switched ON. If a bit is cleared or turned OFF (0), the output is switched OFF.

Most controllers have an area for internal storage bits. These storage bits are also called internal outputs, internal coils, internal control relays, or just internals. The internal output operates just as any output that is controlled by programmed logic; however, the output is used strictly for internal purposes. In other words, the internal output does not directly control an output device. Internal outputs include timers, counters, and internal control relay instructions of various types. Each internal output, referenced by an address in the control program, has a storage bit of that same address. When the control logic is true, the internal (output) storage bit turns ON.

The information stored in the input/output table is ON/OFF status information that is easily represented by 1 or 0. To denote any quantity having a value that cannot be represented by a single 1 or 0, groups of bits or memory words must be used. Memory words that store value-type information are also called storage registers. In general, there are three types of storage registers: input registers, holding registers, and output registers. The total number of registers varies, depending on the controller memory size and according to how the data table is configured. Values stored in the storage registers are in a binary or BCD format. Each register can generally be loaded, altered, or displayed using the industrial terminal or other data entry devices.

Input registers are used to store numerical data received via input interfaces from devices such as thumbwheel switches, shaft encoders, and other devices that provide BCD input. Analog signals also provide numerical data that must be stored in input registers. The current or voltage signal generated by various analog transmitters is converted by the analog interface. From these analog values, binary representations are obtained and stored in the designated input register. The value contained in the input register is determined by the input device and is therefore not alterable from within the controller or via any other form of data entry. Holding registers are required to store variable values that are program-generated by instructions (such as math, timer, or counter) or constant values that are entered via the industrial terminal or some other data entry method.

Output registers are used to provide storage for numerical or analog values that control various output devices. Typical devices that receive data from output registers are alphanumeric LED displays, recorder charts, analog meters, speed controllers, and control valves. Output registers are essentially holding registers that are used to control outputs.

The user program memory is an area reserved in the application memory for the storage of the control logic. All the PLC instructions that control the machine or process are stored here. The addresses of inputs and outputs, whether real or internal, are specified in this section of memory.

When the processor is in the run mode and the program is executed, the processor interprets the user program memory locations and controls the bits of the data table that correspond to real or internal outputs. The interpretation of the user program is accomplished by the processor's execution of the executive program.

The maximum amount of user program memory available is normally a function of the processor size, such as I/O capacity. In medium and large controllers, the user program area is normally flexible by altering the size of the data table so that it meets the minimum data storage requirements. In small processors, however, the user program area is normally fixed. In some cases, the application memory can be expanded to some maximum point by using add-on memory modules.

6.0.0 ◆ SOFTWARE

Software is a general term used to designate the various types of programs used to configure the PLC processor. In practice, there are two types of configuration data entered into each processor. One type is called software and the second type is called firmware. Software is the user-configured program that determines the status of PLC outputs as determined by the inputs. An example of this is a ladder logic program. Firmware is the program that is permanently entered into a processor. Its purpose is to define the operation of the processor and to allow the entering of operational software.

6.1.0 Languages

The following are the five types of programming languages normally used to develop software for programmable logic controllers:

- Ladder logic
- Boolean mnemonics
- English statements
- Functional block
- Machine stage

Figure 17 shows each of these programming languages contrasted for completion of a typical electrical motor starter application. (Some are shown in abbreviated form.)

6.1.1 Ladder Logic

Ladder logic programming is the most widely used form of PLC programming. This is because it is most like conventional electrical relay ladder schematics that represented the initial purpose of a PLC. Ladder logic is programmed by successively entering rungs to the ladder. Each rung is comprised of input and output symbols that are very similar to conventional electrical symbols. Ladder logic programming will be explained in more detail later in this section.

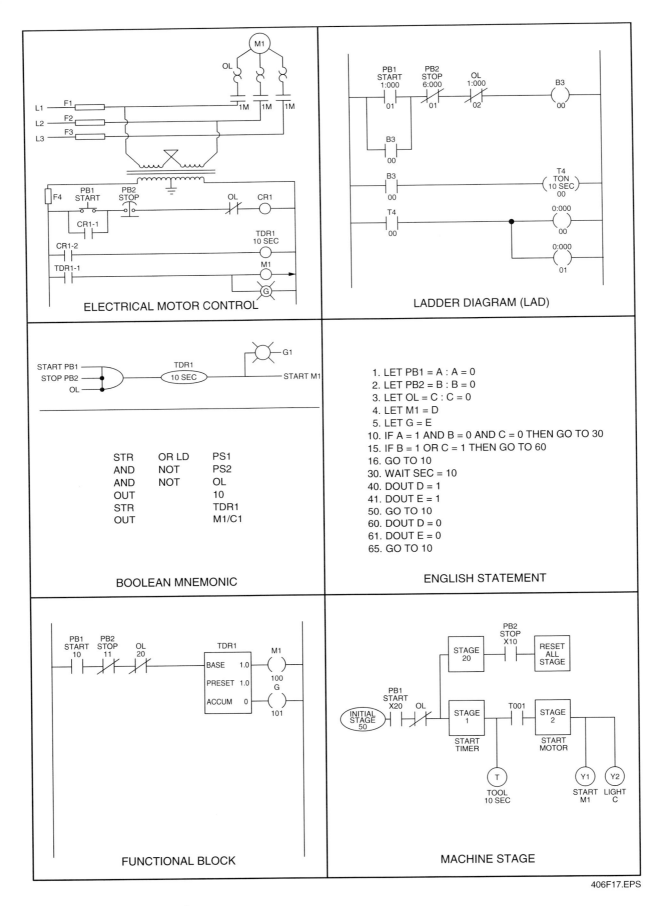

Figure 17 ◆ PLC language comparison.

6.1.2 Boolean

The Boolean language is a basic level PLC language that is based primarily on the Boolean logic gates: AND, OR, and NOT. A complete Boolean instruction set consists of the logic gates and other instructions that will implement all the functions of an electrical ladder diagram. Each instruction is written in an abbreviated form, using three or four letters that generally imply the operation of the instruction.

6.1.3 English Statement

English statement languages for programmable logic controllers can be considered a derivative of computer languages. Compared with the languages that many computers use, the English statement instructions are much easier to understand and more operator oriented.

The high-level control statements provide an advantage in the manipulation of data and numerical I/O because data can be inserted in registers directly through simple English instructions. Peripheral communication for reporting can also be simplified when using these languages because an instruction such as PRINT can be used to output a string of data.

These control statements have been adopted in larger PLCs. Control statements have two main reasons for their support: their simplicity eases the programming of a control task, and the English instructions allows other users to easily interpret the program once it has been written. There are generally a fixed number of instructions available, and they differ from one manufacturer to another; nevertheless, in one way or another, they accomplish the same task.

There are several types of control statements that are presently used in the industry. Most of these languages are similar to the widely used BASIC language used in personal computers.

6.1.4 Functional Block

Functional blocks are high level instructions that permit the user to program more complex functions using the electrical ladder diagram format. The instruction set is composed of blocks that execute or perform a specific function.

When using block instructions, input conditions are programmed using normally open (NO) and normally closed (NC) contacts, which will enable the block operation. There are also several parameters associated with the block that must be programmed. These parameters normally include storage or holding registers used to set preset values, or I/O registers (variables) used to input or output numeric data (analog, BCD, and others).

Functional blocks can be classified into four main types: timer and counter instructions, arithmetic, data manipulation, and data transfer blocks. Each of these classifications is formed by a group of instructions of similar operation. Depending on the block type, there will be one or more control lines and one or more data specifications inside the block. Many features of functional block programming have been incorporated into advanced-level ladder logic programming.

6.1.5 Machine Stage

Machine stage is a relatively recent development in PLC programming. This system combines the flow chart for the process or machine to be controlled with basic relay ladder logic instructions. Machine-stage programming is similar to functional block programming in that the use of extensive interlocking is not necessary, and each block controls one complete operation. It differs from functional block programming because it does not follow the ladder logic format.

6.2.0 Ladder Diagram Instructions

Ladder diagrams are a symbolic instruction set used to create programs for a PLC. The symbols that are used to represent the various operations have been selected to closely approximate common electrical schematic symbols. The following are the six general categories of instructions available for use in ladder diagram programming:

- Relay
- Timer/counter
- Arithmetic
- Data manipulation
- Data transfer
- Program control

Each of these instructions is entered onto a rung within the ladder. When the input to a rung is present, it is said to be true. When a given rung's inputs are true, this causes the output to become energized (also called true). When logic continuity exists across a rung, then the rung is true and the output is energized.

6.2.1 Relay

Relay instructions are comprised of input contacts and output coils. These instructions are the most fundamental of ladder logic instructions and allow the PLC to communicate with external devices.

Inputs into a PLC's input module are shown as contacts on a ladder rung. These contacts can be either normally open (NO) or normally closed

(NC) in their deactivated or 0 state. Examples of possible inputs include switches, pushbuttons, limit switches, and output coil contacts. By convention, input contacts are entered on the left side of a ladder rung and output relay coils are entered on the right. Typical symbols for input contacts are shown in *Figure 18*.

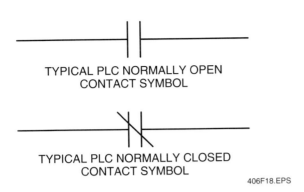

TYPICAL PLC NORMALLY OPEN
CONTACT SYMBOL

TYPICAL PLC NORMALLY CLOSED
CONTACT SYMBOL

406F18.EPS

Figure 18 ◆ Input contact symbols.

Outputs from a PLC's output module are shown as a coil on a ladder rung. These coils can be of two different types: standard or latch. A standard coil will energize when the rung is true and de-energize when a rung is false. A latch coil is shown like a standard coil with the letter L in the center of the coil. When the rung is true, a latch coil will energize. When the unlatch coil (coil with U in the center) with the same address is energized, the latch coil will unlatch. Typical symbols for various output coils are shown in *Figure 19*.

Figure 20 illustrates how the contacts and coils are used to represent logic flow in a ladder diagram. Rung 0 shows a NO contact that is deactivated or off (logic state of 0 or false), resulting in a de-energized coil (logic state of 0 or false). Rung 1 shows a NO contact that is activated or on (logic state of 1 or true), resulting in an energized coil (logic state of 1 or true). Rungs 2 and 3 show the

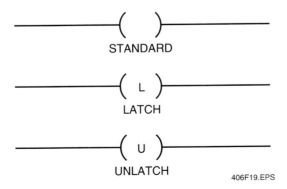

STANDARD

L

LATCH

U

UNLATCH

406F19.EPS

Figure 19 ◆ Output coil symbols.

logic states of NC contacts that result in the exact opposites of the coil logic states in rungs 0 and 1. The logic states correspond to logic bits of 1 or 0 in the ladder program of a PLC.

Figure 21 shows the operation of a latching/unlatching coil and its input contacts. In rung 0 of this figure, output coil 3 will latch on when input contact 1 is closed. If contact 1 then opens, output 3 will remain latched on until unlatched by unlatch coil 3. Unlatch coil 3 will be energized when input contact 2 is closed. Each of these input contacts could be a push button connected to the PLC's input module. The output coils of the ladder control some point on the PLC's output module. When the ladder's coil is energized, so is the device connected to the PLC's output module. Some PLCs only allow internal outputs to be latched.

Observe that in *Figure 21*, the inputs are labeled I:1 and I:2 while the output is labeled O:3. This is an overly simplified example of an addressing method for inputs and outputs. The discrete I/O addressing scheme used in Allen-Bradley PLC5 ladder diagram programming to identify hardware physical location and input and output table memory locations would comprise the I: or O: identifier along with a group of numbers such as 001/12. The first two digits of the number group designate the rack location number, and the third digit specifies the I/O table group number (0 through 7). The two digits after the slash (/) represent the input or output bit memory location/terminal number (00-07 and 10-17). For example, I:001/12 would represent an input device connected to terminal 12 of an input module at rack location 00 and input table group 1, bit memory location 12. This method of addressing makes it very easy to match the input device to the actual hardware terminals and assigned memory locations of the PLC. This system will be used in the examples given in the remainder of this module. Other addressing schemes are used by different manufacturers, and some are more difficult to interpret.

6.2.2 Timer and Counter

Timers and counters are output instructions that provide the same functions as hardware timers and counters. They are used to activate or deactivate a device after an expired interval or count. The timer and counter instructions are generally considered internal outputs. Like the relay-type instructions, timer and counter instructions are fundamental to the ladder diagram instruction set.

The operations of the software timer and counter are similar in that they are both counters. A timer counts the number of times that a fixed

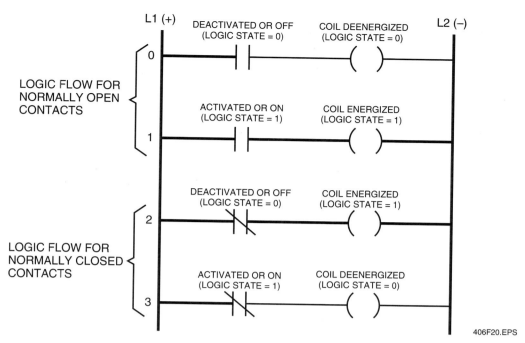

Figure 20 ◆ Logic flow for normally closed and normally open contacts.

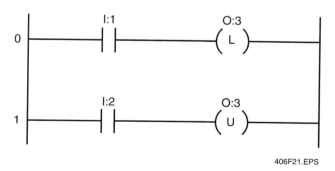

Figure 21 ◆ Latching/unlatching coil operation.

interval of time (such as 0.1 second or 1.0 second) elapses up to a preset value. For example, to time an interval of 3 seconds, a timer counts 3 one-second intervals (time base). A counter simply counts the occurrence of an event. Both the timer and counter instructions require an accumulator register (word location) to store the elapsed count and a preset register to store a preset value. The preset value will determine the number of event occurrences or time-based intervals that are to be counted.

The timer on delay (TON) output instruction is programmed to provide time-delayed action or to measure the duration for which some event is occurring (*Figure 22*, rung 3). If the rung path for the timer has logic continuity (true), the timer begins counting time-based intervals and times until the accumulated time equals the preset value. If the input remains true, then when the accumulated time equals the preset time, the timer done (DN) output is energized (or de-energized), and the timed-out contact associated with the output is closed (or opened), as shown in rung 4. If the rung logic for the TON goes false at any time, the accumulated value is reset to 0.

The timer off delay (TOF) output instruction is programmed to provide time-delayed action (rung 5). If logic continuity is lost, the timer begins counting time-based intervals until the accumulated time equals the programmed preset value. When the accumulated time equals the preset time, the timer done (DN) output is de-energized (or energized), and the timed-out contact (rung 6) associated with the output is opened (or closed). If the input logic to the TOF is restored to true at any time, the accumulated value is set to 0, and the DN output is restored to its original condition.

Timers may also be retentive. A retentive timer on (RTO) output instruction is one that stores its timed value until it has been reset by a reset instruction. The RTO (rung 7) will retain its accumulated count, even if the input logic or power is lost. If the input rung is true and remains true, the timer will count intervals until the accumulated value equals the preset value. If the input becomes false before the timer is timed out, the accumulated value is retained. When the input is restored to true, the timer will resume the count until the accumulated value equals the preset value and the timer done (DN) bit is set. The DN bit can be a

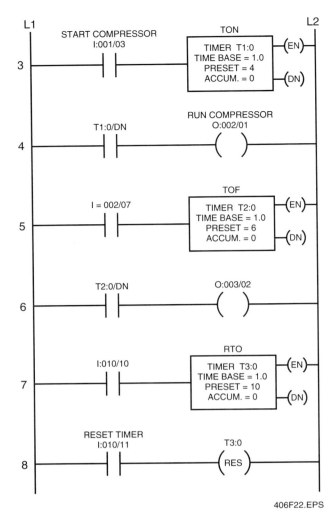

Figure 22 ◆ Timers.

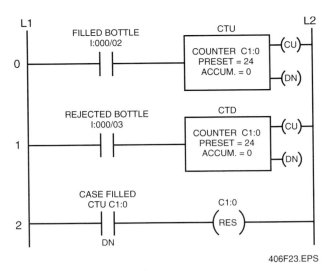

Figure 23 ◆ Counter application.

NO or NC instruction as desired for use in the rest of the program. To clear the RTO accumulator value, a reset (RES) instruction for the timer must be set by a NO reset input (rung 8).

The up-counter (CTU) output instruction will **increment** by one each time the counted event occurs (*Figure 23*, rung 0). The CTU increments its accumulated value each time the up-count event makes an OFF-to-ON transition. When the accumulated value reaches the preset value, the counter done (DN) output is turned ON, and the DN contact associated with the referenced output is closed (or opened).

The down-counter (CTD) output instruction will count down by one each time a certain event occurs (rung 1). Each time the down-count event occurs, the accumulated value is **decremented**. In normal use, the down-counter is usually used in conjunction with the up-counter to form an up/down-counter. For example, while the CTU counts the number of filled bottles that pass a certain point, a CTD with the same reference address would subtract one from the accumulator each time an empty or improperly filled bottle occurs.

Counters require a reset instruction to clear their accumulator. In this example, a CTU is used to determine the completion of a case of filled bottles. The counter done (DN) contacts signal the completion of the case when the accumulator reaches the preset value so that the next case can be moved to the fill point. The DN bit is also used to reset the counter for the next case. A momentary false-to-true transition is required for both counters to increment. Unlike timers, an up-counter will continue to increment the accumulator value after the preset has been reached unless the counter is reset. If the accumulated value exceeds the maximum range of the counter, an overflow (OV) bit will be set. This overflow signal can be used to cascade counters for applications that require counts greater than the maximum value of the counter.

6.2.3 Arithmetic

Arithmetic operations include the four basic operations of addition, subtraction, multiplication, and division. These instructions use the contents of two registers and perform the desired function and store the result in a third register. *Figure 24* shows the four arithmetic operations.

The ADD instruction performs the addition of two values stored in the referenced memory locations. The SUB instruction performs the subtraction operation of two registers.

The MUL instruction performs the multiplication operation. It uses two words in a register to hold the result of the operation between two operand registers. The DIV instruction performs the quotient calculation of two registers. The result of the division is held in two words of a

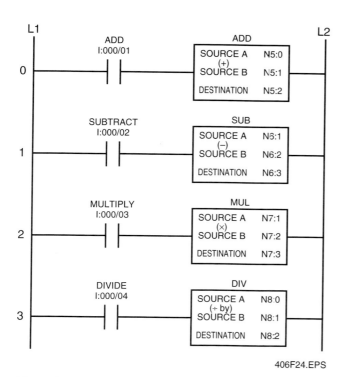

Figure 24 ◆ Arithmetic operations.

406F24.EPS

third register. The first word generally holds the integer while the second word holds the decimal fraction. In multiplication and division, the operand registers can be values or addresses.

6.2.4 Data Comparison

The data comparison instructions are an enhancement of the basic ladder diagram instruction set. Whereas the relay-type instructions were limited to the control of internal and external outputs based on the status of specific bit addresses, the data comparison instructions allow multi-bit operations. In general, the comparison of data using ladder diagram instructions involves simple register operations to compare the contents of two registers. In the ladder language, there are three basic data comparison instructions: equal to (EQU), greater than (GRT), and less than (LES). Based on the result of a greater than, less than, or equal to comparison, an output can be turned ON or OFF, or some other operation can be performed. The comparison can be performed on either addresses or values. *Figure 25* shows typical data comparison instructions.

6.2.5 Data Transfer

Data transfer instructions involve the transfer of the contents from one register to another. Data transfer instructions can address any location in the memory data table, with the exception of areas restricted to user applications. Prestored values can be automatically retrieved and placed in any new location. The common instructions used are bit distribute (BTD), move (MOV), and masked move (MVM). Similar data transfer instructions are used by most PLC manufacturers. *Figures 26* and *27* show typical data transfer operations.

The BTD output instruction is used to move up to sixteen bits of data within or between words. The source of the data is not affected. The destination is overwritten with the data bits being moved. If the length of the moved bit field exceeds the bit field of the destination word, the overflow bits are lost. When the BTD instruction is true, the desired bit field is moved from the source word to the destination word during a scan cycle of the PLC. If bits are to be moved within the source word, as shown in *Figure 26*, the source and destination word address is the same.

The MOV output instruction is used to copy the data in a source to a destination. When the MOV instruction is true, the content of the source is transferred to the destination during a scan cycle of the PLC. The source may be a program constant or a data address that is used by the instruction to read an image of the value. This imaged value is used by the instruction to overwrite any data stored at the destination.

The MVM output instruction is used to copy an image of the source to a destination while allowing portions of the source data to be masked from the transfer. The source data is not changed. This instruction can be used to copy I/O image table, binary, or integer values. For instance, bit data such as status or control bits can be extracted from an address that contains bit and word data. The

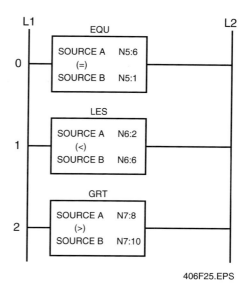

406F25.EPS

Figure 25 ◆ Examples of data comparison instructions.

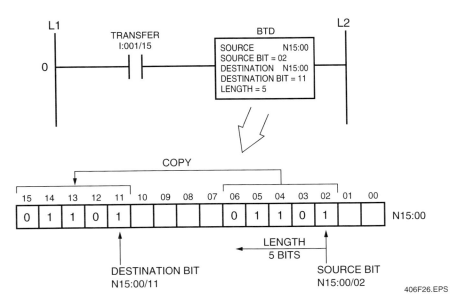

Figure 26 ◆ BTD transfer operation.

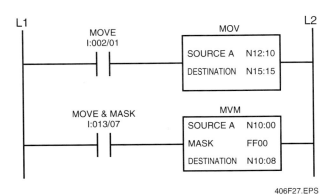

Figure 27 ◆ MOV and MVM transfer operations.

source can be a program constant or data address. The mask can be an address or hexadecimal value that specifies which bits to pass or block. The mask bits must be set to 1 by the programmer to allow the desired data bits to pass to the destination. The moved data overwrites the destination data. Note that in the example, the mask (FF00) used for the source has the higher eight bits set to 1 (two hex Fs) to allow transfer while the lower eight bits are blocked (two 0s).

6.2.6 *Program Control*

Program control operations are accomplished using a series of conditional and unconditional branches and return instructions. These instructions provide a means of executing sections of the control logic if certain conditions are met. The instructions described here are representative of similar program control instructions for other PLCs.

The master control relay (MCR) output instruction is used in pairs (conditional and unconditional) to fence and activate or de-activate the execution of a group of ladder rungs (*Figure 28*). When the MCR rung condition is true, the conditional MCR output is activated, and all rung outputs within the zone can be controlled by their respective input conditions. If the MCR output is turned OFF, all non-retentive outputs within the zone will be de-energized and reset. With the MCR output OFF, PLC scan time is reduced by the false state of the non-retentive outputs, even though they are scanned.

The zone control (ZCL) instruction is similar to the MCR instruction. It determines if a group of ladder rungs will be evaluated or not. If the referenced ZCL output is activated, the outputs within the zone are controlled by their respective rung input conditions. If the ZCL output is turned OFF, the outputs within the zone will be held in their last state.

The jump (JMP) and label (LBL) instructions are used in pairs to skip portions of the ladder logic. The JMP allows the normal sequential program execution to be altered if certain conditions exist. If the rung condition is true, the JMP coil reference address tells the processor to jump forward and execute the rung labeled with the same reference address as the JMP coil. In this way, the order of execution can be altered to execute a rung that needs immediate attention.

The LBL instruction is used to identify a ladder rung that is the target destination of a JMP or JSR (jump to subroutine) instruction. The LBL reference number must match that of the JMP and/or JSR instruction with which it is used. The LBL

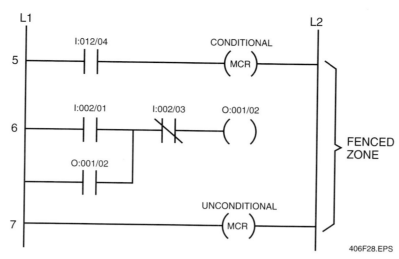

Figure 28 ◆ Typical MCR instructions.

instruction does not contribute to logic continuity and, for all practical purposes, is always logically true. It is placed as the first condition instruction in the rung. An LBL instruction referenced by a unique address can be defined only once in a program.

The jump to subroutine (JSR) instruction, when true, causes the processor to execute a subroutine file outside the main program during a scan and then return to the main program. If required, data passed to and received from the subroutine is defined. An optional subroutine (SBR) instruction can be used to store incoming data. It is used only if the data must be passed to and from the subroutine. If used, it must be the first instruction on the first rung of the ladder for the subroutine program. The subroutine is ended by a return (RET) instruction. If required, the RET instruction causes data that is returned to the JSR instruction in the main program to be stored.

7.0.0 ◆ HARDWARE TO PROGRAM CORRELATION

Some method must be used to designate to you and the processor the correlation between the hardware input and output connections to the PLC and the input contacts and output coils in the ladder logic program. This is called the addressing system. *Figure 29* illustrates the physical PLC connections and software designation correlation for a simple electric motor.

In this simple PLC system, the processor module containing the ladder logic program is shown at the bottom of *Figure 29*. The input module with external connections is shown on the left. The output module with external connections is shown on the right.

Note that in the addressing system for this PLC application, the circled field devices shown in the

electrical motor control circuit are retained and hardwired to the PLC inputs and outputs as shown in the ladder diagram. Also, note that all the other devices in the motor control circuit have been eliminated and replaced with logic elements in the ladder diagram. Their function during a processor scan is represented by internal storage bits in the processor memory.

8.0.0 ◆ GUIDELINES FOR PROGRAMMING AND INSTALLATION

While each manufacturer will recommend specific methods of programming and installing a PLC, some or all of the following guidelines may be applicable, depending on the application. Because of the number of different PLC programs available, detailed programming procedures including solutions to the many types of control problems that can be encountered are beyond the scope of this module.

8.1.0 Programming

Use the following guidelines to program a PLC:

- *Determine the control task* – This task is very important and should be done by persons very familiar with the process to be controlled.
- *Determine the control strategy* – Based on the control task, this is the sequence of detailed actions that will be necessary in the PLC program to produce the desired result from start to finish.
- *Program implementation* – After the control strategy has been thoroughly reviewed, observe the guidelines in *Table 11* when developing the PLC program.

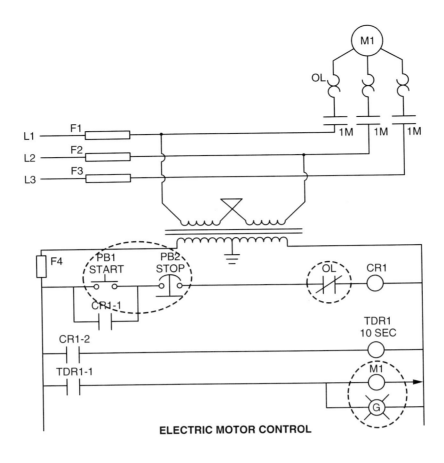

ELECTRIC MOTOR CONTROL

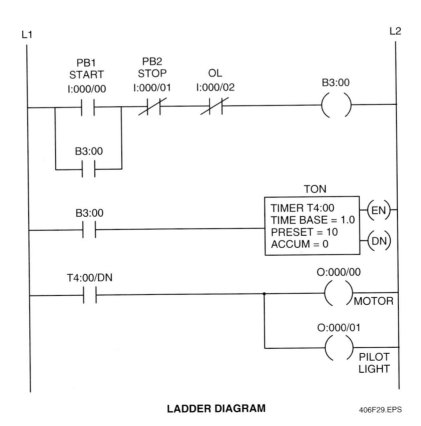

LADDER DIAGRAM   406F29.EPS

Figure 29 ◆ Hardware to software correlation.

Table 11 Programming Guidelines

System Modernizations	New Systems
Understand the process and/or machine function.	Understand the system functions that are required.
Review and optimize existing machine logic. Create and optimize an updated ladder diagram.	Recheck control strategy.
Assign I/O and internal addresses to inputs and outputs.	Create a flowchart of the process operation.
Translate ladder diagrams into PLC coding.	Create ladder diagrams or other logic symbology.
If available, run emulator software to evaluate program operation.	Assign I/O and internal addresses to inputs and outputs.
	Translate ladder diagrams or other symbology into PLC coding.
	If available, run emulator software to evaluate program operation.

8.2.0 Installation

Use the following guidelines to install a PLC:

- Make sure that the PLC selected has 25% to 50% extra capacity for future expansion.
- Select an enclosure as recommended by the manufacturer so that the PLC is easily accessible for wiring, testing, troubleshooting, and local programming.
- The location and enclosure must be tailored to mitigate the effects of the surrounding environmental conditions, such as temperature, humidity, electrical noise, and vibration.
- Group I/O modules in base units according to their type. If possible, physically separate input modules from output modules.
- Incoming and output power wiring, external input power devices, power supplies, output power starters, contactors, relays, and other electromechanical devices should be located at the top of the enclosure to minimize wiring runs and electrical noise.
- Make sure that power wiring in ducts or raceways is separated from all low-level I/O wiring. Make sure that both the PLC I/O wiring and electrical wiring materials and installations conform with the *National Electrical Code®* (NEC®) and local codes as well as the manufacturer's recommendations.
- Make sure to ground the system components in accordance with the manufacturer's recommendations as well as the NEC® and local codes.
- AC power for the PLC and field devices should be appropriately fused and should come from a common source at the PLC location. If necessary, install line filters to remove electrical noise.

- Personnel safety devices should be external to the PLC and hardwired using reliable electromechanical devices to remove power from the controller and/or inputs and outputs in the event of a personal emergency or PLC failure. Master control relay circuits can be used to remove all power. Safety control relay circuits can be used to disable only the inputs or the outputs without shutting down the PLC.
- Check that all primary and PLC power supply wiring is correctly connected to the PLC.

8.3.0 I/O Wiring

Use the following guidelines for I/O wiring:

- Remove and lock out all controller and input/output power.
- Check that all I/O modules are in the correct locations in accordance with the I/O address assignment document.
- Loosen all terminal screws on each I/O module.
- Locate the wire bundle corresponding to each module. Identify each of the wires.
- Start wiring each module from the bottom up, bending the wires at a right angle at the appropriate terminal.
- Cut the wires so that they extend about ¼" beyond the terminal. Then strip approximately ⅜" of insulation from each of the wires.
- Insert each wire under the appropriate terminal pressure plate, and tighten the screw.
- If using shielded wire, follow the manufacturer's recommendations for terminating the shield. In most cases, ground only one end of the shield, usually at the PLC end, to prevent ground loops.

- After all wires are connected, pull gently on the wires to make sure they are securely clamped under the terminals.
- If necessary, install any special leakage suppression for input devices that have known current leakage problems or inductive spike protection for outputs to inductive load devices.

Use the following guidelines to make static input wiring checks:

- Place the PLC in a mode that will inhibit automatic operation.
- Apply power to the PLC and the input modules. Verify that any system indicators show proper operation.
- Activate each emergency stop, and check that the PLC and/or input modules are de-energized.
- Reestablish power to the PLC and the input modules/devices.
- Manually activate each input device, and check that the proper input is registered at the PLC.
- Take precautions to avoid injury or damage when activating input devices connected in series with loads that are external to the PLC.

Use the following guidelines to make static output wiring checks:

- Remove power from the PLC and the input/output modules.
- Locally disconnect all output devices that will result in mechanical motion.
- Apply power to the PLC and the output modules/devices. Verify that any emergency stops remove power from the PLC and/or output modules.
- Reapply power to the PLC and output modules.
- Use the forcing function of the PLC and a hand-held device to force each output to ON by setting the corresponding terminal address to 1. Verify that the proper output indicator lights and the output device energizes (except for the disconnected devices).
- Once all outputs have been activated, reconnect each disconnected device one at a time, and repeat the forcing function for that device to make sure it energizes. Take precautions to make sure that energizing the device will not cause injury or damage when the load operates independently.

8.4.0 Dynamic System Checkout

Use the following guidelines to make dynamic system checks:

> **NOTE**
>
> During dynamic checkout, it is advisable to bring large systems up a section at a time. Usually, large systems have remote subsystems that can be brought up one at a time for checkout before the entire system is brought up.

- Make sure all corrections from the previous checks have been incorporated into the PLC program.
- Load the program into the PLC memory.
- Switch the PLC to a test or emulate mode that will allow running and debugging of the program with the outputs disabled.
- Use a single scan mode or the emulate mode to check each rung of the ladder diagram for proper operation.
- Correct the program and program documentation to reflect all changes.
- Place the PLC in the run mode, and verify that the subsystem or system operation is correct, as applicable.

Summary

A programmable logic controller is a microprocessor-based device that is user configured to meet a specific requirement. The hardware components of the PLC include the power supply, processor, and input/output devices. The processor contains the microprocessor and the memory. The software components of the PLC include the operating configuration, called software, and the program.

Understanding PLCs and their components, power supplies, wiring, and languages is absolutely necessary for being a successful instrumentation technician.

Review Questions

1. Certain large capacity, multi-tasking PLCs can have up to _____ connections.
   a. 50,176
   b. 125,000
   c. 1,000,000
   d. 3,000,000

2. The binary number system is a base _____ system.
   a. 2
   b. 8
   c. 10
   d. 16

3. The decimal numbering system is a base _____ system.
   a. 2
   b. 8
   c. 10
   d. 16

4. A binary code that uses alphanumeric characters is the _____ code.
   a. ASCII
   b. BCD
   c. Gray
   d. hexadecimal

5. Rotational encoders use the _____ code.
   a. ASCII
   b. BCD
   c. Gray
   d. hexadecimal

6. Discrete I/O modules are used for _____ interfaces.
   a. special
   b. analog
   c. true/false
   d. arithmetic

7. Discrete I/O modules that use isolation circuits are _____ interfaces.
   a. AC/DC
   b. analog
   c. true/false
   d. arithmetic

8. I/O modules that convert a varying voltage or current to a digital value or vice versa are _____ interfaces.
   a. AC/DC
   b. analog
   c. true/false
   d. arithmetic

9. A special I/O that sends out a train of pulses to a motor is a(n) _____ module.
   a. encoder/counter
   b. stepper control
   c. smart transmitter interface
   d. numerical data

10. HART®-compliant field devices are used with a(n) _____ module.
   a. encoder/counter
   b. stepper control
   c. smart transmitter interface
   d. numerical data

11. The process by which a PLC polls the inputs, executes the program, and updates the outputs is called a scan.
   a. True
   b. False

12. The internal storage memory of a PLC can include _____.
   a. input data
   b. output data
   c. constants
   d. timer and counter instructions

13. The most widely used PLC language is _____.
   a. Boolean mnemonics
   b. ladder logic
   c. English statement
   d. functional block

14. PLC output coils are of two different types: standard or latch.
   a. True
   b. False

15. A TON is a _____.
   a. timer on delay
   b. timer off delay
   c. retentive timer
   d. constant run timer

Trade Terms Introduced in This Module

American Standard Code for Information Interchange (ASCII): A system for converting alphanumeric characters into numerical designations for computer processing.

Baud rate: The rate at which data is passed from one computer to another.

Binary coded decimal (BCD): A way to encode binary numbers in a decimal format.

Binary: The number system based on the two digits 0 and 1 that is used by computers.

Bit: In computers, a bit is the smallest unit of data.

Decimal: The number system based on the 10 digits 0 through 9 that is used by people in everyday applications.

Decrement: A decrease.

DIN: Abbreviation for Deutsches Institut für Normung. An international standard developed in Germany.

Discrete: Having unique and separate parts. In PLCs, discrete outputs and inputs only have two possible states: true or false.

Hexadecimal: The number system based on the 16 digits 0 through 9 and A through F that is typically used in addressing very large numbers, such as in a computer addressing scheme.

Increment: An increase.

Octal: The number system based on the eight digits 0 through 7 that is often used in programmable logic controllers.

Parity: In PLCs, this refers to how data is checked for errors on transfer.

Programmable read only memory (PROM): A type of ROM.

Random access memory (RAM): Memory which can be read from and written to.

Read only memory (ROM): Memory which can only be read from.

Scan: The process of a programmable logic controller evaluating all inputs and outputs. Normally referred to as scan time and expressed in milliseconds.

Sinking: An inputting device that must receive current when its output is in a true or ON state.

Sourcing: An inputting device that provides current when its output is in a true or ON state.

Stop bit: In computers, this refers to the bit that designates the end of a data transfer.

Additional Resources

This module is intended to be a thorough resource for task training. The following reference works are suggested for further study. These are optional materials for continued education rather than for task training.

Instrumentation, 1975. F.W. Kirk and N.R. Rimboi. New York, NY: American Technical Society.

Basic Instrumentation, 1966. New York, NY: McGraw-Hill.

Process Control Instrumentation Technology, 1977. John Wiley and Sons. New York: NY.

Figure Credits

Rockwell Automation   406F01, 406F02, 406F03, 406F04

NCCER CURRICULA — USER UPDATE

NCCER makes every effort to keep its textbooks up-to-date and free of technical errors. We appreciate your help in this process. If you find an error, a typographical mistake, or an inaccuracy in NCCER's curricula, please fill out this form (or a photocopy), or complete the online form at **www.nccer.org/olf**. Be sure to include the exact module ID number, page number, a detailed description, and your recommended correction. Your input will be brought to the attention of the Authoring Team. Thank you for your assistance.

Instructors – If you have an idea for improving this textbook, or have found that additional materials were necessary to teach this module effectively, please let us know so that we may present your suggestions to the Authoring Team.

NCCER Product Development and Revision

13614 Progress Blvd., Alachua, FL 32615

Email: curriculum@nccer.org
Online: www.nccer.org/olf

❑ Trainee Guide   ❑ AIG   ❑ Exam   ❑ PowerPoints   Other _____

Craft / Level: _____   Copyright Date: _____

Module ID Number / Title: _____

Section Number(s): _____

Description: _____

_____

_____

_____

_____

Recommended Correction: _____

_____

_____

_____

_____

Your Name: _____

Address: _____

_____

Email: _____   Phone: _____

Distributed
Control Systems

COURSE MAP

This course map shows all of the modules in the fourth level of the Instrumentation curriculum. The suggested training order begins at the bottom and proceeds up. Skill levels increase as you advance on the course map. The local Training Program Sponsor may adjust the training order.

INSTRUMENTATION LEVEL FOUR

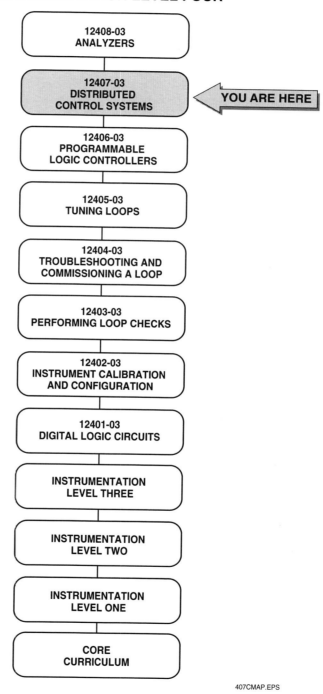

407CMAP.EPS

1.0.0 INTRODUCTION 7.1

2.0.0 MANUFACTURING 7.1

    2.1.0   Wet and Dry Processes 7.2

    2.2.0   Plant Management 7.2

3.0.0 DEFINITION OF DISTRIBUTED CONTROL SYSTEM 7.2

4.0.0 EVOLUTION 7.2

    4.1.0   Traditional Control Loop 7.3

    4.2.0   Central Monitoring 7.3

    4.3.0   Central Control 7.4

    4.4.0   Programmable Logic Controllers (PLCs) 7.4

    4.4.1   Architecture 7.5

    4.4.2   Extended Control 7.5

    4.5.0   Distributed Control 7.6

    4.5.1   Multiple Control Boxes at the Process Area Level 7.6

    4.5.2   Overall Process Level Control 7.6

    4.5.3   Fieldbus Design 7.7

    4.5.4   Redundant Control 7.7

5.0.0 HUMAN INTERFACE 7.8

    5.1.0   Data Highway 7.8

    5.2.0   Topology 7.8

    5.2.1   Star 7.8

    5.2.2   Ring 7.8

    5.2.3   Bus 7.9

    5.2.4   Hybrid 7.9

    5.3.0   Protocols 7.10

    5.4.0   Workstations 7.11

    5.5.0   Local Area Network 7.11

    5.6.0   Remote Communications 7.12

    5.6.1   Computer to Computer 7.12

    5.6.2   Wide Area Network (WAN) 7.12

    5.6.3   Supervisory Control and Data Acquisition (SCADA) 7.12

    5.6.4   The Internet 7.12

6.0.0 MAINTENANCE 7.14

    6.1.0   Instruments 7.14

    6.1.1   Staff Capabilities 7.15

    6.1.2   History Files 7.15

    6.2.0   Calibration 1.16

    6.3.0   Methods 7.16

    6.4.0   Acquiring Expertise 7.16

    6.4.1   Equipment Knowledge 7.16

    6.4.2   Repair Tool Skills 7.17

SUMMARY 7.17

REVIEW QUESTIONS 7.18

GLOSSARY 7.19

Figures

Figure 1   Typical DCS 7.3
Figure 2   Traditional control loop 7.3
Figure 3   Central monitoring 7.4
Figure 4   Central control 7.4
Figure 5   PLC control loop 7.4
Figure 6   Early PLC design 7.5
Figure 7   PLC with extended control 7.5
Figure 8   Control box distributed control 7.6
Figure 9   Overall process level control .7.7
Figure 10   Fieldbus 7.7
Figure 11   Redundant control 7.8
Figure 12   Star network 7.9
Figure 13   Ring network 7.9
Figure 14   Bus network 7.9
Figure 15   Ring-wired star network 7.10
Figure 16   Transfer frame format 7.10
Figure 17   Local area network 7.11
Figure 18   Computer-to-computer communication 7.13
Figure 19   Wide area network communication 7.13
Figure 20   SCADA system 7.14
Figure 21   The Internet 7.15

Distributed Control Systems

Objectives

When you have completed this module, you will be able to do the following:

1. Define a distributed control system (DCS).
2. Identify the typical components associated with a DCS.
3. Identify the various network configurations used with a DCS.
4. Describe uses for a DCS.
5. Explain how an instrument technician interfaces with a DCS.

Prerequisites

Before you begin this module, it is recommended that you successfully complete the following: Core Curriculum; Instrumentation Levels One through Three; Instrumentation Level Four Modules 12401-04 through 12406-04.

Required Trainee Materials

1. Pencil and paper
2. Appropriate personal protective equipment

1.0.0 ◆ INTRODUCTION

Microprocessor-based control systems became available in the early 1970s. Over time, they have become dominant in the process control industry. During the same period, the use of pneumatic controls, which once dominated industrial instrumentation, declined.

The seventies saw the advent of **analog** electronic controls, but embedded in this control strategy was the concept of computer-based control systems. Computers manufactured for businesses by IBM were adapted to process control. These machines incorporated analog to digital converters to centralize control of the process variables in the field. Better informed and safer decisions could be made because of this centralized concept. The engineers wanted more; for example, why not use a DAC to send pulses to a stepping motor to supervise the board **controllers** by controlling their **setpoints**.

The period from 1970 to 2002 saw the rapid growth of many new microprocessor-based control companies and many advancements in control and sensing hardware, application software, interconnecting cables, and **protocol** standards. This evolving computer-based technology has not only impacted the way process control is achieved but also how plants are managed.

Microprocessor-based control systems evolved from a centralized system to a **distributed control system (DCS)** in the process control industry. This module focuses on the evolution of DCS components, the application of DCS in industrial instrumentation control, and your role in maintaining a DCS.

Now, instead of evaluating process data at a computer console then walking over to manipulate an electronic controller, an operator can stay at the console and remotely optimize the process. The 1980s saw the advent of personal computing, which quickly refined the computer processor to the microprocessor.

2.0.0 ◆ MANUFACTURING

Control instrumentation expanded from mechanical to include hydraulic, pneumatic and electronic control systems to manage industrial processes. DCS has had a major impact, not only on process control methods but also on the plant organizations that support the control methods. Therefore, this study begins by looking at the generic types of manufacturing processes and traditional process control plant organization.

2.1.0 Wet and Dry Processes

Traditionally, manufacturing processes are divided into two distinct categories by the process control industry: wet and dry.

A **wet process** deals with fluids that flow through conduits and valves and require many analog control functions. Examples include the use of modulating control valves to manage such fluids as liquid chemicals, liquid ice cream flavorings, and petroleum products. A wet process tends to be continuous in nature and may also be referred to as an analog process.

A **dry process** deals with materials that are transported or carried by conveyors or robots and require predominantly start/stop or motion type of control. Examples include management of conveyer systems to move machined parts used to manufacture automobiles, appliances, and boats, and in the manufacturing of food and beverage products. Because a dry process tends to use multiple starts, it is commonly referred to as a **discrete process**.

2.2.0 Plant Management

In the early 1970s, wet (continuous) and dry (**discrete**) processes were managed by devices that were, in turn, maintained by different departments. Instruments that had meters and gauges managed continuous processes and were maintained by a plant's instrument department. Similarly, discrete processes were managed by electrical controls that used relay logic and feedback lights. A plant's electrical department maintained these controls.

Acquisition of data for record keeping and quality control was typically part of the business operation and not an integral part of the process control system. In this environment, adjustments to the process took coordination between three departments and time to execute. Operators had to walk from control station to control station, and record keeping involved physically acquiring charts that ran on recorders close to the process. Acquired data was good for determining what had happened but not so useful for making real-time adjustments to the process.

The emergence of DCS blurred the distinctions between these three plant departments and resulted in cross-training personnel to keep pace with the advancements in control technology.

3.0.0 ◆ DEFINITION OF DISTRIBUTED CONTROL SYSTEM

DCS is a term used by the process control industry to describe a control system consisting of microprocessor-based sensors (**primary elements**), actuators (**final elements**), associated supportive devices, and a communication network that uses three network levels. A typical DCS (*Figure 1*) has an overall supervisory network level that provides human interface, a second-level network that distributes control execution across multiple microprocessor-based devices, a series of third-level communicating networks that link smart sensors and smart actuators to their PLCs, and dedicated localized computer PID control.

DCS has evolved with computer technology and has provided the process industry with the ability to more closely manage and monitor increasingly complex processes.

Continuous processes once used controllers that specialized in analog control functions. Discrete manufacturing processes used controllers that relied mainly on relay logic. The advancements in technology have resulted in controllers that are a mixture of analog and discrete control—readily adaptable to either continuous or discrete process control. This new technology has blurred the traditional functions of the plant instrumentation and electrical departments.

4.0.0 ◆ EVOLUTION

Lessons learned from early uses of computers in process control included the need for reliability and redundancy. As computer technology became cheap and available, a new concept for control emerged. Instead of one central computer doing everything, why not distribute the control of the process to many computers, and leave the central computer to do housekeeping chores like history logging and optimizing setpoints. This strategy would provide reliability if each computer-controlled area were made to work independently.

The evolution of DCS in the process industry has gone through distinct stages as computer technology has developed and as understanding of computer applications has grown. Plants all around the country have a mixture of installed control equipment that reflects each of these evolutionary stages. To understand this evolution, you should begin by looking at the operation of a traditional process **control loop**.

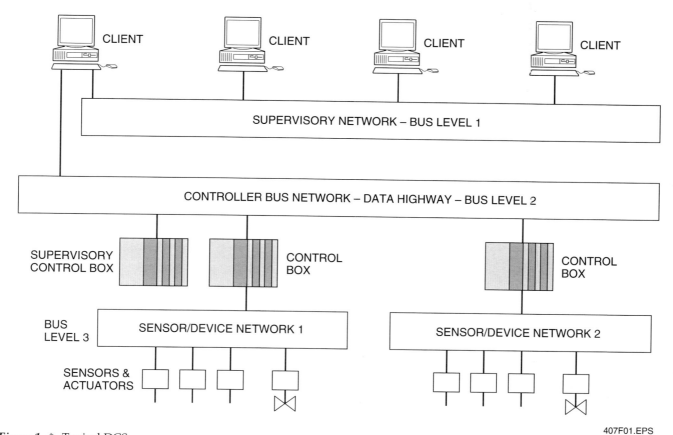

Figure 1 ◆ Typical DCS.

407F01.EPS

4.1.0 Traditional Control Loop

Figure 2 shows a traditional control loop. In traditional process control, one pneumatic or electronic controller managed a single control loop. The controller was located first near the process then in a central plant control room. It continually monitored the process.

Sensors, referred to as primary elements, measure process variables, such as pressure, flow, level, and temperature, and are connected to signal conditioning devices called transmitters. Each transmitter sends a converted, standardized signal to the controller. Based upon a setpoint established by a setpoint knob, the controller compares an actual process variable, such as temperature, against the setpoint value and calculates a corrective response. To initiate the corrective response, the controller sends a control signal to a process control element—a control valve in this instance. The control valve is commonly referred to as the final element in the control loop. Powered pneumatically or electronically, the valve opens or closes to the degree prescribed by the control signal. As the process variable changes, the controller continually adjusts the control signal so the control valve can maintain the variable at or very near the required setpoint.

4.2.0 Central Monitoring

Computers first found their way into the process industry to solve business related problems such as payroll and accounting. However, understanding the computer's capability, instrument engineers began to apply computers to process control. Early control applications of the computer involved data acquisition for engineering management. Signals from process sensors and actuators were connected back to a central computer (*Figure 3*). Process temperatures, pressures,

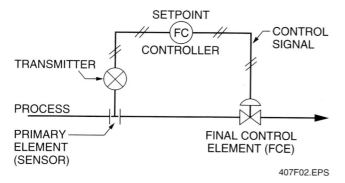

Figure 2 ◆ Traditional control loop.

407F02.EPS

flow rates, levels, valve positions, and other parameters were monitored continually. Reports were generated and trends were plotted. The availability of this information at a central location made it possible for management to closely monitor the process, quickly respond when the process was out of normal range, prepare reports, **archive** information, and engage in studies aimed at process optimization.

This approach also required instrument engineers to expand their computer understanding and skills. They had to merge their control technology understanding with computer operational and programming skills.

4.3.0 Central Control

Recognizing that computers could be used to control the process, engineers created computerized control strategies (PID algorithms) to calculate the required control signal value and send that value directly to the controlled devices. As *Figure 4* shows, the central computer now took direct control of final elements in the process—in this case the valve—while the existing controller remained in place as a backup. This provided 100% redundancy and improved process reliability.

Because control valves were typically equipped with pneumatic actuators, current-to-pressure transducers were used to convert the computer's electronic control signal to a pneumatic signal. This approach allowed pneumatic control valves to be used without disrupting piping and made it easy to use the original analog controller's pneumatic control signal as a backup signal.

4.4.0 Programmable Logic Controllers (PLCs)

Central computers provided a major improvement in process control, but such installations were quite expensive. Computers were costly, and programming efforts were labor intensive. Wire runs between the process and the central computer were large in number and long in distance. One significant drawback became apparent: if the central computer should fail, and no redundant controllers were in place, the entire plant could be shut down.

With awareness of these disadvantages in mind, control companies began manufacturing a small microprocessor-based multi-loop controller called a **programmable logic controller (PLC)** to take the place of the traditional pneumatic and electronic single-loop controller (*Figure 5*). Designs simply duplicated the functions performed by the controllers they replaced. In order

to physically look just like the controllers they were replacing, the early multi-loop controllers and associated components were physically mounted in a cabinet out of sight of the operator. Multi-loop controller interface displays and buttons were created to look just like those associated with the original controller except they were virtual on a screen.

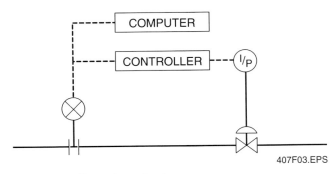

Figure 3 ◆ Central monitoring.

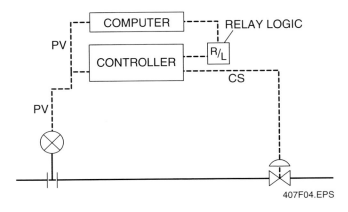

Figure 4 ◆ Central control.

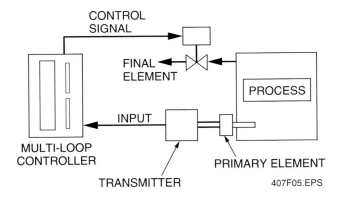

Figure 5 ◆ PLC control loop.

INSTRUMENTATION LEVEL FOUR — TRAINEE MODULE 12407-03

4.4.1 Architecture

Early PLCs were developed to duplicate relay logic circuits that existed in discrete process control. Later versions of the controller also contained analog output control routines to control continuous single-loop processes. The PLC (*Figure 6*) was a single piece of hardware that consisted of the following:

- A power supply
- A processor
- Input terminals
- Output terminals
- A local interface keyboard

The processor contained a microprocessor chip, a control **program**, and associated circuitry to interpret the input signals and produce an output signal. A local interface keyboard was used to program the PLC for the single loop it was to control. A power supply transformed distribution voltage in the plant to 24 volts to power the PLC components.

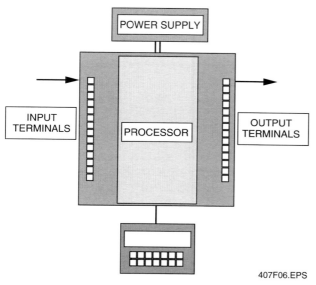

Figure 6 ◆ Early PLC design.

4.4.2 Extended Control

As the speed and computing power of the microprocessor grew, the multi-loop controller became capable of controlling more than a single control loop. Later models allowed the control of 4, 8, 32, 64, 128, and up to 512 control loops with distributive technology improvements (*Figure 7*). Thus, small processes could be controlled by a single multi-loop control box—in short, a DCS.

Input and output signals between the multi-loop controller and the remote I/O modules were transmitted along a manufacturer specific sensor communications bus. Because the same manufacturer produced the DCS and its associated I/O modules, the communication language used on the sensor bus, called a protocol, varied with each manufacturer. Mixing one manufacturer's DCS with another manufacturer's I/O modules was not feasible.

With the enhanced microprocessor capability, programming of the DCS also became more elaborate. Now, the DCS could not only control multiple loops but could link control loops by cascading the results of one process step into another. Thus, if one control loop lost control of a variable in an upstream process area, the downstream control loops could alter their setpoints or loop responses without having to wait for their sensors to measure the downstream impact. Internally, information from the upstream control loop was passed through software to the downstream control loops—all within the same local controller box. This allowed control of the process to be more timely and accurate.

The only downside of this arrangement was that a single computer contained all the control programming for the entire process or for many steps of the process. Failure of the DCS would result in shutdown of the process.

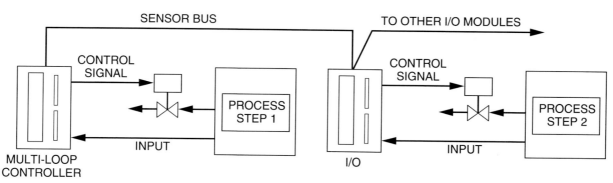

Figure 7 ◆ PLC with extended control.

4.5.0 Distributed Control

As the number of controlled areas increased, and the distance between them increased, the next evolution in microprocessor-based control used multiple control boxes. In this evolution, control boxes were used at two levels. First, multiple control boxes were used at the process area level, particularly when large distances were involved between areas. A good example of this would be a mining conveyer system where large distances exist between each process area. The second application of the control box was at the overall process level. This was done to coordinate actions between individual control boxes that were applied at the process level.

4.5.1 Multiple Control Boxes at the Process Area Level

Figure 8 shows the use of multiple control boxes at the process area level. Each control box controls one or more areas of an overall process. Each controller is designed to be a stand-alone controller capable of controlling its portion of the process. A communications bus, sometimes referred to as a **data highway**, ties the individual controllers together. An optional central computer is connected to the data highway and can automatically extract information from the process for management use, or allow management to manually intervene with the process. In this control approach, the central computer does no real-time control. Real-time control is accomplished at the control box level. The computer can be off line, and the control system will still function as usual.

Using multiple control boxes allows the programs for process control loops to operate close to the process. It also minimizes the length of wire between the controller and the sensors. The response time of loop control programs is reduced because multiple control boxes take the place of the central computer or a single control box with extended I/O capability. Thus, failure of a single control box can be inconvenient but will not cause the entire plant to shut down.

4.5.2 Overall Process Level Control

With each control box controlling one or more areas of the overall process, coordination between control loops is still necessary, particularly to cascade the impact of one control loop onto other downstream control loops. With the enhanced programming capability of the control boxes, overall process coordination (*Figure 9*) can be handled by a control box operating at the network level. Using communications on the interconnecting data highway, the host control box for the overall process can exchange information with control boxes operating at the process area level. Thus, control of the overall process is further distributed. The use of default or average values by the control box can keep the process going during system repair.

All control functions take place at the control box level; the central computer does no real-time control. The central computer acts as a human interface to the process control system by receiving alarms, archiving data, and generating reports. It also provides for human monitoring and human intervention with the process.

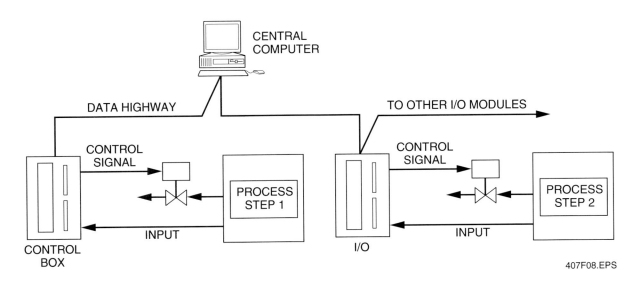

Figure 8 ◆ Control box distributed control.

4.5.3 Fieldbus Design

Innovations in distributing control throughout a DCS have resulted in the development of a third-level communicating network bus (*Figure 10*) that connects remote I/O sensors and actuators to a control box processor without the use of I/O modules. Using smart sensors and actuators that are equipped with microprocessors, data is exchanged with the control box in a transmitted digital packet format. This bus design eliminates the traditional wiring needed between I/O devices and the control box I/O modules. It also allows an actual variable, such as a temperature value in degrees, to be exchanged instead of voltage or current signals, such as 4–20mA, which would have to be translated at the DCS processor.

This third level bus has come to be known as a **fieldbus**. In reality, however, there are many fieldbus designs that depend upon the type of information being exchanged and on the protocol used to transmit this information.

4.5.4 Redundant Control

In some critical processes, totally redundant control boxes are installed in parallel to control the same process area. *Figure 11* shows two controllers connected to remote I/O modules through redundant sensor network wires. One controller is a backup to the other. If one control box or control loop fails, control is shifted to the backup control box, and an alarm is sent via the data highway to alert the appropriate personnel. In fact, this type

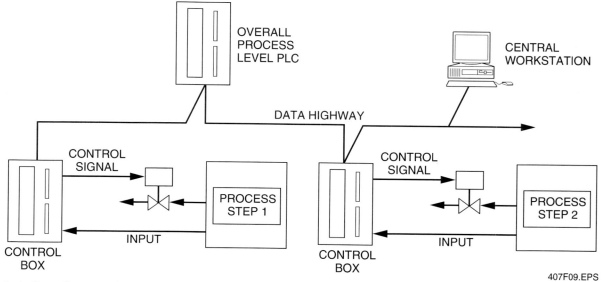

Figure 9 ◆ Overall process level control.

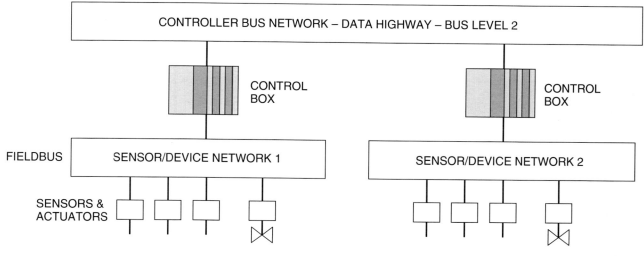

Figure 10 ◆ Fieldbus.

of application is also likely to have a redundant data highway bus. The redundant data highway bus would be run in a different physical path through the plant to reduce the risk of catastrophic failure.

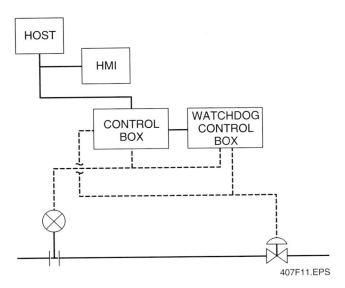

Figure 11 ◆ Redundant control.

5.0.0 ◆ HUMAN INTERFACE

The use of multiple control boxes is the heart of the process control approach known as a distributed control system (DCS). However, DCS also involves the ability to provide human interfaces that allow manual intervention, configuration, and monitoring of the process. Human/machine interfaces (HMI) exist at the DCS sensor/actuator bus level, data highway level, and supervisory LAN bus level.

With these various human interface devices, information can be extracted from the system where and when it is needed. This module has already covered control boxes and their interaction with the industrial process. The following sections will cover the data highway, local interface devices, **workstations**, and **local area network (LAN)**.

5.1.0 Data Highway

Information exchange between control boxes, and information exchange between control boxes and a central workstation take place along the data highway. Typically, one workstation is connected to the data highway and is called the **server**. The server contains communication software that allows it to communicate with the data highway.

Physically, the data highway network bus consists of two or three copper cables that are twisted, stranded, and shielded. The data highway is con-

nected to the communication terminal of each control box and the appropriate communications (COMM) port of the host computer. In some installations, the data highway may be fiber-optic cables or a combination of stranded copper and fiber optics.

5.2.0 Topology

The manner in which the data highway is connected to the control box or the manner in which multiple work stations are connected together on the local area network is called a **topology**. Typical topologies include star, ring, bus, and hybrid configurations.

5.2.1 Star

One of the most prevalent topologies is the star topology. *Figure 12* shows a simple seven-node system with the seventh node serving as the hub of the star configuration. Each node represents a connection to a communicating component (control box or workstation) in the network. This topology was first used in the telephone industry and is the basis for many commercial and industrial control systems. Because the hub is the central point through which all communications flow, the hub simplifies data management in the network. No two nodes can communicate simultaneously through the hub. This topology is used successfully when the physical layout of nodes favors a central connection point.

One disadvantage of the star configuration is the large cost of running individual wires from the hub to each node when long distances are involved. For example, if the hub is located on the first floor of a 20-story building, and 100 nodes are located on the twentieth floor, 100 sets of wires will have to be run from the first to the twentieth floor.

Another disadvantage of the star configuration is its vulnerability to the failure of the hub. Any failure of the hub eliminates communications among the nodes.

5.2.2 Ring

The use of ring topology (*Figure 13*) is an attempt to circumvent the problems of wire length and hub failure associated with the star topology. In a ring, each node is connected to two adjacent nodes, and the communications message is passed in a closed loop. If a node fails, the ring is broken, and message flow is impeded but not prevented. When communications cannot be passed in one direction, the message flow is reversed. However, trying to constantly reverse the message flow can lead to communication problems.

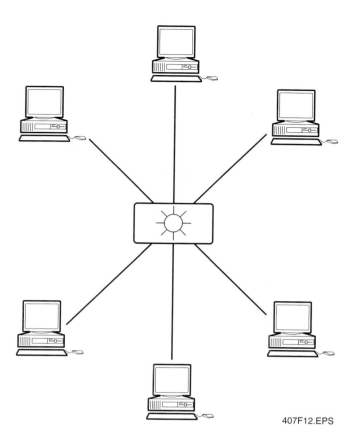

Figure 12 ◆ Star network.

5.2.3 Bus

Bus topology, also referred to as multi-drop topology, has become more prevalent, especially in process control networks. As shown in *Figure 14*, a bus acts as a common conductor among the nodes. Connections from the bus to the nodes are made in a T or a daisy chain manner. This wiring system promotes easy connections and low susceptibility to complete network failure. If any one node fails, the other nodes can still communicate. This layout also minimizes wire lengths and reduces installation costs.

5.2.4 Hybrid

No single wiring topology meets all application needs. The most common approach is to use hybrid wiring schemes. One of the most common hybrid schemes is the ring-wired star network shown in *Figure 15*. This hybrid topology uses a ring as a backbone wire to connect localized hubs. The localized hubs are, in turn, wired in a star topology to individual nodes. This layout minimizes wiring distances by minimizing the star wire lengths yet provides reliability by allowing communication reversal along the ring if needed.

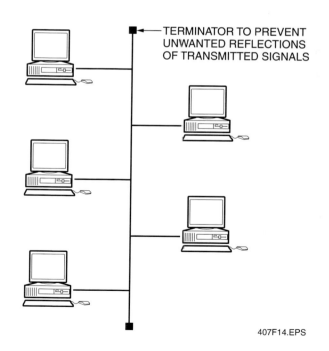

Figure 13 ◆ Ring network.

Figure 14 ◆ Bus network.

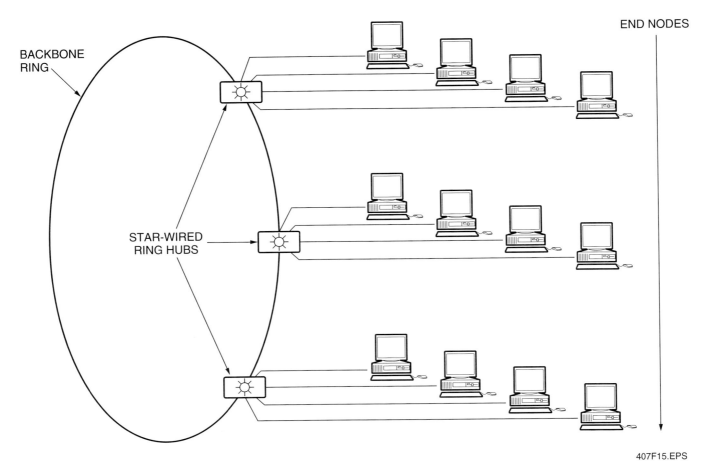

BACKBONE RING

STAR-WIRED RING HUBS

END NODES

407F15.EPS

Figure 15 ◆ Ring-wired star network.

5.3.0 Protocols

In order for the DCS to communicate internally and with the host, all network components must share a common protocol. That is, they must share and understand the same communications language. A protocol is a set of rules that govern the arrangement and transmission of binary data. Binary data is transmitted as a string of ones and zeros. A protocol defines how the string of ones and zeros is segregated into ordered groups and the ordered groups into frames. The ordered serial strings of data represent an intelligent statement or set of commands between devices on the data highway or bus. Each serial transmission is called a **transfer frame** and contains a start and stop group, addresses, command statements, and a communication verification group.

To illustrate the protocol architecture, *Figure 16* shows the format of information to be sent between two devices that use the *Institute of Electrical and Electronics Engineers* (IEEE) *802.3* protocol. A message (transfer frame) sent by one DCS, using the *IEEE 802.3* protocol, would be made up of eight parts. The **data field** portion of the message is the actual process information that needs to be sent, and it is imbedded in the transfer frame. The receiving DCS has to be able to receive the message packet and interpret the data field in order to complete the communication.

IEEE 802.3 FRAME

PREAMBLE	START DELIMITER	DESTINATION ADDRESS	SOURCE ADDRESS	LENGTH	DATA FIELD	PAD	FRAME CHECK SEQUENCE
7 OCTETS	1 OCTET	2 OR 6 OCTETS	2 OR 6 OCTETS	2 OCTETS	0–n OCTETS	0–P OCTETS	4 OCTETS

407F16.EPS

Figure 16 ◆ Transfer frame format.

The problem is that not all component and software manufacturers use the same protocol. In fact, some manufacturers use unique protocols in an attempt to tie the customer to them, assuring their equipment is used in the process.

Protocols also include methods that govern how communication between nodes on the network topology occurs and the communications frequency. Three popular communication management methods are as follows:

- *Contention* – In this method each node on the network waits for quiet time in milliseconds on the data highway. When the highway is quiet, the node sends a data packet and waits for a confirmation that it has arrived at its destination. This system works well when the amount of communications on the highway is limited. When too many nodes try to send messages (contend) simultaneously, communications collisions (failures) will occur on the highway.

- *Polling* – In this method, one node is the master and regularly polls all other nodes in successive order. Current information is limited to the length of time required for one complete polling sequence.

- *Token passing* – This method is similar to polling in that only one node communicates on the highway at a time. A token is passed, not necessarily in sequence, from one node to another along the data highway. Only the node with the token can communicate on the data highway. Like polling, current information is limited to the length of time required for one complete passage of the token along the highway.

5.4.0 Workstations

Human interface into the overall industrial process is accomplished through workstations. Each workstation is located on the supervisory local area network and is a computer equipped with an **operating system** like Windows NT®, Windows XP®, LINUX®, and UNIX®, as well as **DCS software** that allows communication with the control boxes. Because of the harsh environment in which the computer may be required to work, the workstation computer is often housed in a protective console. Workstations accomplish no real-time control. That is, workstations could be disconnected from the data highway bus, and the DCS would continue to maintain control of the industrial process.

The workstations are programmed with multiple access levels to manage interactions with the industrial process. Operators, for instance, might be allowed to change setpoints and respond to alarms. Engineering management and instrument technicians, on the other hand, would be allowed a much wider interaction with the process, such as PID tuning and optimizing the control scheme. This is typically accomplished through a series of user names and passwords assigned to the operating software and to the DCS software. Another method of accessing workstations is through the use of magnetic cards that contain the access definition for each user or keys.

Workstations also serve as the recipients of alarms generated by the DCS when something in the process is not performing within specifications or when emergencies occur. Alarms can be routed to individuals or workstations so that the proper individual can be reached to solve a specific problem. Workstations that are the recipient and storage device for process alarms are typically called alarm servers.

Workstations called **historians** serve as storage areas for accumulated historical process monitoring information. Storing historical information is called archiving. Historical data can be archived over large time periods to satisfy historical queries and investigations.

5.5.0 Local Area Network

Workstations are physically connected together by communications cabling in a local area network (LAN). In the LAN (*Figure 17*), one workstation is called the server, and the other workstations are called **clients**. With this layout, it would be possible for the server to be located in the engineering management office while several operator workstations might be located at strategic points along the process that they are monitoring

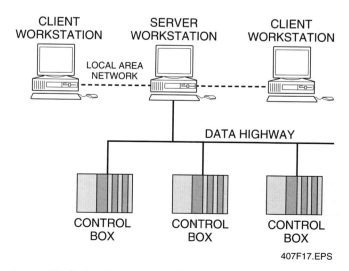

Figure 17 ◆ Local area network.

The server contains special software that allows it to act as the central database for the control boxes controlling the industrial process. A complete duplicate of the current programming for each control box is kept on the server. Client computers do not contain their own industrial process database. Rather, each individual who signs on to a client computer has access to the server through the LAN. Thus, the individual has access to the data highway through the server computer's physical connection to the data bus. Changes made to any of the control boxes from any workstation on the LAN are recorded in the server's database.

Communication between computers on the LAN also requires a common protocol. The workstation operating systems are all programmed to use a common protocol designed for computer networks, such as **transmission control protocol / Internet protocol (TCP/IP)**, Ethernet, or synchronous optical network (SONET).

5.6.0 Remote Communications

Communications with the industrial process can be accomplished from a remote location in several ways:

- A remote client can dial up the modem of a local workstation computer connected to the data highway of the process.
- The remote and central server can be part of a **wide area network (WAN)**.
- The remote server can be accessed through the Internet or World Wide Web.

5.6.1 Computer to Computer

Figure 18 shows a remote connection to a local process using modems that establish a computer-to-computer connection. The connection between computers is accomplished using the computer's operating system software. Once connected, the remote operator activates the DCS software on the local server workstation. The operator can interact with the local industrial process as if he or she were physically located in the local plant. The dial-up connection from the remote location is manual and does not occur automatically.

5.6.2 Wide Area Network (WAN)

As remote usage of dial-up services into local industrial process sites became more frequent, and the location of sites became more distributed, permanent communication connections were

needed. Thus, the development of wide area networks (WANs) evolved. However, chemical industrial operations tend to resist the usage of WANs, probably due in large part to proprietary concerns. A WAN is a client/server computer network spread over a much larger geographical area than that covered by a LAN. The WAN is unlimited in geographical size and usually connects several LANs together through the use of dedicated telephone lines, satellite links, and/or a data packet carrier service. All computers are continually linked and exchange information as required. *Figure 19* shows a WAN spread across the United States connecting sites in many states from Florida to California.

5.6.3 Supervisory Control and Data Acquisition (SCADA)

A method of remote connection that uses WAN technology is called **supervisory control and data acquisition (SCADA)**. SCADA systems (*Figure 20*) use a large mainframe computer at a central location. This computer is connected through dedicated telephone communication to monitoring computers at remote industrial process sites. The remote locations are usually unattended. However, the remote locations are equipped with an intelligent computer sometimes referred to as a **remote server unit (RSU)**. The RSU monitors the remote process, gathers data, and transmits gathered information on a regular basis back to the central mainframe computer. The frequency of transmission depends on what part the central computer plays in real-time control of the remote process. The mainframe computer analyzes the information gathered from the remote location and sends corrective commands back to the RSU to be used in adjusting the remote process operating parameters.

Alarms that occur in the remote locations are transmitted by the RSU according to their priority. Critical alarms are sent as they occur; less critical alarms are held until a regular transmission back to the central location occurs.

SCADA systems tend to be used by industries that are connected over vast distances. Some connections may be thousands of miles apart. Offshore drilling as well as gas and oil pipeline applications fit these criteria.

5.6.4 The Internet

The Internet (*Figure 21*) has become another way to span large geographical distances and connect multiple remote process sites into a central hub.

INSTRUMENTATION LEVEL FOUR — TRAINEE MODULE 12407-03

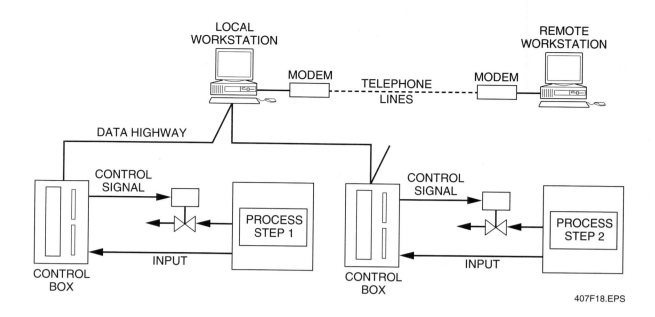

Figure 18 ◆ Computer-to-computer communication.

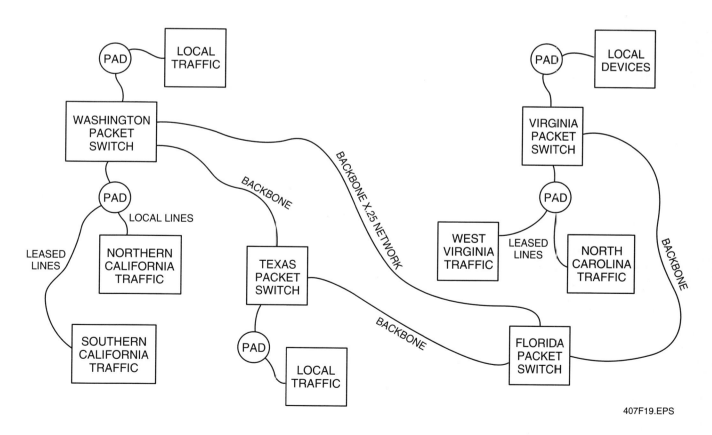

Figure 19 ◆ Wide area network communication.

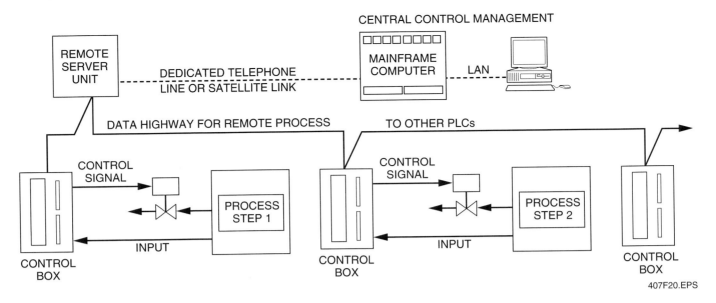

Figure 20 ◆ SCADA system.

The Internet is a vast global network developed by the U.S. Department of Defense project called the Advanced Research Projects Agency Network (ARPANET). ARPANET was originally designed to connect several local area networks at research institutions involved in government funded projects. Over time, the Internet has evolved into the use of TCP/IP to handle end-to-end message integrity functions and routing instructions. Now operators and engineers, using computers equipped with browser technology and web server functions, can gain access to remote process sites to view and control remote plant processes and equipment. Actual common usage of this procedure is futuristic, however.

6.0.0 ◆ MAINTENANCE

Many technicians and engineers are needed to maintain a DCS. Normally, your function as a technician is to service the instrumentation so that production downtime is minimized or eliminated, and product quality is maintained. Servicing the instrumentation includes equipment maintenance, calibration, and failure repair. Keeping the production line running is of paramount importance. Therefore, efforts aimed at failure repair take priority over routine maintenance and calibration activities.

6.1.0 Instruments

Production downtime can be minimized by servicing instruments according to a scheduled maintenance program. Some operations feel it is less costly to perform breakdown maintenance. In other words, as long as the instruments are working properly, don't work on them except during a general overhaul when operations are in a shutdown mode. However, it is generally accepted that equipment that is serviced regularly will fail less often and provide more accurate control of the process. A scheduled maintenance program is typically planned around the following elements:

• The operation's production schedule
• Maintenance staff capabilities
• History files containing the frequency and reason for past instrument failures

The production schedule dictates what instruments are critical and when service can be performed with minimal interruption to production. Many maintenance activities are planned around scheduled downtimes or when the process is running at less than optimum levels. Ideally, the technical capabilities of individuals will be known to allow the right staff personnel to be scheduled to service the appropriate instruments in a timely manner. Finally, maintenance department instrument history files provide valuable information about the types of instruments used, their failure frequency, and the reasons they failed. You might find yourself, though, working in an operation where there is no production schedule per se; the only schedule is to make as much product as possible for as long as possible. However, you might be in a situation where all personnel are considered to have the same capabilities. You might even be in an operation where there are no instrumentation histories except the original documentation from when the plant was built. Be prepared for it.

Figure 21 ◆ The Internet.

407F21.EPS

6.1.1 Staff Capabilities

Most plants have a combination of pneumatic, hydraulic, electrical, electronic, and digital instruments in service. Because the facilities and expertise required to deal with these instruments vary, a logical division of personnel generally falls into three areas:

- Electric/electronic instruments
- Pneumatic/hydraulic/mechanical instruments
- Digital instruments

Technicians working in any one of the these areas typically have general all-around instrument repair knowledge and, depending upon their training and expertise, should be assigned primary responsibility in one or two areas. However, some operations expect you, an instrument technician, to have expertise in all areas of instrumentation, which implies extensive and continual training.

6.1.2 History Files

The format of history files varies from one operation to another. The following list contains essential information that is usually available either in written or digital format or both:

- The instrument name
- Name of the manufacturer
- Model number or serial number
- Location in the plant
- Date when first placed into service
- Instrument tag number
- A diagram showing the process loops
- Associated logic ladder or wiring diagrams
- Associated installation drawings
- File showing the instrument's operating history
- File showing each control loop's history

The operating history file typically includes the following information:

- A description of all troubleshooting and maintenance work
- The date when a failure occurred
- The date the instrument was repaired or maintained
- A description of the repair or maintenance work done
- An identification of the technician who performed the work.

The loop history file provides all related data for each control loop and a record of historical changes made to the loop. This information is typically stored on a network workstation in two files: the working file and the master file. The master file is a backup to the working file and is updated when changes to the working file have been made. This redundant system insures that historical data is never lost and is always available to the repair technician or engineering management.

6.2.0 Calibration

If instruments are not kept calibrated, control of production quality will degrade, and some product may have to be thrown away. Therefore, regular instrument calibration is typically scheduled either on site or at the instrument shop. On-site calibration is typically done when the instrument is too large, has complex connections to the process, or when time constraints prohibit the instrument being taken out of service for any longer than necessary to calibrate it on the spot. On-site calibration requires portable calibration equipment. It also requires that calibration be traceable to the National Institute of Standards and Technology (NIST). If connections to the process are minimal, and the instruments are small enough, they are typically inspected and repaired at the shop, where the work can be performed in a controlled environment. However, many engineering guidelines and specifications require a final check be made with the instrument installed.

6.3.0 Methods

In order to minimize process downtime, the most common method of repairing or maintaining instruments involves replacing the instrument on the line and then repairing or servicing the instrument later in the shop. This method minimizes the impact on production and allows work to be done in an environment equipped with appropriate tools. Thus, work can be done without the pressure of the production line and in a more efficient manner.

On-site repair or maintenance is normally done when it is critical to production or if no spare in-kind instrument is available. Before performing on-site work, the first thing you must do is consult with the operator in charge of the area in which you will be working. Remember that operators are typically like ship captains: they must be aware of and approve all activity within the areas for which they have responsibility. The operator in charge will know if the proposed work will directly affect production downtime, whether or not there is other work going on in the area that could affect the technician's access to the instrument, if any extra safety precautions need to be taken, whether or not the work is likely to affect other production processes, and whether or not operation can safely proceed in a bypass mode. After being given the all clear from operations, the technician must follow proper lockout/tagout procedures before proceeding with the work.

On-site maintenance in most operations is typically restricted to performance checks, routine adjustments, and cleaning. Calibration is typically and more efficiently performed in the instrument shop. Instruments are brought to the shop for routine maintenance based upon the manufacturer's recommended maintenance intervals, manufacturer's service bulletins, or failure mode analysis reports from the maintenance department. The shop should be equipped with necessary spare parts to perform the necessary work. Maintenance work, which is normally scheduled to avoid interference with production, has a lower priority than repair work.

6.4.0 Acquiring Expertise

The cornerstone of repairing and maintaining instruments revolves around having knowledge of the equipment to be serviced, the skills necessary to operate test equipment and repair tools, and the ability to think logically. In support of these primary skills, you must have access to process instrument maintenance and repair records (history files) and supportive product literature.

6.4.1 Equipment Knowledge

In order to repair any instrument, you must have an in-depth knowledge of how it works; you must know its theory of operation. A cursory knowledge of the equipment will not be satisfactory.

Because there are many types of instruments used in a production process, the acquisition of instrument knowledge is a constant challenge. There are, however, several sources for acquiring such knowledge, and plant management recognizes the need and normally supports their use.

The following sources should be used to obtain product-specific knowledge:

- Manufacturer's service training programs
- Self-study service training programs provided by the manufacturer
- Manufacturer's operating and troubleshooting literature
- Manufacturer's service technicians
- Independently published training material like the material in this module

6.4.2 Repair Tool Skills

What has been said about acquiring equipment knowledge also applies to acquiring the skills and knowledge necessary to operate mechanical, power, and electronic tools used to repair instruments. Training in the use of tools must also be pursued, and practice makes perfect. The same resources exist to acquire the necessary skills and knowledge for operating repair tools.

Summary

This training module has provided an overview of distributed control systems. All DCSs use multiple microprocessor-based control boxes, along with their associated sensors and actuator devices, to distribute control responsibility over the many areas of an industrial process. Beyond this common core of devices, the variations and complexity of a DCS expand with the use of sensor/actuator buses, data highways, local area networks, and wide area networks.

In this ever-expanding digital world, your primary role as instrumentation technician is to maintain, calibrate, and repair instrumentation to keep a production process up and running. However, developments in computer technology continually present a variety of new digital devices. Continual learning will be a way of life for you to keep pace with evolving digital technology.

Review Questions

1. A system that uses multiple microprocessor-based controllers, sensors, actuators, associated supportive devices, and a communications network that links controllers to a central computer is called a _____ control system.
   a. dial-up
   b. distributed
   c. discrete
   d. device

2. In some critical processes, _____ are installed _____.
   a. redundant control boxes; in parallel
   b. redundant control boxes; in series
   c. workstations; in parallel
   d. workstations; in series

3. The workstation that contains communications software that allows it to communicate with the data highway is called the _____.
   a. client
   b. historian
   c. server
   d. monitor

4. The _____ wiring system promotes easy connections and low susceptibility to complete network failure.
   a. ring
   b. bus
   c. star
   d. hybrid

5. The most commonly used wiring scheme uses _____ topology.
   a. ring
   b. bus
   c. star
   d. hybrid

6. The set of rules that govern the arrangement and transmission of binary data along a bus is called _____.
   a. token passing
   b. polling
   c. topology
   d. protocol

7. This communications management method sends a data packet and waits for a confirmation that it has arrived at its destination.
   a. Polling
   b. Contention
   c. Token passing
   d. Transfer frame

8. A system that physically connects workstations together with communications cabling is called a _____.
   a. data highway
   b. wide area network
   c. local area network
   d. supervisory control and data acquisition system

9. A scheduled plant instrument maintenance program is typically planned around the operation's production schedule, maintenance staff capabilities, and _____.
   a. instrument lab schedule
   b. history files of past failures
   c. management meeting schedule
   d. plant layout

10. To minimize downtime, the most common method of repairing and maintaining instruments involves _____.
    a. installing extra loops and rerouting the process while equipment is serviced
    b. scheduling all equipment to be serviced on weekends and holidays
    c. using portable calibration equipment while the line is running
    d. replacing the instrument on the line and then servicing later in the shop

Trade Terms Introduced in This Module

Analog: A signal that is continuously variable over time.

Archive: The process of storing historical data on a workstation.

Client: A computer located on a LAN that does not contain a central database but uses the server database.

Control loop: The process used by a controller to continually monitor and control the quality (pressure, flow, level, and temperature) of a medium such as current, air, or water.

Controller: A digital hardware device containing input channels, a microprocessor, software application algorithms, output channels, and communications software.

Data field: The information to be transferred between devices on a network. The data field is a bit stream, whose encryption (assemblage of information) varies from manufacturer to manufacturer.

Data highway: A term used to define a network bus (wire) that connects PLCs and their associated devices with computer workstations in an industrial process control system.

DCS software: Application software located on a workstation that is used to communicate with controllers in a distributed control system.

Discrete: A signal that has only two positions such as on/off or in/out.

Discrete process: An industrial batch process that transports materials with conveyers or robots.

Distributed control system (DCS): An industrial process control system consisting of microprocessor-based controllers, primary elements, final elements, associated support devices, and a communication network that links controllers to central computers.

Dry process: A dry process deals with materials that are carried by conveyors or robots and tends to use multiple starts and stops. It is also referred to as a discrete process.

Fieldbus: A communicating network bus that connects smart sensors and actuators with a PLC processor without using traditional I/O modules and their associated I/O wires.

Final elements: A term used in the process industry to refer to throttled components, such as valves, dampers, and drives.

Historians: Workstations located on a LAN that are used to store historical operating information for an industrial process control system.

Local area network (LAN): A combination of computers connected together by a physical communications bus.

Microprocessor: A programmable integrated circuit that processes digital data in a computer or industrial controller.

Operating system: The computer program that runs continuously on a computer to manage the basic services and execute other application programs.

Primary elements: A term used in the process industry to refer to the devices used to measure process variables such as temperature and pressure. Also known as sensors.

Program: A set of software instructions running in a controller that executes a control loop to continually monitor and control the temperature, pressure, flow, and level of a medium such as air, water, or gasoline.

Programmable logic controller (PLC): A micro-processor-based industrial controller consisting of input channels, a microprocessor, software application algorithms, output channels, and communications software.

Protocol: The pre-defined set of rules that govern how information is shared between controllers in a network.

Remote server unit (RSU): An industrial computer located at a remote industrial process site that monitors and supervises the process.

Server: A computer located on a LAN that contains a central database used by all other computers on the LAN.

Setpoint: The desired condition required for a medium being maintained by a control loop.

Supervisory control and data acquisition (SCADA): An industrial control system that uses a large mainframe computer at a central location connected to monitoring computers at remote industrial process sites through dedicated telephone communications.

Topology: The physical wiring architecture used to connect multiple computing devices (nodes) together into a single communicating network and the pattern created.

Transfer frame: The architecture of a data bit stream that includes all information required to successfully transmit a data field from a source to a destination address on a network.

Transmission control protocol / Internet protocol (TCP/IP): The transport and network protocols of the Internet that connect millions of dissimilar services around the world.

Wet process: A wet process deals with fluids that flow through conduits and valves and tends to be continuous in nature. It may also be referred to as an analog process.

Wide area network (WAN): A client/server type computer network spread over a much larger geographical area than that covered by a LAN.

Workstation: A computer loaded with operating system and application software and is connected to a data highway and/or computer LAN.

NCCER CURRICULA — USER UPDATE

NCCER makes every effort to keep its textbooks up-to-date and free of technical errors. We appreciate your help in this process. If you find an error, a typographical mistake, or an inaccuracy in NCCER's curricula, please fill out this form (or a photocopy), or complete the online form at **www.nccer.org/olf**. Be sure to include the exact module ID number, page number, a detailed description, and your recommended correction. Your input will be brought to the attention of the Authoring Team. Thank you for your assistance.

Instructors – If you have an idea for improving this textbook, or have found that additional materials were necessary to teach this module effectively, please let us know so that we may present your suggestions to the Authoring Team.

NCCER Product Development and Revision

13614 Progress Blvd., Alachua, FL 32615

Email: curriculum@nccer.org
Online: www.nccer.org/olf

❏ Trainee Guide   ❏ AIG   ❏ Exam   ❏ PowerPoints   Other _____

Craft / Level: _____   Copyright Date: _____

Module ID Number / Title: _____

Section Number(s): _____

Description: _____

_____

_____

_____

_____

Recommended Correction: _____

_____

_____

_____

_____

Your Name: _____

Address: _____

Email: _____   Phone: _____

Analyzers

COURSE MAP

This course map shows all of the modules in the fourth level of the Instrumentation curriculum. The suggested training order begins at the bottom and proceeds up. Skill levels increase as you advance on the course map. The local Training Program Sponsor may adjust the training order.

INSTRUMENTATION LEVEL FOUR

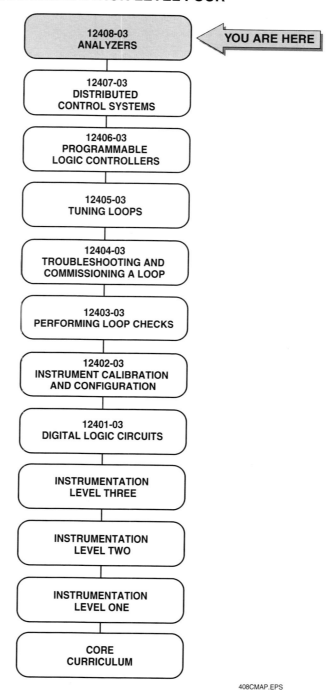

12408-03
ANALYZERS

YOU ARE HERE

12407-03
DISTRIBUTED
CONTROL SYSTEMS

12406-03
PROGRAMMABLE
LOGIC CONTROLLERS

12405-03
TUNING LOOPS

12404-03
TROUBLESHOOTING AND
COMMISSIONING A LOOP

12403-03
PERFORMING LOOP CHECKS

12402-03
INSTRUMENT CALIBRATION
AND CONFIGURATION

12401-03
DIGITAL LOGIC CIRCUITS

INSTRUMENTATION
LEVEL THREE

INSTRUMENTATION
LEVEL TWO

INSTRUMENTATION
LEVEL ONE

CORE
CURRICULUM

408CMAP.EPS

1.0.0 INTRODUCTION 8.1

    1.1.0   Classification 8.1

    1.2.0   Calibration 8.1

2.0.0 DENSITY AND SPECIFIC GRAVITY 8.2

    2.1.0   Air Bubble Measurement 8.2

    2.2.0   Displacement Measurement 8.2

    2.3.0   Densitometer 8.3

    2.4.0   Nuclear Detectors 8.3

3.0.0 VISCOSITY 8.3

    3.1.0   Viscometers 8.4

4.0.0 TURBIDITY 8.5

    4.1.0   Jackson Turbidimeter 8.5

    4.2.0   Transmission Analyzer 8.6

    4.3.0   Reflection Analyzer 8.7

    4.4.0   Ratio Analyzer 8.7

5.0.0 FLASH POINT 8.8

    5.1.0   Standardized Systems 8.8

    5.2.0   *OSHA 1910.106(a)* 8.9

6.0.0 OXIDATION-REDUCTION POTENTIAL (ORP) 8.10

    6.1.0   Probe Calibration 8.11

    6.2.0   Probe Maintenance 8.11

7.0.0 pH 8.12

    7.1.0   pH-Sensitive Electrodes 8.12

    7.2.0   Reference Electrodes 8.12

    7.2.1   *Errors* 8.13

    7.2.2   *Calibration* 8.13

8.0.0 CONDUCTIVITY (OF A LIQUID) 8.13

    8.1.0   Electrodes 8.14

    8.2.0   Inductive Probes 8.14

9.0.0 OXYGEN (O^2) 8.15

    9.1.0   High-Temperature Electrochemical Sensors 8.15

    9.2.0   Paramagnetic Analyzers 8.16

    9.3.0   Galvanic Sensors 8.17

10.0.0 CARBON MONOXIDE (CO) 8.17

11.0.0 CARBON DIOXIDE (CO_2) 8.18

    11.1.0 Emissions 8.18

    11.2.0 Monitoring 8.19

12.0.0 HYDROGEN SULFIDE (H₂S) 8.19
  12.1.0 Personnel Protection Indicators 8.19
  12.2.0 Semiconductor Sensors 8.20
  12.3.0 Electrochemical Sensors 8.20

13.0.0 TOTAL HYDROCARBON (THC) 8.20
  13.1.0 Flame Ionization Detector 8.20

14.0.0 PARTICULATES 8.21
  14.1.0 Optical Microscopy 8.21
  14.2.0 Discrete Particle Counters (DPCs) 8.21

15.0.0 CHEMICAL COMPONENTS 8.22
  15.1.0 Gas Chromatography 8.22

16.0.0 INFRARED RADIATION 8.24
  16.1.0 Basic Theory 8.24
  16.2.0 Affecting Factors 8.25
  16.3.0 Sensing Equipment 8.25
  16.4.0 Spectrometry 8.26

17.0.0 ULTRAVIOLET LIGHT WAVE ABSORPTION 8.27
  17.1.0 Analysis 8.28
  17.1.1 Applications 8.28
  17.1.2 Calibration 8.28
  17.1.3 Advantages 8.29
  17.2.0 Flame Detectors 8.29

SUMMARY 8.29

REVIEW QUESTIONS 8.30

GLOSSARY 8.33

REFERENCES & ACKNOWLEDGMENTS 8.35

Figures

Figure 1   Dual bubblers used to measure density 8.2
Figure 2   Displacement measurement of density 8.2
Figure 3   Measurement by the radiation detector 8.3
Figure 4   Basic falling ball viscometer 8.4
Figure 5   Basic rotating disc viscometer 8.4
Figure 6   Basic capillary viscometer .8.5
Figure 7   The Jackson turbidimeter .8.6
Figure 8   Principle of a transmission-type turbidity analyzer . . .8.6
Figure 9   Colorimeter calibration curve for chlorine 8.7
Figure 10   Typical reflection-type, in-line turbidity analyzer 8.7
Figure 11   Light beam traveling through a ratio
            turbidity analyzer 8.8
Figure 12   ORP probe assembly 8.10
Figure 13   Piping design including an ORP in-line system 8.11
Figure 14   pH/millivolt relationship at three different
            temperatures 8.12
Figure 15   pH-sensitive electrodes .8.13
Figure 16   Reference electrodes 8.13
Figure 17   Simple method to measure conductivity
            in a liquid 8.14
Figure 18   Principles of inductive conductivity measurement . . .8.14
Figure 19   High-temperature electrochemical
            oxygen detector 8.15
Figure 20   Magneto-dynamic oxygen analyzer 8.16
Figure 21   Principles of a galvanic oxygen sensing cell 8.17
Figure 22   Bacharach Monoxor® III handheld CO analyzer 8.17
Figure 23   Personnel protection indicator 8.19
Figure 24   Optical particle counter (OPC) 8.22
Figure 25   Operation of a condensation particle
            counter (CPC) 8.23
Figure 26   Simplified chromatography analyzer 8.23
Figure 27   Electromagnetic spectrum 8.24
Figure 28   Characteristics of infrared radiation 8.25
Figure 29   Basic thermography equipment 8.26
Figure 30   Simple single-beam infrared spectrometer 8.27
Figure 31   Simple ultraviolet analyzer block diagram 8.28
Figure 32   Ultraviolet flame detector 8.29

Tables

Table 1   Standardized Systems for Flash
           Point Determination 8.9

Table 2   Carbon Dioxide Emissions by Industry 8.18

Table 3   Common Particulates and Their Micron Sizes 8.21

Table 4   Emissivity Table 8.25

Table 5   Light Spectrum 8.27

Analyzers

Objectives

When you have completed this module, you will be able to do the following:

1. Define the following properties in a process or environment, and identify methods used to analyze them:
   - Density
   - Specific gravity
   - Viscosity
   - Turbidity
   - Flash point
   - Oxidation-reduction potential (ORP)
   - pH
   - Conductivity of a liquid
   - Oxygen (O_2)
   - Carbon monoxide (CO)
   - Carbon dioxide (CO_2)
   - Hydrogen sulfide (H_2S)
   - Total hydrocarbon content
   - Particulates in a clean room
2. Describe chromatography and its uses.
3. Describe ultraviolet analyzers and their uses.
4. Describe infrared analyzers and their uses.

Prerequisites

Before you begin this module, it is recommended that you successfully complete the following: Core Curriculum; Instrumentation Levels One through Three; Instrumentation Level Four Modules 12401-03 through 12407-03.

Required Trainee Materials

1. Pencil and paper
2. Appropriate personal protective equipment

1.0.0 ◆ INTRODUCTION

A process analyzer may be defined as an unattended instrument that monitors the chemical presence in, a particular property of, or the environment of a process.

Most process analyzers operate on a **laboratory principle** but with added mechanisms and circuitry to perform the required sampling and present the resulting data as required. Likewise, process analyzers must be housed in a manner that complies with electrical standards and requirements and protects from weather, ambient environment, and physical abuse.

1.1.0 Classification

Process analyzers may be classified in various ways, depending upon the purpose of classification. Some classifications are determined by the following:

- Operating principle, such as infrared, ultraviolet, and **chromatography**
- Type of analysis, such as oxygen and carbon dioxide
- Selection (infrared may be sensitized to monitor only one component, while a chromatograph may monitor several components)

1.2.0 Calibration

Process analyzers are almost always calibrated by applying standard samples prepared and analyzed by a laboratory. For this reason, process calibration can be no better than the laboratory analysis because any errors of both units are compounded. **Repeatability** is essential in process analyzers to eliminate the human errors due to variances and ambient conditions.

2.0.0 ◆ DENSITY AND SPECIFIC GRAVITY

Density is defined as mass per unit volume and is most commonly expressed in grams per cubic centimeter, pounds per cubic foot, or pounds per gallon.

Specific gravity is a term often used interchangeably with density. Specific gravity is expressed as the ratio of the density of the liquid to the density of water or the density of gas to the density of air at a specified temperature, with 60°F being the standard temperature used in this ratio calculation.

Because liquids are essentially uncompressible, their densities are generally unaffected by operating pressure. If temperature compensation is applied, the specific gravity and density of liquids are interchangeable. However, this rule does not hold true for gases because gases are compressible, and specific gravity (or molecular weight) is no longer equal to the density at operating conditions. Gases are generally referenced to their specific gravity relative to air or to their direct density at operating conditions when determining mass flow measurement.

2.1.0 Air Bubble Measurement

The simplest and perhaps most widely used method of measuring density and specific gravity in processes is using bubble tubes and a reference chamber in conjunction with a differential pressure transmitter. The two bubble tubes are installed in a vessel containing the process liquid to be measured but at different levels in the vessel, as shown in *Figure 1*.

The pressure required to cause bubbles to escape into the liquid is a measure of the pressure at the level of that tube's outlet. The difference between the two pressures required to cause bubbles to escape out of each tube is measured by the differential pressure transmitter and is equal to the weight of a constant-height (reference chamber) column of the liquid. The change in differential pressure caused by changing density is proportional to the density of the liquid. The adjustable **rotameters** are used to keep a calibrated continuous purge flowing through the two bubbler lines and can also be used to calibrate the transmitter.

2.2.0 Displacement Measurement

A displacer is sometimes mistakenly called a float. A true displacer submerses into the liquid based on its calculated buoyancy, whereas a true float rides on top of the liquid.

Figure 2 illustrates the application of displacement in measuring density. The force acting on the

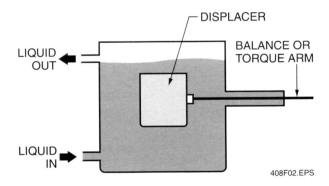

Figure 2 ◆ Displacement measurement of density.

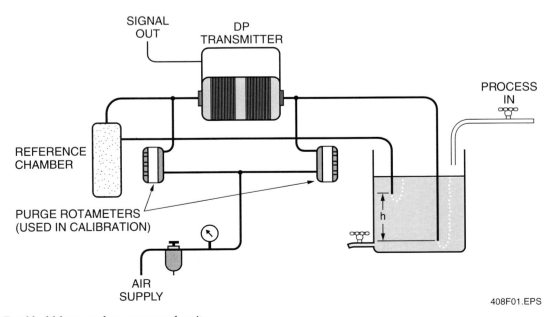

Figure 1 ◆ Dual bubblers used to measure density.

balance, or torque, arm is directly related to the density of the fluid displaced by the displacer or float. Because the displacer is in direct contact with the process liquid, a continuous flow in the process is necessary to purge the unit with fresh liquid. Low flows are necessary through the process lines, and the temperature of the process liquid must remain constant. The process liquid must be clean, to prevent settling of solids in the chamber and buildup of material on the displacer.

2.3.0 Densitometer

Measuring the density of gases is important in the gas transmission and petrochemical industries. This measurement can be accomplished using a densitometer with a probe. The probe contains a vane that is symmetrically positioned across the supporting cylinder. When the unit is to be put into operation, the probe is installed in the pipeline that contains the flowing process fluid. The vane oscillates in a simple, harmonic motion. This causes an acceleration of the flowing process fluid. The frequency of the oscillation varies with the density of the fluid. As the density increases, the oscillation frequency decreases. The relationship is expressed by the following:

$$Density = f^2[(A - B)/(f + C)]$$

In this equation, f is the frequency of the oscillation, and A, B, and C are constants related to the size of the probe and the properties of the fluid. The signal is amplified by the transmitter and energizes a driver with the probe. The driver sustains the oscillation. The transmitter converts the frequency to a 4–20mA current. This densitometer is also used with liquid. It is capable of measuring density even if the fluid is not flowing.

2.4.0 Nuclear Detectors

The density of a liquid can also be measured using the radiation from a suitable radioactive isotope. A source of radioactivity, such as radium or cobalt 60, is placed at one side of a vessel. Its radiated particles are directed across the vessel. A radiation detector is placed on the opposite side of the vessel or tube so that it receives the amount of radioactive energy remaining after the radiation passes through the vessel walls and the liquid, as shown in *Figure 3*. The amount of radiation absorbed by any material varies directly with its density. Such a method is best suited for liquids that do not permit a sensing element to be immersed because of threat of corrosion, abrasion, or other limitation.

3.0.0 ◆ VISCOSITY

Viscosity is a comparative measure of the ease with which particles in a fluid can change their relative positions and yield to an external force. For example, a thick liquid like honey offers more resistance to flow than does water. More specifically, the viscosity of a liquid or gas is the physical property that determines the magnitude of the resistance of the fluid to a **shearing force**. It is very important to both understand and accurately measure viscosity.

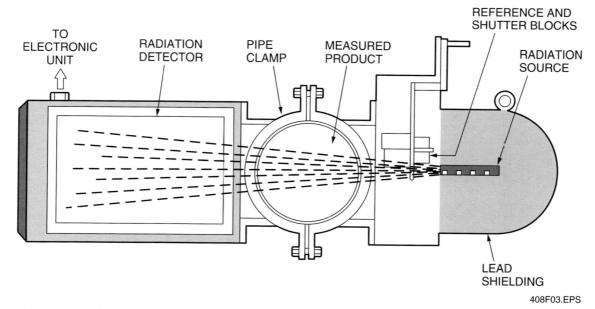

408F03.EPS

Figure 3 ◆ Measurement by the radiation detector.

3.1.0 Viscometers

Viscometers are instruments used for measuring the viscosity and flow properties of fluids at ambient or other specified temperatures. There are many types and models of viscometers available for determining liquid viscosities, most of which are designed for use with specific liquids and viscosity ranges. Commonly used viscometer designs typically fall into the three following categories:

* Falling ball or piston
* Rotating disc
* Capillary

A falling ball-type viscometer (*Figure 4*) measures viscosity by measuring the time required for a sphere (ball) to travel through a known distance of the liquid under test. The ball used for the test has a specific density and is typically made of glass or stainless steel. Some are made of tantalum. Balls of different density are used for testing different liquids. The precision ball is contained in a precision-diameter (-bore), temperature-controlled glass tube that contains the calibrated quantity of the liquid under test. The speed of ball movement through the calibrated distance of the liquid is typically measured using a stopwatch.

A rotating disc viscometer (*Figure 5*) requires a certain amount of torque to rotate a disc at a constant speed when immersed in a sample of the fluid being tested. The torque is a measure of the test fluid's **dynamic viscosity** and is proportional to the viscous drag on the immersed disc. A synchronous induction motor is typically used to drive a container coupled through a calibrated spring (torsion wire) to the disc. During measurement, the spring tends to wind up until its force equals the viscous drag on the disc that is immersed in the fluid under test. When this occurs, the container and disc both rotate at the same speed but with an angular difference between them. This angular difference is proportional to the torque on the spring and is converted into a viscosity reading. Depending on the device, the viscosity indication can be accomplished electronically or by a pointer. Other versions of a rotating viscometer that operate in a similar manner rotate a cone, sphere, or spindle in the fluid under test.

Capillary viscometers (*Figure 6*) measure the flow rate of a fixed volume of fluid at a controlled temperature through a small orifice. The rate of shear can be varied by changing the diameter and length of the capillary and the applied pressure. The time it takes for a specific volume of fluid to pass through the capillary orifice is proportional to the **kinematic viscosity** of the fluid. It should be pointed out that it also depends on the density of the fluid because the denser the fluid, the faster it will flow through the capillary orifice.

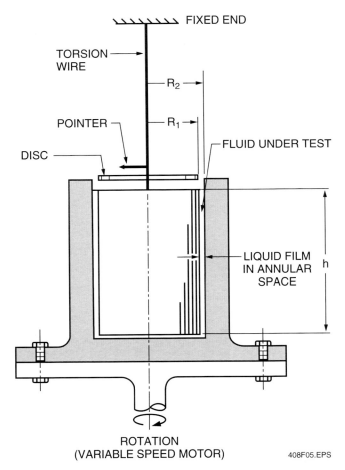

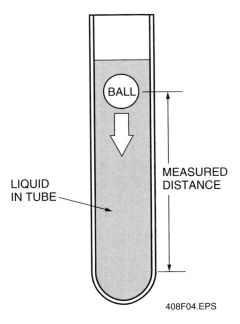

Figure 4 ◆ Basic falling ball viscometer.

Figure 5 ◆ Basic rotating disc viscometer.

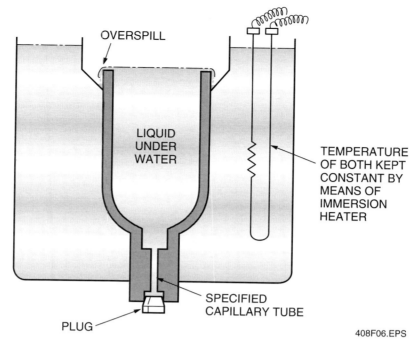

OVERSPILL

LIQUID UNDER WATER

TEMPERATURE OF BOTH KEPT CONSTANT BY MEANS OF IMMERSION HEATER

SPECIFIED CAPILLARY TUBE

PLUG

408F06.EPS

Figure 6 ◆ Basic capillary viscometer.

Viscometer manufacturers often describe their instruments as being used to measure Newtonian or non-Newtonian fluids. A Newtonian fluid is defined as a fluid that has a viscosity that is independent of shear. Oils, liquid hydrocarbons, beer, and milk are some examples of a Newtonian fluid. A non-Newtonian fluid is defined as a fluid that has a viscosity that is dependent on shear. This means as the shear rate changes, the fluid velocity changes. Some examples of non-Newtonian fluids are grease, paint, clay slurries, and candy compounds. Note that with an increase in shear rate, the viscosity of some non-Newtonian fluids will increase, while for others, it will decrease. Obviously, the type of viscometer selected for a specific use must be compatible for use with the type of fluid being tested.

Viscometers can be portable instruments, fixed instruments, or in-line devices. In-line instruments can be inserted into and retracted out of a process line without disturbing the flow of the fluid in the line. Viscometers are available that have analog, digital, video, or cathode ray tube (CRT) displays. Some have no local display and are made to send the measured viscosity data to a remotely located display or device. Operating controls can also be analog or digital. Some also have the capability of being operated under the control of a host computer. Some models are pre-programmed and have no user controls. Other features available for viscometers include temperature compensation, temperature sensing, and data storage such as on a hard drive or floppy drive.

4.0.0 ◆ TURBIDITY

Turbidity is defined as an optical appearance property of liquids caused by the presence of suspended particles. The particles cause a scattering of the light energy passing through the liquid, and the turbidity is influenced by the concentration, size, shape, and optical properties of the particles in addition to the optical properties of the fluid.

The theory of the scattering of light by particles was developed late in the nineteenth century. A theoretical understanding of the light-scattering phenomena by particles is well established, but the general theories are so complex that they require computer technology for analytical solutions. However, this level of complexity is not generally required in the analysis of turbidity in industrial applications.

4.1.0 Jackson Turbidimeter

The Jackson turbidimeter, represented by *Figure 7*, and the Jackson turbidity unit are the standard instrument and unit of turbidity measurement and represent the birth of instrumentation designed to measure turbidity. The Jackson turbidimeter consists of a special candle and a flat-bottomed glass tube graduated in Jackson turbidity units. You gradually pour the sample into the tube as you observe the image of the candle from the top. At the point of pouring when the image of the candle burning disappears into a uniform glow, you take the reading at the mark on the tube.

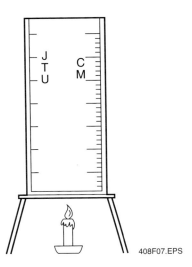

Figure 7 ◆ The Jackson turbidimeter.

The Jackson turbidimeter compares the strength of the transmitted light with that scattered or reflected. However, you cannot use the Jackson turbidimeter for very fine particles because the yellow-red candle flame is at the long wavelength end of the visible spectrum and is not effectively absorbed or scattered by very fine particles. Therefore, the candle flame will not disappear. Also, black particles absorb most if not all the light, and the liquid will become dark before enough of the sample is poured into the tube to reach an image extinction point.

In modern instrumentation, you can largely avoid these difficulties by using light sources that provide a wider light spectrum or using comparison rather than extinction techniques for measurement.

4.2.0 Transmission Analyzer

A transmission turbidity analyzer contains a light source, lens, sample cell, photocell, and a control unit containing a readout device. The lens directs a portion of the light from the source through the sample cell and at the photocell, which acts as a receiver of the light. The light that is not received by the photocell is either absorbed or reflected by the solids in the sample, while the light that is received by the photocell varies inversely with the concentration of solids in the sample. The light that is received by the photocell is converted into electrical energy, amplified in the control unit of the analyzer, and transmitted to a readout device. Turbidity, measured in percent of light transmitted, is calibrated by checking zero and then 100% transmission of light with a clear sample.

The illustration of the principle associated with a transmission-type analyzer is shown in *Figure 8*.

It indicates an adjustable shutter that can be used to regulate the amount of light from the source, depending on the degree of turbidity in the sample. As long as the shutter is set before calibrating the analyzer with a clear sample, the amount of light will not affect the reading because the reading is based on a percentage of light available.

The transmission-type turbidity analyzer may also be used as a colorimeter to measure darkness or **absorbance** due to the presence of color. Occasionally, however, it is desirable to monitor one specific component in the stream based on the presence of color. The light wavelength of the transmission-type turbidity analyzer may be so calibrated and limited, based on a colorimeter calibration curve like the one shown in *Figure 9* for chlorine, so that the instrument responds only to changes in particle quantity of a particular component in the stream.

For example, if the light wavelength is limited to a narrow band near 340 millimicrons, the instrument will be only sensitive to changes in chlorine content. This is because chlorine absorbs strongly at that wavelength. The transmission-type turbidity analyzer can be sensitized in this manner by selecting the proper light source, filters, and photocell. Sensitivity may be increased or decreased within limits by varying the wavelength or sample cell length or both.

Gauging color in a lab is usually accomplished by holding a container filled with liquid in front of a light and comparing its color to a group of color standards. The transmission-type process analyzer uses the same method for gauging color, but it substitutes more reliable and calibrated photocells for the unreliable human eye.

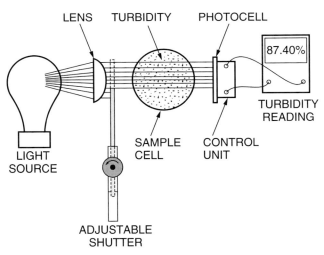

Figure 8 ◆ Principle of a transmission-type turbidity analyzer.

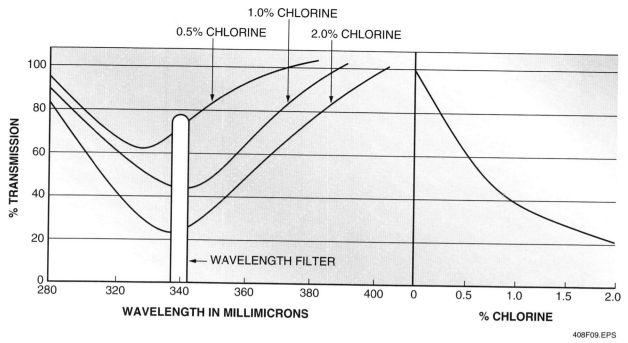

Figure 9 ◆ Colorimeter calibration curve for chlorine.

4.3.0 Reflection Analyzer

Another analyzer that is similar to the transmission type is the reflection-type turbidity analyzer, which measures the amount of light reflected by the suspended particles in a sample stream. A simplified reflection analyzer (*Figure 10*) has a shield between the light source and photocell to prevent the transmission of light directly to the cell. Particles in the stream reflect light around the shield to the photocell. The quantity of light reaching the photocell is directly proportional to the concentration of the particles present in the stream. The electrical signal from the photocell is typically connected externally to either an indicating or recording instrument that is calibrated in parts per million of turbidity.

The turbidity sensor is designed with a self-contained calibration feature consisting of a push button that actuates a slide plate, exposing a precisely machined opening in the light shield. This allows an amount of light exactly equivalent to a turbidity of 5 ppm to be transmitted directly to the photocell. When you push this button, it changes the output of the sensing cell by an amount that is equal to 5 ppm turbidity. This signal change permits the system to be quickly and accurately calibrated.

4.4.0 Ratio Analyzer

A ratio turbidity analyzer uses a more complex system to measure turbidity, sharing technology from both the transmission-type and reflection-type analyzers. See *Figure 11*. Light from a common light source is passed through a rotating chopper disc that alternately blanks out one light beam at a time. The transmitted beam is then passed through the sample cell to photocell A, followed by the scattered beam at right angles to each other. Photocell A sees the incident beam during one half of the chopper disc's cycle and the reflected beam during the last half of the cycle, causing its output signal to be digital in nature. This type of detector is usually associated with a microprocessor-based turbidity analyzer.

Figure 10 ◆ Typical reflection-type, in-line turbidity analyzer.

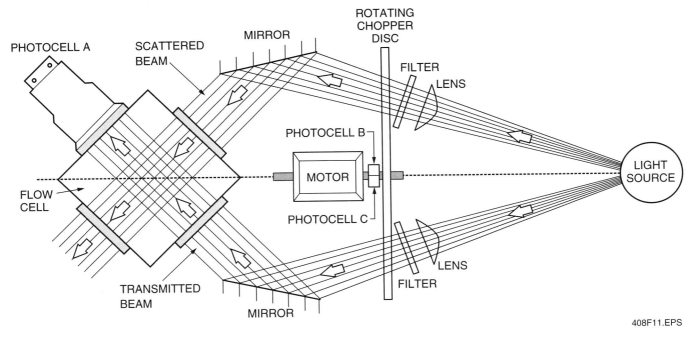

Figure 11 ◆ Light beam traveling through a ratio turbidity analyzer.

Photocells B and C generate pulses that are synchronized with the transmitted and scattered beams and are only used as gating pulses in the electronic circuit associated with this analyzer. The two light beam signals are amplified, switched to the output circuit, and transmitted to the readout device. The readout device is usually calibrated in parts per million.

5.0.0 ◆ FLASH POINT

OSHA 1910.106(a) defines flash point as the lowest temperature at which a liquid gives off vapor within a test vessel in sufficient concentrations to form an ignitable mixture with air near the surface of the liquid. A closely related and less common term associated with flash point is the fire point. This is defined as the temperature at which the flame becomes self-sustained and continues burning the liquid. At the flash point, the flame does not need to be sustained. The fire point is usually a few degrees above the flash point.

Materials with a flash point below 100°F (38°C), such as solvents and solvent-borne coatings, are considered dangerous. Manufacturers provide flash point data about their products in the related material safety data sheet (MSDS). For flash point testing purposes, a test sample flashes when a large flame appears and instantly spreads across the surface of the sample. It should be pointed out that a blue halo does not constitute a flash point. The U.S. Department of Transportation requires that all transported substances must have their

flash point determined and that any materials with flash points lower than 140°F (60°C) be handled with extra caution.

5.1.0 Standardized Systems

You can determine the flash points of various materials by performing flash point tests using the correct instruments and procedures. You must follow correct sampling methods and standardized procedures precisely. Flash point testing methods have become standardized by a number of national and international organizations over the years. In general, two methods for determining flash points are the closed-cup and open-cup methods. The closed-cup method prevents vapors from escaping and, therefore, usually results in a flash point that is a few degrees lower than in an open cup. Because the two systems give different results, you must always identify the method you use for testing.

There are several types of commonly used flash point analyzing systems, including the following:

- Abel
- TAG
- Rapid
- Pensky-Martens
- Cleveland
- Auto-ignition

You determine the point of ignition with the Abel, TAG, Pensky-Martens, and Cleveland systems of flash point testing by increasing the temperature of the sample and applying an igniter. You can use the rapid flash test system to quickly determine if a flash point occurs below or above a specific temperature. However, it does not find the actual flash point temperature. The auto-ignition system works at extremely high temperatures. The hot surfaces and air inside the container cause the sample to ignite. When performing flash point testing, always operate the flash point test instrument in accordance with the manufacturer's instructions, and perform the test in accordance with the applicable standard.

Flash point testers typically operate by one of the four following methods:

- Standard instruments with gas heating and gas ignition
- Electrically heated instruments with gas or electric ignition
- Partially automated instruments with electric heating and automated gradient control
- Fully automatic instruments with automatic flash point detection

WARNING!

Follow all safety procedures when using a flash tester. Fire extinguishers, safety visors, and breathing apparatus must be close at hand and should be used as needed. Proper ventilation is extremely important because heating can produce toxins, such as PCBs.

Table 1 shows the applicable standards associated with each of the most common standardized systems used to determine flash point.

NOTE

The Deutsches Institut für Normung, or DIN, standard is an international standard developed in Germany.

5.2.0 *OSHA 1910.106(a)*

As mentioned earlier, *OSHA 1910.106(a)* defines flash point. It also defines the methods of testing flash point in some detail. Regarding the methods of testing, the standard states that if a liquid has a viscosity of less than 45 **SUS** at 100°F (37.8°C), does not contain suspended solids, and does not have a tendency to form a surface film while

Table 1 Standardized Systems for Flash Point Determination

Low Temperatures (−20°C to +80°C)	
System	**Standard**
Abel	*Institute of Petroleum (IP) 33, IP 170*
Tag	*American Society for Testing and Materials (ASTM) D56, IP 304*
Rapid	*ASTM D3228, ASTM D3243, ASTM D3278, IP 303, International Standards Organization (ISO) 3679, ISO 3680*
Pensky-Martens	*ASTM D93*
Medium Temperatures (60°C to 360°C)	
Pensky-Martens	*ASTM D93, IP 34, Deutsches Institut für Normung (DIN) 51 758, ISO 2719, ASTM D3228, ASTM D3243*
Rapid	*ASTM D3228, ASTM D3243, ASTM D3278, IP 303, ISO 3679, ISO 3680*
High Temperatures	
Cleveland	*ASTM D92, ISO 2592*
Highest Temperatures	
Auto-ignition	*DIN 51 794, ASTM E659*

under test, then the test procedure specified in *ASTM D56-70* must be used. However, if a liquid has a viscosity of 45 SUS or more at 100°F (37.8°C), or contains suspended solids, or has a tendency to form a surface film while under test, the standard method of test for flash point by Pensky-Martens closed tester specified in *ASTM D93-71* must be used. There are some exceptions to this rule, which may be found in *ASTM D93* Notes.

WARNING!

Know the flash point of any material you work with by referring to the MSDS. Always avoid heat, open flame, sparks, or other sources of ignition when a material is near, at, or above its flash point. A common error is failure to pay attention to flash points when using a heating bath in a lab or on a calibration bench.

6.0.0 ◆ OXIDATION-REDUCTION POTENTIAL (ORP)

Oxidation may be defined as the loss of electrons by one molecule and **reduction** defined as the absorption of electrons by another. Every liquid has both oxidizing and reducing ions, but their balance varies from liquid to liquid. Those liquids that tend to reduce have an excess of electrons while those that tend to oxidize have a shortage of electrons.

The oxidation-reduction potential (ORP) of a process liquid indicates by its voltage polarity whether the process liquid has an oxidizing potential or a reducing potential. A liquid with an oxidizing potential has a positive polarity relative to an applied reference voltage; a liquid with a reducing potential has a negative polarity relative to the reference voltage. The unit of ORP measurement, which is the voltage potential measured between two electrodes submerged in the sample, is the millivolt.

An ORP probe, as shown in *Figure 12*, is made up of two separate electrodes—a reference electrode and a measurement electrode—which are housed in an outer shell. The reference electrode provides a constant voltage potential. The lead wire of a reference electrode is positioned in an inner tube inside the shell. The inner tube typically contains silver metal and silver chloride paste. The paste is in contact with a saturated solution of potassium chloride, which acts as an electrical bridge to the solution being measured. The potential of the reference electrode depends on temperature and on the concentration of potassium chloride. The potassium chloride slowly migrates from the reference electrode to the solution being measured by means of a liquid junction consisting of a porous ceramic material near the bottom of the electrode. Crystals of solid potassium chloride in the bottom of the electrode ensure that the solution stays saturated. A portal is provided for replenishment of the potassium chloride, and a rubber sleeve protects the portal from contamination.

The ORP measurement electrode is a platinum wire that is exposed at its bottom end to the process liquid. This electrode reacts to variations in voltage. Voltage variations caused by changes in the activity of the oxidizing and reducing ions in the process liquid are amplified and displayed as the ORP of the liquid.

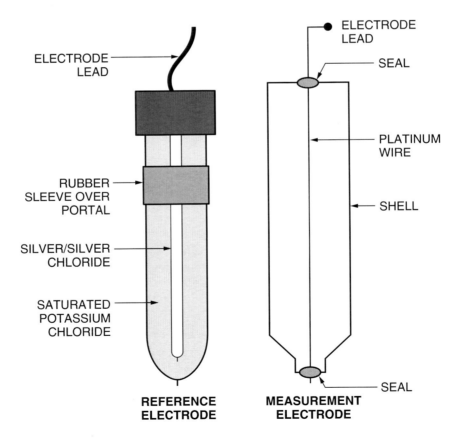

REFERENCE ELECTRODE

MEASUREMENT ELECTRODE

408F12.EPS

Figure 12 ◆ ORP probe assembly.

6.1.0 Probe Calibration

You can calibrate an ORP probe by immersing both the measurement and reference electrodes in liquids that have a known oxidation-reduction potential. These liquids are used specifically for calibration and are known as standards.

You can calibrate the ORP probe with standards in one of two ways—by removing the electrodes from the process and immersing them in a suitable, safe container holding the standard or by putting the standard directly into the sample tubing leading to the installed electrodes. The first method is the easiest and the most accurate. A piping design that allows you to remove the electrodes easily without shutting off the process flow is recommended. If you use the second method, you must be extremely careful of the following:

• Do not allow the process liquid to come into contact with the standard, which would change its ORP value.

• Use a sufficient quantity of the standard to rinse the tubing to the electrode to wash out any leftover process liquid that would cause an inaccurate reading.

• Flush the standard to a chemical sewer or waste dump after the test. Standards and process liquids are frequently different chemical solutions, so the system must be flushed thoroughly using either the process liquid itself or demineralized water.

Figure 13 is a one-line diagram of a typical sampling installation for ORP analysis while the electrodes are still installed in the system. The block and bleed valves are installed to keep the process liquid from leaking into the sample system during the calibration process. The rotameter regulates the flow of sample liquid through the sampling system. One of the two drain valves allows the flow to return to the process or to an external chemical drain system. The other drain valve is used for obtaining a quick sample, which is referred to as a grab sample. This system also includes an in-line demineralized flush system to flush the lines before and after calibration.

6.2.0 Probe Maintenance

ORP probes usually require little maintenance as long as the process liquid does not contain contaminants that could coat or erode the electrodes or liquid junctions. One important maintenance procedure is the replacement of potassium chloride solution in the reference electrode. Although most reference electrodes are refillable, some cannot be refilled, and they must be replaced when they lose their potassium chloride solution.

You must periodically clean or replace the porous tip of the reference electrode that forms the liquid salt bridge between the silver–silver chloride electrode and the process liquid. A very small amount of potassium chloride must flow from the electrode into the process liquid to keep the process liquid from contaminating the reference electrode solution.

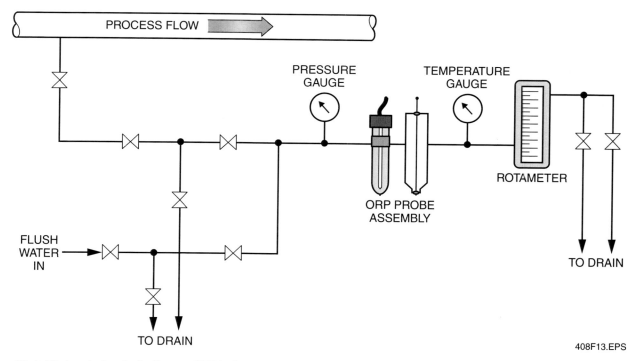

Figure 13 ◆ Piping design including an ORP in-line system.

7.0.0 ◆ pH

The pH analysis is one of the most widely used analytical measurements in industrial applications. In the refining and petrochemical industry, pH measurement is applied extensively for the following purposes:

- Quality control, where pH control determines product uniformity
- Corrosion inhibition, where pH monitoring is used in conjunction with oxygen control to minimize corrosion in high-pressure boilers and other water systems
- Effluent pH control, in the neutralization of liquid discharges into public streams

The measure of the **acidity** or **alkalinity** of a solution, pH, is defined as the negative **logarithm** of the hydrogen ion activity. The pH range covers both acid and alkaline solutions. In water, the equilibrium (balanced) product of the hydrogen (H^+) and hydroxyl-ion (OH^-) concentration is a constant, 10^{-14}, at 22°C. Because the H^+ and OH^- concentrations are equal at equilibrium, the H^+ ion concentration must be 10^{-7}, and the pH = $\log(1/10^{-7})$, or 7.0. Acid solutions increase in strength as the pH value decreases below 7. The scale is not linear because a change of one unit of pH represents a change by a factor of 10 in the effective strength of the acid or base.

Figure 14 shows how errors due to temperature changes can occur in pH measurements. As the temperature rises, more hydrogen ions are dissociated. Therefore, a change in the millivolt output occurs to either lower the apparent pH in the case of acid solutions or raise the apparent pH in the case of base solutions. A basic pH assembly consists of a pH-sensitive electrode, a reference electrode, a temperature-compensating electrode, and a potential-measuring device.

7.1.0 pH-Sensitive Electrodes

A pH-sensitive electrode consists of a pH-sensitive glass membrane joined to an insulating envelope or tube, as shown in *Figure 15*. When you immerse the electrode in an **aqueous solution**, a potential develops across the membrane that is proportional to the concentration of hydrogen ions in the solution. Because the glass membrane offers such a high resistance to ion flow (approximately 100 megohms), this potential must be measured from a point inside the envelope to a point in the solution that is outside the envelope. A silver/silver chloride element is immersed in a **buffer solution** to act as the inner contact. An adjacent reference electrode contacts the outer surface through the measured solution, and an electrical measuring circuit then measures the potential across the glass membrane.

7.2.0 Reference Electrodes

The voltage measurement must be made in reference to some stable electrode that does not change as the pH changes. This is accomplished by using a reference electrode like one of those shown in *Figure 16*.

There are four types of liquid junctions commonly used to accomplish electrical continuity in the reference electrode. Each type is best-suited for a particular type of liquid or a particular set of operating conditions. The four types of junctions are as follows:

- *Quartz fiber* – Maintains a rapid flow rate for fast response and minimum tendency to clog. It is the best general-purpose junction.
- *Ceramic frit* – Is recommended for minimal potassium and chloride contamination of the sample.
- *Inverted sleeve* – Provides a very high flow rate for troublesome samples. May be disassembled for cleaning. Recommended for pastes, creams, colloids, sludge processes and highly viscous samples.
- *Double* – Has the same characteristics as the ceramic frit junction. It is recommended where the flowing **electrolyte** must be matched to the sample and/or the sample must be completely isolated from the filling solution. The inner chamber of the double-junction electrode is not refillable. The outer chamber can be refilled with a custom filling solution to match the sample.

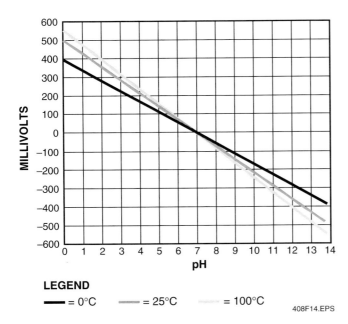

LEGEND

━━ = 0°C   ▩▩ = 25°C   ░░ = 100°C

408F14.EPS

Figure 14 ◆ pH/millivolt relationship at three different temperatures.

INDICATING SPHERICAL BULB

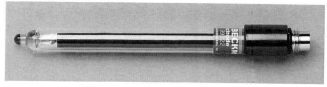

INDICATING RUGGED BULB

408F15.EPS

Figure 15 ◆ pH-sensitive electrodes.

CALOMEL QUARTZ

CALOMEL CERAMIC

CALOMEL INVERTED

SILVER CHLORIDE QUARTZ

SILVER CHLORIDE INVERTED

SILVER CHLORIDE DOUBLE JUNCTION

408F16.EPS

Figure 16 ◆ Reference electrodes.

7.2.1 Errors

Temperature fluctuations can affect the calibration of a pH analyzer. Small errors occur in the reference electrode voltage and in the internal voltage of the sensing electrode. These errors are minimized, however, by using electrodes that are made from similar materials. Solubility of chemicals in cells may vary with temperature changes and cause a period of signal drift until a new state of equilibrium is reached after the change in temperature.

Modern pH process analyzers are direct-reading types because any change in pH is more easily read over the wide range of 0 to 14 pH. Accuracy has been improved with the more modern pH analyzers and may be as accurate as ±0.03 pH in the dual or folded scales (0–8 and 6–14 pH).

7.2.2 Calibration

The electronic measuring circuit includes a feedback amplifier in which the output is applied across a precision resistor. The resulting current may be applied as an analog output signal or read using a milliammeter. The unit zeros itself automatically every second for a more stable and accurate operation.

pH-sensitive electrodes require a buffer solution in the approximate range at which the analyzer operates. Standard buffer solutions that cover the entire pH range are available from the manufacturers of the analyzer. Calibration using a standard buffer solution should be made at two points—one below and one above the expected normal pH value of the sample.

8.0.0 ◆ CONDUCTIVITY (OF A LIQUID)

The conductivity of a liquid is a measure of the ability of a solution to carry an electrical current. It is also sometimes referred to as specific conductance and is defined as the **reciprocal** of resistance in ohms of a one centimeter cube of the solution at a specific temperature. The symbol for conductance is the letter G. It is represented by the following formula:

$$G = \frac{1}{R}$$

The former unit of measurement of conductivity was the mho/cm, which represents the reciprocal of ohm/cm, a resistance measurement. However, the **siemens/cm** is now the accepted term of conductivity, and it has replaced the mho/cm in many applications. Mho/cm is still used by many to represent the measurement of conductivity.

The flow of current in liquid solutions is different from that in metals. It is accomplished by the drift of positive and negative ions, compared to the movement of free electrons in metallic conductors.

8.1.0 Electrodes

One simple method used to measure conductivity in a liquid is by inserting two plates in the liquid, as shown in *Figure 17*. These plates are referred to as the electrodes. In order for conductivity to be equal to the reciprocal of the resistance, the plates must be one cm square in size and separated from each other by one cm of space. If any other dimensions are used in the electrode installation, these dimensional differences are considered as multipliers in the calculation and are referred to as cell constants when figuring the calculation. A voltage of 1V is applied across the electrode plates, and the current in the 1V circuit is read. By applying Ohm's law, which states that the voltage (E) is equal to the current (I) times the resistance (R), you can find the conductivity because conductivity is always the inverse of the resistance.

For example, if 1V is applied to the electrode plates and a current of 0.8 amperes flows in the circuit, you can calculate the conductivity using Ohm's law as follows:

$$E = IR$$

$$1 \text{ volt} = 0.8A \times R$$

$$R = \frac{1V}{0.8A}$$

$$R = 1.25 \text{ ohms}$$

$$G = \frac{1V}{R}$$

$$G = \frac{1V}{1.25 \text{ ohms}}$$

$$G = 0.8 \text{ siemens/cm or mho/cm}$$

If the size of the electrode plates changes, the amount of current that flows through the plates will change. Therefore, the indicated conductivity changes although the voltage supply has not changed. This is referred to as changing the cell constant of an instrument. Each cell with electrode plates has its own cell constant. Cell constants range from 0.01 to 100 and vary by multiples of ten. An instrument with a cell constant of 0.01 has a range of 1 to 10 microsiemens, an instrument with a cell constant of 0.1 has a range of 1 to 100 microsiemens, and an instrument with a cell constant of 1 can measure 1 to 1,000 microsiemens, and so on. Normally, the instruments with lower ranges are used for liquids of lower conductivity, and the higher ranges are used for liquids of higher conductivity.

8.2.0 Inductive Probes

Inductive conductivity probes operate on the principle of a magnetic field and usually consist of two enclosed electrical coils. The instrument transmits an AC current of known value through the primary coil, as shown in *Figure 18*. As this current passes through the primary coil, it creates a magnetic field and induces a current into the process liquid. This current in the process liquid

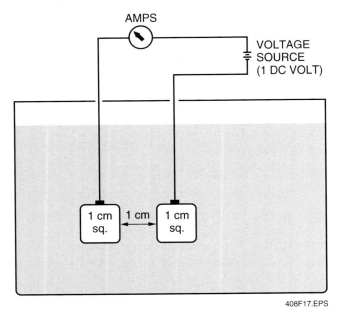

408F17.EPS

Figure 17 ◆ Simple method to measure conductivity in a liquid.

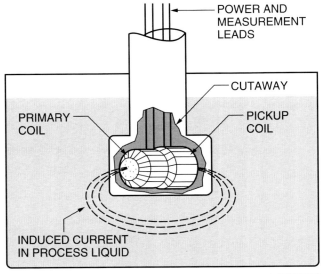

408F18.EPS

Figure 18 ◆ Principles of inductive conductivity measurement.

then induces a current in the pickup coil. The value of the current is directly proportional to the conductivity of the liquid.

No direct contact is required between the coils and the solution; thus, maintenance problems are minimized. The inductive conductivity analyzer typically transmits a 4–20mA signal that is proportional to the measured conductivity.

Calibration of the inductive conductivity cells is performed in much the same way as any probe in that you use a liquid containing a known standard measurement of conductivity. When using standard liquids for conductivity calibration, you should first immerse the probe in the liquid of known conductivity, and then adjust the instrument to that conductivity. It is recommended that you calibrate the instrument using two or more known liquid standards that are within the instrument's range. If only two standards are used, it's best to use one that is at the low end and one that is at the high end of the instrument's range. The low end is often at or near 0.0 microsiemens/cm and is referred to as the zero standard measurement. The high end of the range is referred to as the span standard.

9.0.0 ◆ OXYGEN (O$_2$)

The demand for oxygen analysis is due in part to the role oxygen plays in combustion, oxidation, and other industrial process applications. However, in industry the major contributor to oxygen analysis is, by far, combustion.

Several sensor cells are available for measuring the concentration of oxygen, including the following three common types:

- High-temperature electrochemical
- **Paramagnetic**
- Galvanic

9.1.0 High-Temperature Electrochemical Sensors

High-temperature electrochemical sensors are widely used in combustion control applications to measure exhaust gas oxygen concentrations. One such application is the measurement of the oxygen concentration in the exhaust gases of steel-producing blast furnaces. This type sensor is also widely used in exhaust stack analyzer applications by manufacturers of chemicals and petro-chemicals, ceramics, and glass.

A type of high-temperature electrochemical sensor commonly used in combustion control applications to determine combustion efficiency is a zirconium oxide oxygen sensor. This sensor

consists of a cell (*Figure 19*) made of ceramic zirconium oxide combined with an oxide of either **yttrium** or calcium. A coating of porous platinum applied to the inside and outside walls of the cell acts as a pair of electrodes. The cell is then heated to maintain it at a constant temperature.

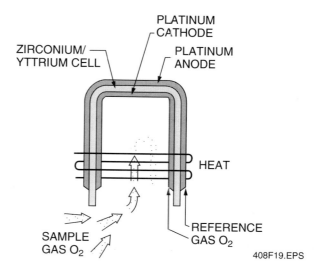

Figure 19 ◆ High-temperature electrochemical oxygen detector.

At high temperatures (typically above 1,200°F), minuscule openings in the cell walls permit the passage of oxygen ions from a sampled gas through the cell. When the partial oxygen pressures are equal on both sides of the cell, only a random movement of oxygen ions occurs within the cell. This results in no voltage being generated across the platinum electrodes. However, when a sample gas is present on one side of the cell, and a referenced gas (typically air) is present on the opposite side, oxygen ions will pass through the cell from one electrode to the other.

The rate at which oxygen ions pass through the cell is determined by the temperature of the cell and by the difference in the oxygen partial pressures of the sample gas versus the reference gas. The passage of oxygen ions causes a voltage to be produced across the platinum electrodes. The magnitude of this voltage is a function of the ratio of the sample gas and reference gas oxygen partial pressures. Because the partial pressure of the reference gas is known, the voltage produced by the cell indicates the oxygen content of the sample gas.

The major advantage of using a zirconium oxide oxygen analyzer for high-temperature combustion gas analysis is that the measuring probe can be placed directly into a flue with the probe mounted in virtually any position. One disadvantage is that the zirconium oxide cell has a relatively short life that is typically less than 18 months. Frequent thermal cycling can cause stress cracks that can greatly shorten cell life and overall reliability. Another disadvantage is that, as the cell ages, it becomes increasingly more difficult to calibrate.

Zirconium oxide oxygen analyzers are not recommended for use in trace oxygen analysis applications. This is because the presence of gases, such as hydrogen, hydrocarbons, or carbon monoxide, will chemically consume oxygen in the sample at the high temperatures necessary for operation and will result in lower than actual readings of oxygen content.

9.2.0 Paramagnetic Analyzers

The paramagnetic oxygen analyzer works on the principle that oxygen is a highly magnetically susceptible material. Magnetic susceptibility is a measure of the intensity of the magnetization of a substance when it is placed in a magnetic field. Oxygen has an exceptionally high magnetic susceptibility when compared to other gases. Paramagnetic types of oxygen analyzers are most commonly used in the following applications:

- Analysis of combustion efficiency
- Testing the purity of breathing air and protective atmospheres
- Laboratory instruments
- Medical applications
- Selected industrial process monitoring and control applications

The magneto-dynamic oxygen analyzer described here is the most widely used type of paramagnetic oxygen analyzer. It consists of a small dumbbell-shaped body made of glass and charged with a low magnetic susceptibility gas (such as nitrogen), a light source, a photocell, a mirror, and a calibrated indicating unit (*Figure 20*). The dumbbell-shaped body is suspended within the magnetic field of a permanent magnet and is free to rotate in the space between the poles of the magnet. Because of its nitrogen charge, the dumbbell body has negative magnetic susceptibility causing its balled ends to naturally deflect slightly away from the point of maximum magnetic field strength.

When you introduce a test sample containing oxygen into the test cell, the oxygen in the sample is attracted to the point of maximum field strength, which causes the dumbbell to move. Its displacement is proportional to the amount of oxygen in the sample. The movement of the dumbbell is detected by a light beam from the light source outside of the test cell. The light beam is reflected to an exterior photocell by a mirror attached to the dumbbell body. The amount of light reflected to the photocell is determined by the amount of dumbbell movement. The output of the photocell is then amplified and transmitted to an indicating unit that is calibrated to read out the percentage of oxygen content in the test sample.

An advantage for using a magneto-dynamic oxygen analyzer is that it provides a direct measurement of magnetic susceptibility that is not influenced by the thermal properties of the

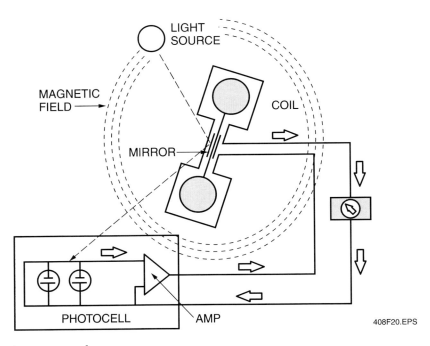

Figure 20 ◆ Magneto-dynamic oxygen analyzer.

background gas. It also has excellent speed of response. Its main disadvantage is that it is a delicate instrument with moving parts. This makes it sensitive to vibration and, therefore, requires precise positioning. It should be pointed out that the magnetic susceptibility of the sample is a function of temperature. This characteristic can result in significant errors if the sample temperature deviates from the calibrated temperature. In addition, the accuracy of a dumbbell-type analyzer can be affected by the magnetic susceptibility of the prevailing background gases.

9.3.0 Galvanic Sensors

The galvanic oxygen sensor, also called a **coulometric** oxygen sensor, is a chemical cell whose oxidation-reduction potential (ORP) is changed by the introduction of the sample gas. The primary element of the galvanic oxygen sensor is an electrochemical cell consisting of two electrodes in contact with a basic electrolyte, usually potassium hydroxide (*Figure 21*). The cell electrodes are made of dissimilar metals, typically silver and lead.

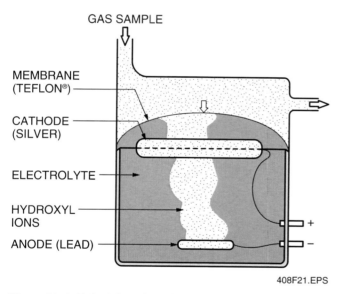

408F21.EPS

Figure 21 ◆ Principles of a galvanic oxygen sensing cell.

When a gas sample enters the cell, it diffuses through a membrane usually made out of Teflon®. The oxygen in the sample contacts the silver cathode and is chemically reduced to hydroxyl ions. These ions then flow toward the lead anode, where an oxidation reaction occurs with the lead. This reaction results in a flow of electrons that is proportional to the oxygen concentration of the sample. An external metering circuit connected to the cell electrodes measures the electron flow (current) between the electrodes. This current is proportional to the

rate of consumption of the oxygen and is indicated on a meter either as a percentage or as parts per million of oxygen in the sample.

10.0.0 ◆ CARBON MONOXIDE (CO)

The gases used as fuel for furnaces and the various heated vessels in industry, manufacturing, and power generating facilities are very dangerous. They can cause potential health problems or even death in confined areas where their byproducts can accumulate.

One such byproduct of combustible gas is carbon monoxide, which is a colorless, tasteless and odorless gas that combines with the **hemoglobin** of the human blood and inhibits its oxygen-carrying capacity. It is the same deadly gas that is expelled from the exhaust system in your car and from gas space heaters in your home.

CO sensors are used in industrial applications to warn personnel of dangerous levels of this lethal byproduct of combustion. Handheld carbon monoxide analyzers are now available for personnel who must enter confined spaces to work. One such example is shown in *Figure 22*. This particular handheld analyzer is the Bacharach Monoxor® III. It is a continuous sampling instrument that quickly and accurately measures levels of CO between 0 and 2,000 ppm. It has a stainless steel probe with 5' of tubing to sample CO found in flue gases of residential furnaces, combustion appliances, and commercial and industrial boilers.

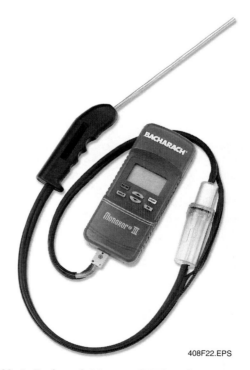

408F22.EPS

Figure 22 ◆ Bacharach Monoxor® III handheld CO analyzer.

By adding an accessory module to this particular unit, it can be used to determine the presence and level of CO in the bloodstream. It is also available in a high-range version (0–85,000 ppm) for vehicle emissions testing and forklift tuning.

CO analyzers are primarily used in the following industrial and manufacturing applications:

- CO safety checks on combustion stack gas and ambient air levels
- Analyzing indoor air quality and safety
- Furnace and boiler servicing or replacement

11.0.0 ◆ CARBON DIOXIDE (CO₂)

Carbon dioxide (CO_2) is a nonflammable, colorless, odorless gas somewhat heavier than air. Carbon dioxide is one of the gases that form the earth's atmosphere, along with nitrogen, oxygen, and argon. Carbon dioxide is evenly distributed at a concentration of about 0.033% over the earth's surface. Because of its low concentration in the atmosphere, it is not practical to commercially obtain carbon dioxide gas by extracting it from the atmosphere. Commercial quantities of carbon dioxide gas are normally produced as byproducts of manufacturing processes such as the combustion of coal and natural gas, the production of ethanol, and the manufacture of ammonia. Bulk quantities of carbon dioxide are usually stored and shipped as liquid under elevated pressure and refrigeration.

Carbon dioxide is non-reactive (**inert**) with many materials and is commonly used for applications such as blanketing and purging of tanks and reactors and as a shielding gas in arc welding applications. It is also used in carbonated beverages and to fill certain types of fire extinguishers. In addition to its inert properties, at extremely low temperatures (–100°F) carbon dioxide solidifies into dry ice, which is often used to freeze and preserve foods.

11.1.0 Emissions

Manufacturing is the single largest source of carbon dioxide emissions in the U.S. industrial sector. It accounts for approximately 85% of industrial energy-related carbon dioxide emissions, and it also accounts for approximately 84% of industrial energy consumption. *Table 2* shows current estimates of energy-related carbon dioxide emissions from the manufacturing industry.

Of the 405.2 million metric tons carbon equivalent emitted by manufacturers in 1998, about 43% (174.5 million metric tons carbon equivalent) was emitted by two industry groups—petroleum and coal products (21.6%) and chemicals (21.5%). Four other groups account for roughly 37% of the industries' total: primary metals (16.9%), paper (8.0%), food (6.1%), and stone, clay, and glass (5.6%). The other manufacturing group consisting of 14 industries, which range from apparel to fabricated metals to computer manufacturing, accounts for the remaining 20%. None of those 14 industry groups separately accounts for more than 3% of total energy-related carbon dioxide emissions from the manufacturing industries.

Because both the petroleum/coal and chemical industries use energy for non-fuel purposes, both have lower carbon emissions than the manufacturing average. For example, in 1998 the petroleum/coal industry averaged 11.54 million metric tons carbon equivalent per quadrillion Btu while the chemical industry averaged 12.07 million metric tons carbon equivalent per quadrillion Btus.

Table 2 Carbon Dioxide Emissions by Industry

Industry Group	Carbon Dioxide Emissions (Mega-Metric Tons Carbon Equivalent)	Share of Total Manufacturing Emissions (%)	Carbon Intensity of Energy Use (Mega-Metric Tons Carbon Equivalent per Quadrillion Btus Consumed)
Petroleum and coal products	87.4	21.6	11.54
Chemicals	87.1	21.5	12.07
Primary metals	68.4	16.9	18.69
Paper	32.3	8.0	9.90
Food	24.6	6.1	15.64
Clay and glass	22.6	5.6	18.15
Other manufacturing	82.8	20.4	14.95
Total	405.2	100.0	13.48

(Data courtesy of Energy Information Administration, Form EIA-846, Manufacturing Energy Consumption Survey, and Form EIA-810, Monthly Refinery Report, 1998.)

The paper industry, which uses wood byproducts extensively, yielded carbon intensities of 9.90 million metric tons carbon equivalent per quadrillion Btu in 1998. Carbon dioxide emissions from wood consumption are considered to be zero because the carbon emitted has been recently **sequestered,** and the replanting of trees will re-sequester the emitted carbon.

The primary metals industry, however, is a heavy user of energy sources with high carbon content, such as coal. As a result, the overall carbon intensity for the primary metals industry was 18.69 million metric tons carbon equivalent per quadrillion Btu in 1998.

11.2.0 Monitoring

Combustion processes that use boilers, furnaces, and kilns require proper mixtures of oxygen and fuel in order to burn properly. Because carbon dioxide deprives any environment of oxygen, carbon dioxide levels must be monitored in these types of combustion processes to achieve ideal combustion conditions.

Additionally, there is much scientific discussion about global warming and the **greenhouse effect.** When the sun's energy reaches the earth, some of this energy is reflected back to space and the rest is absorbed. The absorbed energy warms the earth's surface, which then emits heat energy back toward space as radiation. This outgoing radiation is partially trapped by greenhouse gases, such as carbon dioxide, methane, and water vapor, which then radiate the energy in all directions, warming the earth's surface and atmosphere. Scientists are concerned that higher greenhouse gas concentrations will lead to an enhanced greenhouse effect, which may lead to global climate change.

Carbon dioxide analyzers and monitoring are available in many forms, including stack-mounted, portable, and **in-situ**, depending on the environment and application. Carbon dioxide can be measured by infrared absorption and by electrochemical methods similar to those used for oxygen analysis. The process of electrochemical analysis was discussed earlier, and infrared analysis will be investigated later in this module.

12.0.0 ◆ HYDROGEN SULFIDE (H₂S)

Hydrogen sulfide is a colorless, very flammable gas. In low concentrations, it smells like rotten eggs. It is heavier than air and is considered to be a very toxic gas. After only 2 to 15 minutes of exposure at 100 ppm, you lose your sense of smell, which makes it impossible to either smell more

dangerous concentrations or to detect when concentration levels have decreased to a safe level. When H_2S burns, it produces another very toxic gas—sulfur dioxide (SO_2).

H_2S is used in metallurgy, the preparation of phosphorous and oil additives, and as a reagent in chemical analysis. During the recovery and processing of crude oil, H_2S can contaminate the atmosphere and become a major health hazard. The gas can be detected at a level of 2 parts per billion. To put this into perspective, 1 ml of the gas distributed evenly in a 100-seat lecture hall is about 20 ppb. The **OSHA ceiling** for H_2S is 20 ppm, and the peak is 50 ppm for 10 minutes. The evacuation point is 65 ppm.

12.1.0 Personnel Protection Indicators

Because hydrogen sulfide is very toxic and can cause your sense of smell to be lost after brief exposure, it's very important that those of you who could potentially be exposed to H_2S are always equipped with some type of personal H_2S detection device.

One very simple device is the wearable personnel protection indicator (*Figure 23*) that changes color when exposed to toxic gases, including hydrogen

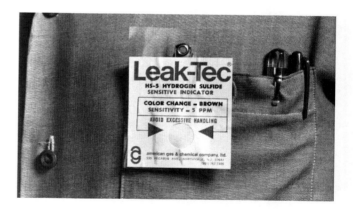

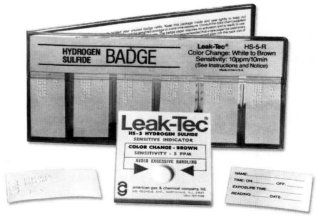

408F23.EPS

Figure 23 ◆ Personnel protection indicator.

sulfide. Badges must be ordered according to the gas being monitored. In the case of H_2S, the badge's sensitivity is 10 ppm/10 minutes, and the color strip will change from white to brown. Extra color strips are shipped with the badge and may also be separately ordered for H_2S as well as other gases, such as ammonia, carbon monoxide, chlorine, hydrazine, nitrogen dioxide, and ozone. Coincidentally, the ozone strip will also change from white to brown when exposed to ozone, but has a much smaller sensitivity—0.1 ppm/15 minutes.

Natural gas contains several percentage points of H_2S, and because of this content, natural gas wells are often called sour gas wells from their offensive odor. Volcanoes also discharge hydrogen sulfide. **Anaerobic decay**, aided by bacteria, produces hydrogen sulfide, which in turn, produces sulfur. This process accounts for much of the sulfur found in nature.

Hydrogen sulfide has few important commercial uses. However, it is used to produce sulfur, which is one of the most commercially important elements. About 25% of all sulfur is obtained from natural gas and crude oil. Hydrogen sulfide dissolves in water to make a solution that is weakly acidic.

12.2.0 Semiconductor Sensors

Semiconductor sensors work well in H_2S gas measurements. Although the quality of sensors varies widely from manufacturer to manufacturer, semiconductor sensors are among the best for H_2S gas monitoring where sensitivity to low concentrations is required.

For this method of H_2S gas monitoring, a semiconductor material is applied to a nonconducting substrate between two electrodes. The substrate is heated to a temperature such that the gas being monitored can cause a reversible change in the conductivity of the semiconductor material. Under zero gas conditions, it is believed that O_2 molecules tie up free electrons in the semiconductor material thereby inhibiting electrical flow. As H_2S gas or vapor molecules are introduced, they replace the O_2, releasing the free electrons and decreasing the resistance between the electrodes. This change in resistance is measured electrically and is proportional to the concentration of the gas being measured.

12.3.0 Electrochemical Sensors

As described previously, the electrochemical sensor is widely used in other analytical functions involving oxygen and carbon monoxide. In addition, it is often used to detect hydrogen sulfide gas because it is simple, reliable, and inexpensive.

The following is a review of how the electrochemical sensor operates: The electrochemical sensor is a self-powered micro fuel cell. The cell consists of a casing containing a gel or electrolyte and two active electrodes—the working electrode (anode) and the counter-electrode (cathode). The top of the casing has a membrane that can be permeated by the gas sample. Oxidization takes place at the anode and reduction at the cathode. A current is created as the positive ions flow to the cathode and the negative ions flow to the anode.

The main difference between the detection of H_2S and the detection of oxygen in the electrochemical sensor is that oxygen is sensed at the cathode in the cell, but hydrogen sulfide is sensed at the anode.

13.0.0 ◆ TOTAL HYDROCARBON (THC)

Hydrocarbons in atmospheric air are a complex mixture of many substances. Among these substances, methane has a concentration of approximately 1.7 ppm, which accounts for the majority of the concentration of total hydrocarbons. Methane is important as one of the greenhouse effect gases, but it causes no photochemical reactions. Therefore, it is necessary to measure nonmethane hydrocarbons (NMHC), which can cause the production of photochemical oxidants.

Many types of organic gases and vapors can potentially be present in air at any one time. Volatile organic contaminants can be manmade (such as gasoline vapor, exhaust fumes, cleaning solvents, and lube oil vapor) or can result from biogenic activities (such as marsh gas, mold, and mildew).

Some of the hydrocarbons are hazardous and may have irritating odors. A common feature of these volatile organic materials is that they all contain at least one hydrogen atom and at least one carbon atom. Because it is impractical to measure each type of organic contaminant present, they are measured as a group and described as total hydrocarbon (THC) content. Any instrument used to measure THC responds to the total amount of H+C-containing molecules present.

13.1.0 Flame Ionization Detector

The instrument most often used to measure vapors for total hydrocarbon content uses the principle of flame ionization detection (FID). For this type of detector, the air sample and fuel (hydrogen) are sent to the burner nozzle at a controlled flow rate. The concentration of hydrocarbon is obtained by ionizing hydrocarbon in the hydrogen flame, and then measuring the generated ion current. In principle, the FID method shows responses in proportion to the number of carbon atoms in hydrocarbons. Therefore, concentrations are shown as equivalent to those of methane in ppm by using units called ppmC.

Because the sensitivity of an FID is dependent on the flow rates of the sample and hydrogen, care must be taken to control these flow rates. The readings and relative sensitivity of the FID method are also adversely affected by oxygen interference. If oxygen is introduced, an error will occur if there is a difference between the oxygen concentrations of the calibration standard gas and those of the sample air. The degree to which these problems affect the results varies depending on the type of organic material combustion conditions. This means that the THC value varies depending on the composition of hydrocarbons and the monitoring system. The FID method should be used bearing these things in mind.

14.0.0 ◆ PARTICULATES

Federal Standard 209E establishes standards for cleanliness for airborne particulate levels in clean rooms. This standard also describes methods for monitoring the air and procedures for verifying the classification level of clean rooms. These classifications are established by the number of particulates that are one micron (μm) or larger in a cubic foot of space per minute. One micron is equal to one millionth of a meter, which is also 1/1,000 millimeter. *Table 3* shows three examples of common objects that are measured in microns. The classes are 1, 10, 100, 1,000, 10,000, and 100,000.

Clean rooms are specially built enclosed areas that are specifically designed to control airborne particulates, temperature, humidity, airflow patterns, air motion, and lighting. These sealed facilities use specialized air handling and filtration systems designed to minimize static electricity as well as concentrations of particles and other contaminants that may interfere with manufacturing.

Typically, clean rooms produce a vertical laminar flow of air throughout a large area of the space. The air is filtered and contaminants are

Table 3 Common Particulates and Their Micron Sizes

Particulate	Size (in microns)
Cigarette smoke particles	0.01–1.0
Bacteria	0.3–40
Household dust	0.5–20

purged through the large airflow. Because air velocity and environmental factors must be controlled within tightly prescribed limits, clean rooms are large users of energy.

14.1.0 Optical Microscopy

The method used in monitoring and counting particulates in a clean room depends on the standards and policies of the clean room. If the clean room is involved in any government work, chances are that many of the standards required to be followed will have originated within a governmental agency.

A common, yet tedious, method often used to determine particulate content in a clean room is to physically count the particulates in a specified sample. Usually, the method for counting is determined by the size or size range of the particulates specified. For example, one method specified in *Federal Standard 209E* for counting and sizing the concentration of particles 5μm and larger in clean rooms is to collect the particles on a membrane filter then to look through a microscope to count them.

In this method, a vacuum is used to draw a sample of air through a membrane filter. The rate of flow of the sample is controlled by a limiting orifice or by a flowmeter; thus, the total volume of air sampled is determined by the sampling time. The membrane filter is examined microscopically to determine the number of particles 5μm and larger collected from the sample of air.

Image analysis or projection microscopy may replace direct optical microscopy for the sizing and counting of particles, provided the accuracy and reproducibility equal or exceed that of direct optical microscopy.

14.2.0 Discrete Particle Counters (DPCs)

Discrete particle counters (DPCs) provide data on the concentration and size distribution of airborne particles within the approximate range of 0.01 to 20μm on a near real-time basis. A DPC will correctly size only those particles within the limits of its dynamic range.

Air in the clean room or clean zone to be verified or monitored is sampled at a known flow rate from the sample point(s) of concern. Particles in the sampled air pass through the sensing zone of the DPC. Each particle produces a signal that can be related to its size, either directly or with reference to the operation of a pre-detection sample processing program. An electronic system sorts and counts the pulses, registering the number of particles of various sizes that have been recorded within the known volume of air sampled. The concentration and particle size data can be displayed, printed, or processed.

Differences in the design of DPCs that can lead to differences in counting include dissimilar optical and electronic systems, pre-detection sample processing systems, and sample handling systems. Potential causes of difference such as these should be recognized and minimized by using a standard calibration method and by minimizing the variability of sample acquisition procedures for instruments of the same type. Two types of DPC instruments are the optical particle counter (OPC) and the condensation particle counter (CPC).

In the OPC (*Figure 24*), the output from a plasma laser or laser diode light source is **collimated** in a beam used to illuminate a sample volume stream of aerosol flowing through the OPC. The aerosol sample enters the OPC via an input nozzle, and it is drawn out via a vacuum source. Light scattered by **refraction**, reflection, and/or **diffraction** from the various single particles contained within the volume of aerosol is detected by a photo detector via an optical network positioned off-axis from the light beam. Both the size and the number of particles are measured simultaneously with the size of each particle being determined by the intensity of the scattered light.

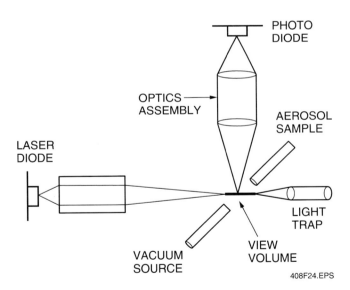

Figure 24 ◆ Optical particle counter (OPC).

Although this example illustrates an aerosol sample OPC, liquid-type sample OPCs are also available. In the case of a liquid OPC, the fluid is constrained to a channel inside a transparent cell, such as a quartz cell.

The condensation particle counter (CPC) operates by saturating the heated sample particles with **butanol**, which is evaporated into the air stream, and then cooling the sample in a condenser. The supersaturation of the vapor increases as it is cooled. The vapor condenses on the particles, which causes them to grow to sizes that are easily detected. The resulting droplets are passed through an optical detector immediately after leaving the condenser.

The detector counts individual pulses produced by the smaller concentrations of particles (10^4 particles per cubic centimeter) as they pass through the sensing zone during the single-count mode. During the photometric mode, higher concentrations (up to 10^7 particles per cubic centimeter) are measured by detecting light scattered by all particles in the sensing zone at any one time and comparing the intensity of the scattered light with calibration levels. *Figure 25* shows the operation of a CPC.

15.0.0 ◆ CHEMICAL COMPONENTS

Some processes require a chemical breakdown for quality control purposes or to meet EPA standard requirements, for example. Chromatography is a technique used to separate and analyze many different kinds of complex mixtures.

15.1.0 Gas Chromatography

A typical application of gas chromatography is to separate and analyze atmospheric mixtures containing volatile organic compounds. Chromatography works on the principle that when a test mixture is processed through a chromatography analyzer, the different components that form the sampled mixture separate and exit the chromatography process at different rates. This separation allows the individual components in the sampled mixture and their concentrations to be identified.

Figure 26 shows a simplified diagram of a gas chromatography analyzer. As shown, it consists of a sample mixture injection point, gas cylinder, a chromatography column and oven, and a detector. The oven can produce a fixed temperature, which means it is **isothermal,** or it can have temperature controls that allow you to program a temperature into the instrument.

The chromatography process involves two phases—mobile and stationary. In the mobile phase, the sample mixture to be analyzed is first vaporized and combined into a moving stream of carrier gas, such as helium, for subsequent passage through the chromatography column. The stationary phase relates to the material used in the column. In gas chromatography, this material is typically an adsorbent or liquid that is distributed over the surface of a porous, inert support column.

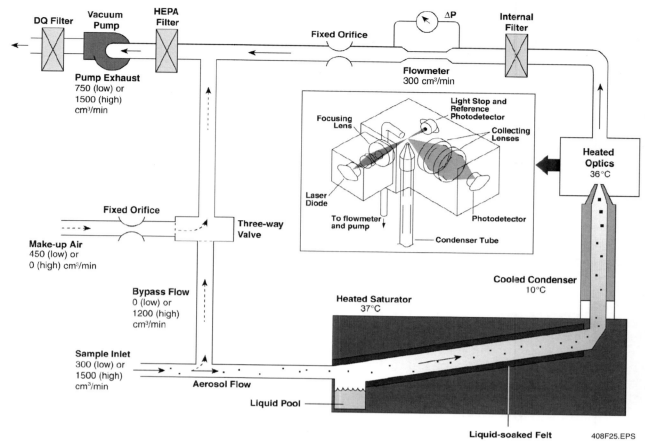

Figure 25 ◆ Operation of a condensation particle counter (CPC).

This adsorbent or liquid material acts to separate the components of the sampled mixture based on the varying attraction or affinity of each of the different mixture components to the column material. As the sample mixture flows through the column, the different components are adsorbed by the column in varying degrees.

The different components of the sampled mixture adsorbed by the column exit the column at different rates. Those with a weak attraction (less adsorption) exit first and are followed by those with a stronger attraction. This action results in the separation of the mixture's components. Following this, the detector gathers and analyzes the different components as they exit from the column to produce an electrical output signal.

The numerous types of detectors used in gas chromatograph analyzers are specifically designed for particular types of applications. Two widely used types of detectors are the thermal conductivity and flame ionization detectors. The output from a gas chromatograph detector can be in the form of computer-readable data or an analog voltage, current, or frequency. Often the detector output is used to drive a printer or plotter to produce chromatograms. Chromatograms are graphical representations that show a series of peaked waveform

responses, each of which corresponds to a different component in the sampled mixture. Another option is to have the output trip an alarm or cause a change in state of a switching device. Some models of gas chromatographs are programmable and have data storage options.

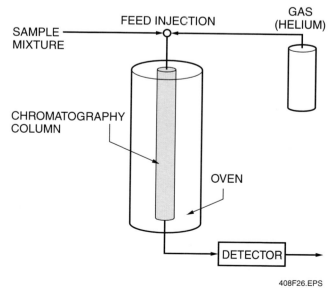

Figure 26 ◆ Simplified chromatography analyzer.

16.0.0 ◆ INFRARED RADIATION

Infrared sensing of heat radiation was initially developed by the military to allow people to see in the dark. In 1958, the technology was introduced to the public sector. Since that time, infrared thermography inspection has developed into a state-of-the-art technique used to monitor equipment performance. It is a nondestructive, noncontact testing method used to monitor energy losses and identify potential failures. Data collected over a period of time can indicate impending failure of a piece of electrical equipment. The equipment can then be repaired before more costly corrective maintenance is required.

16.1.0 Basic Theory

Thermography is the technique of using an optical system that converts invisible infrared (heat) radiation into visible light. Thermography is based on the premise that any object having a temperature above **absolute zero** will emit infrared radiation. The magnitude of infrared radiation emitted by a source is proportional to its temperature. Infrared radiation is part of the electromagnetic spectrum, whose wavelengths range from 0.75 micrometers (the long-wavelength limit of visible red light) to 1,000 micrometers (the shortest microwaves). Commercial thermography uses the 2 to 14 micrometer infrared range (*Figure 27*).

Because infrared radiation is invisible to the human eye, infrared sensing equipment is required. This equipment collects and focuses infrared energy on sensitive infrared detectors that convert the infrared energy into electrical impulses. The electrical impulses are then amplified and relayed to light-emitting diodes (LEDs), which illuminate a light intensity display of the temperatures on the targeted object. With a rotating tilted mirror, a television-like image consisting of multiple scan lines is produced. Some models have 60 scan lines that produce a contrasting light intensity picture of the object. The light intensity indicates temperature differences on the object.

Infrared imaging allows you to see the heat image pattern of the object. This image is called a thermogram. The lighter areas of a thermogram indicate higher temperature regions while the darker areas indicate lower temperature regions. Infrared imaging can also measure temperatures at any point on the object. Other infrared sensing devices use similar principles. Equipment costs are minimized by limiting the output to a digital single-point temperature display or single line scan and not the thermal image of the object.

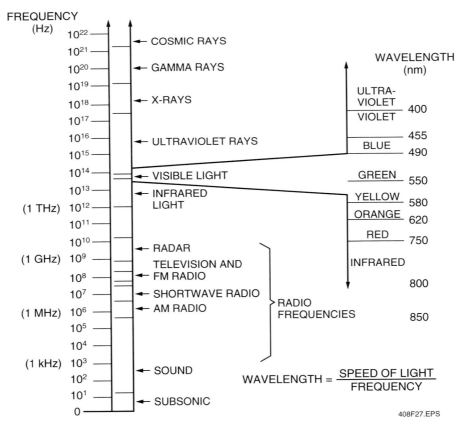

Figure 27 ◆ Electromagnetic spectrum.

16.2.0 Affecting Factors

Infrared sensing works by measuring the amount of infrared radiation emitted from an object. The emitted infrared radiation must pass through a medium, generally air, and be absorbed by the infrared detectors. Objects will appear to be cooler or warmer because of the following factors:

- **Emissivity** of the object
- Direct sunlight or other background infrared sources
- Atmospheric gases in higher than normal concentrations

With two notable exceptions, all matter emits electromagnetic radiation. The exceptions are matter at absolute zero and inert gases. In order to describe thermal radiation, first consider an idealized object. In *Figure 28*, the block to the left represents a hot source that is radiating infrared energy. The block to the right is the target material. Because the block on the left is hotter, energy will transfer from left to right. The block to the right can do one of three things with this transferred energy. First, the energy can be absorbed and converted to heat. Second, the energy can pass through uninhibited and can be measured as the material's **transmittance**. The third interaction that the block and the transferred energy can have is for the energy to be reflected. This, too, can be measured and is known as the material's **reflectivity**. All of these processes can occur at the same time; however, the sum of the reflected energy, the absorbed energy, and the transmitted energy must be equal to the initial transferred energy. If you consider the ability of an object to radiate energy, or its emissivity, it is exactly equal to its coefficient of absorption in the idealized object. The idealized object is given the name black body and is defined as a material that emits the maximum amount of radiation at a given temperature.

The term black body is misleading; its color is not as important as the type of material and the surface finish of the material. Materials that are good radiators and display black body characteristics are carbon and rubber. Highly polished metals are good reflectors but act as poor black bodies. Materials are rated for their emissivity by comparing the amount of energy they reflect to the amount of energy reflected by a black body. Numerically, emissivity is a unitless number between 0 and 1, where 1 represents the emissivity of a black body.

$$\text{Emissivity} = \frac{\text{total radiation from a non-black body}}{\text{total radiation from a black body}}$$

Table 4 lists the approximate emissivity of some common materials.

Table 4 Emissivity Table

Material	Emissivity
Silver (polished)	0.01
Brass (polished)	0.03
Molten steel	0.28
Asbestos cloth	0.90
White paint (ThO$_2$)	0.90
Red brick	0.93
Carbon	0.95
Black paint (CuO)	0.96

16.3.0 Sensing Equipment

The basic infrared sensing equipment consists of a detector and display unit. Most are battery-operated portable units (*Figure 29*). Depending on the purpose of the infrared viewer, numerous options are available to improve analysis capability. Options for the infrared viewers include black and white photography, color photography, wide angle and telephoto lens, and color video displays for enhanced temperature differential detection.

The infrared radiation can generally be detected from –40°C to 1,500°C with a temperature resolution as fine as 0.1°C. Some models have an adjustable focus for increasing the range from 0.2 meters to infinity. As the range is increased, the view field is also increased. By combining the sensitivity of the equipment with the color video display and photographic capabilities, subtle temperature differences can be detected thus increasing the application of thermography.

In addition to thermography infrared viewers, other infrared sensing equipment can be used to

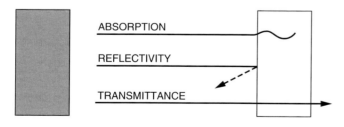

ABSORPTION + REFLECTIVITY + TRANSMITTANCE = 1
ABSORPTION = EMISSIVITY
EMISSIVITY + REFLECTIVITY + TRANSMITTANCE = 1

408F28.EPS

Figure 28 ◆ Characteristics of infrared radiation.

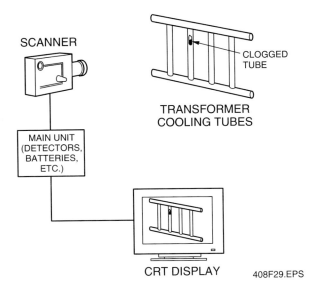

Figure 29 ◆ Basic thermography equipment.

detect temperature differences. These include single infrared line scanners and infrared thermometers. Both of these devices use the same infrared sensing principles as thermography, but the sensing and display components are much simpler. When looking through a single infrared line scanner, the target object can be seen with a thermal line scan superimposed over it. The thermal line scan appears as a plot of temperature versus the horizontal view of the targeted object.

Infrared thermal inspection can be used to detect potential failures in electrical equipment by sensing differences in infrared radiation. While limited use of infrared detection for temperature monitoring is not new to the utility industry, the application of thermography as a diagnostic predictive maintenance technique has gained universal acceptance. Infrared thermal inspection is used in industrial and commercial facilities to locate sources of energy losses and potential failures. The following are examples of infrared thermal inspection applications:

- *Electrical equipment* – Poor connections, corrosion of connections, faulty materials and equipment
- *Electronic circuits* – Faulty components and connections
- *Motors* – Hot spots and worn bearings
- *Power transmission equipment* – Bad connections
- *Power factor capacitors* – Overheating
- *Switchgear and breakers* – Overheating
- *Three-phase circuits* – Unbalanced loads
- *Transformers* – Hot spots
- *Bearings* – Hot spots and wear

Infrared inspection can be used to determine if a bearing is not receiving adequate lubrication and is overheating. Bearings in inaccessible areas are sometimes overlooked. If the bearing can be seen, infrared inspection can be used to determine its temperature and locate hot spots.

Infrared inspection is particularly useful for electrical inspections. Overheating in electrical equipment is a major cause of failures. Infrared inspection of electrical equipment can locate problems that may go undetected using conventional methods. In a plant electrical distribution system and major components, electrical connections that are loose, corroded, or deteriorated overheat. Typical examples are substations, bus bars, motor controllers, motor switching gear, power transformers, power factor capacitors, and lighting circuitry. Failures in these areas are critical to plant operation and safety. Infrared inspection is a fast, reliable method to identify these problem areas. Infrared inspection can view through cabinets and panels, minimizing the inspection time. In addition to locating overheated connections, infrared inspection can locate overheated conduit, clogged transformer cooling tubes, and defective electrical components.

16.4.0 Spectrometry

Another use of infrared detectors is in the science of **spectrometry**. Spectrometry is used to determine the composition of an object. If an unknown sample is analyzed using a spectroscope, the chemical makeup of the sample can be determined.

In absorption spectrometry, radiations of a particular wavelength in the appropriate region of the spectrum are employed. An infrared analyzer responds to the absorption of infrared radiation. When light is passed through a light-transmitting solid or fluid, absorption of radiant energy takes place, depending on the chemical identity of the absorbing medium. The number of molecules per unit volume (concentration) and the length of light path within the medium (thickness traversed) influence the amount of absorption in terms of the ratio between incident and transmitted light. Thus, each hydrocarbon has a characteristic absorption spectrum so that a graph of wavelength versus percentage absorption (transmission) enables the hydrocarbon to be identified.

In operation, first the proper wavelength must be selected for each component from an examination of the spectrum of each component. Then, by setting the instrument successively at each of these wavelengths and comparing the absorption of the unknown with that of a set of standards of known concentrations, the amounts of each component can be determined.

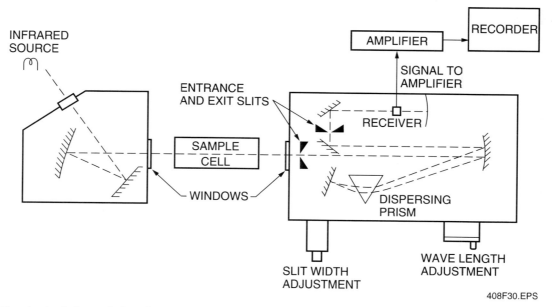

Figure 30 ◆ Simple single-beam infrared spectrometer.

408F30.EPS

The basic components of an infrared absorption spectroscope are as follows:

- A source of infrared radiation
- A detector sensitive to infrared radiation
- An amplifier to enable a record to be obtained
- A comparison cell having infrared-transparent windows and located between the source and the detector through which the sample passes

A simple single-beam infrared spectrometer is shown in *Figure 30*.

Infrared spectrometry is useful in the analysis of a number of organic gases and liquids. However, oxygen, hydrogen, nitrogen, chlorine, as well as the inert gases do not absorb infrared radiation and, therefore, cannot be measured by this method.

17.0.0 ◆ ULTRAVIOLET LIGHT WAVE ABSORPTION

Ultraviolet radiation is also commonly called ultraviolet light waves, and it operates at a much higher frequency than the infrared detectors mentioned earlier. Because of this higher frequency and the resulting shorter wavelength, ultraviolet radiation interacts with materials differently than infrared light waves. *Table 5* shows the relationship of ultraviolet light to infrared light and the visible light spectrum.

Table 5 Light Spectrum

LIGHT SPECTRUM				
REGION	WAVELENGTH (Angstroms)	WAVELENGTH (Centimeters)	FREQUENCY (Hz)	ENERGY (eV)
RADIO	$>10^9$	>10	$<3 \times 10^9$	$<10^{-5}$
MICROWAVE	10^9–10^6	10–0.01	3×10^9–3×10^{12}	10^{-5}–0.01
INFRARED	10^6–7000	0.01–7×10^{-5}	3×10^{12}–4.3×10^{14}	0.01–2
VISIBLE	7000–4000	7×10^{-5}–4×10^{-5}	4.3×10^{14}–7.5×10^{14}	2–3
ULTRAVIOLET	4000–10	4×10^{-5}–10^{-7}	7.5×10^{14}–3×10^{17}	3–10^3
X-RAYS	10–0.1	10^{-7}–10^{-9}	3×10^{17}–3×10^{19}	10^3–10^5
GAMMA RAYS	<0.1	$<10^{-9}$	$>3 \times 10^{19}$	$>10^5$

408T05.EPS

17.1.0 Analysis

Process stream analyzers based on the measurement of ultraviolet radiation absorption are used throughout the process industries to monitor and control the concentrations of components in both gas and liquid streams. Like the infrared detectors discussed earlier, the UV spectrophotometer reads absorbance versus wavelength to identify the components in the mixture as well as the concentration.

The UV absorption pattern of a compound is not so distinctive a fingerprint as its infrared counterpart, and fewer compounds absorb in the ultraviolet region than in the infrared. However, several important classes of compounds absorb strongly in the ultraviolet region, whereas water and the usual components of air do not absorb in this region. As a result, ultraviolet absorption analyzers may be more selective and sensitive than infrared and other types of analyzers.

The basic UV absorption analyzer consists of a radiation source, optical filters, a sample cell, a detector, and an output meter. A simple block diagram of a typical UV detector is shown in *Figure 31*. A transmittance measurement is made by calculating the ratio of the output reading with a sample in the cell to the reading obtained with the cell empty. The concentration can be calculated from the known absorptivity of the substance by means of Beer's law, or it may be obtained by comparison to known samples. Beer's law is stated as follows:

$$A = abc = \ln(I_o/I) = \log(1/T)$$

Where:

A = absorbance

a = molar absorptivity, 1/(mole)(cm)

b = path length, cm

c = concentration, moles/1

I_o = intensity of radiation striking detector with non-absorbing sample in light path

I = intensity of radiation striking detector with concentration c of absorbing sample in light path b

T = transmittance = I/I_o

The source must provide the desired ultraviolet wavelength and may be either a line source or a

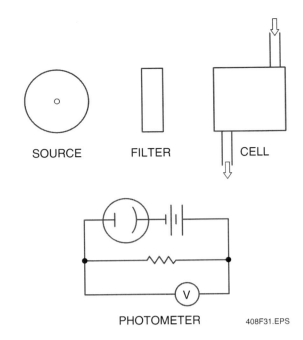

Figure 31 ◆ Simple ultraviolet analyzer block diagram.

continuous source, such as a hydrogen arc. Tungsten and tungsten-iodine lamps also may be used at longer ultraviolet wavelengths. Optical filters or spectral dispersing systems are used to screen out radiations of unwanted wavelengths emitted by the source. The sample cell has windows that are transparent at the chosen wavelengths. The path length between windows must be fixed.

17.1.1 Applications

Because of their high sensitivity, accuracy, reliability, and precision, UV absorption analyzers are suited to thousands of process stream analysis applications. UV absorption analyzers are particularly suited for the detection of halogens such as chlorine; benzene compounds; sulfur compounds; oxidizing agents such as hydrogen, peroxide, and ozone; and sulfur and other pollutants.

17.1.2 Calibration

UV absorption analyzers are calibrated in a similar manner to other gas and liquid analyzers. Two methods can be used—absorbance calibration based on the measured absorbance range of the analyzer and the known absorptivity of the sample, or chemical calibration based on analyzer readings of samples of known concentration. The absorbance method is easier than the chemical method for UV absorption analyzers; however, this method usually acts as a secondary standard with the chemical method as the primary standard.

17.1.3 Advantages

Compared with infrared analyzers, UV analyzers are generally more sensitive yet provide a comparable level of selectability. UV analyzers are less sensitive to temperature variations, are more applicable on liquid streams, and can monitor several important chemicals to which infrared analyzers are completely insensitive. On applications where both infrared and ultraviolet absorption analyzers are of equal sensitivity, such as on sulfur dioxide monitoring, the lack of ultraviolet absorption of commonly expected components from stack gas, such as moisture and carbon dioxide, is an important advantage.

17.2.0 Flame Detectors

In addition to chemical analyzers, UV detectors are also used in furnaces as flame detectors. UV flame detectors use a different type of cell than the device previously mentioned. The cell is a sealed tube containing an ultraviolet-sensitive gas and two electrodes that are connected in an AC circuit. When the gas is exposed to ultraviolet radiation from the flame, it becomes an electric conductor and can carry electrons from one electrode to the other. When the gas conducts the electrons, the AC circuit is completed and electricity flows through the circuit. This process occurs with abrupt starting and stopping of the current, which is known as the avalanche effect. Each avalanche, or period of electricity flow, is counted to provide a flame signal. When the number of counts exceeds a preset minimum, the presence of a flame is indicated rather than background noise from the electronic circuitry. Ultraviolet detectors are considered to be superior to infrared detectors in this area because they are not affected by the radiation emitted from hot surfaces such as the furnace refractory. *Figure 32* shows a typical ultraviolet flame detector.

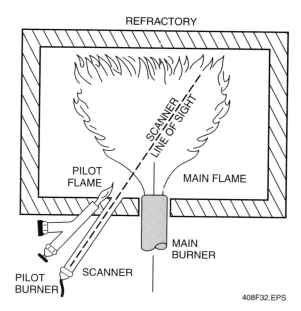

Figure 32 ◆ Ultraviolet flame detector.

Summary

Analytical instrumentation is sometimes considered a specialty trade within the instrumentation industry because of its complexity. A vast amount of knowledge and equipment are required in order to monitor, detect, and analyze the thousands of variables present in various industries.

However, many instrumentation technicians do not have a specialized crew to install, calibrate, and maintain the analyzer instruments within their responsibility. It is for these technicians that this module is presented.

This module introduced the various properties, chemicals, gases, and particles that require monitoring, detecting, and analyzing. These include density, specific gravity, viscosity, turbidity, flash point, oxidation-reduction potential, pH, conductivity, oxygen, carbon monoxide, carbon dioxide, hydrogen sulfide, total hydrocarbon content, and particulates. In many cases, detection and analysis are needed to protect workers from exposure to harmful substances. In other cases, the analysis is used to maintain product quality.

Analytical instrumentation opens up a field of expertise. Each application of analytical instrumentation must be thoroughly researched and learned before assuming the right equipment has been selected for a particular situation.

Review Questions

1. Most process analyzers operate under the _____ principle.
   a. theory of sampling
   b. laboratory
   c. degree of error
   d. span of instrument

2. Process analyzers can be more accurate than the laboratory analysis used to develop their calibration standard.
   a. True
   b. False

3. _____ is a word or phrase often used interchangeably with density and expressed as a ratio of density.
   a. Turbidity
   b. Conductivity
   c. Velocity
   d. Specific gravity

4. Using _____ is considered the simplest method for measuring density or specific gravity in processes.
   a. displacement
   b. a densitometer
   c. bubble tubes
   d. a nuclear detector

5. Because a displacer is in direct contact with the process liquid, _____ must be maintained in order to purge the unit with fresh liquid.
   a. a continuous flow
   b. the blow-down line pressure
   c. the back pressure
   d. a restricted outlet pressure

6. Density of a liquid can be measured using radiation from a _____.
   a. radioactive isotope
   b. hydrogen atom
   c. magnetic field
   d. laser beam

7. A fluid whose viscosity is constant at any shear rate is classified as _____.
   a. temperature-stable
   b. non-Newtonian
   c. Newtonian
   d. equilibrated

8. A falling ball is part of a type of _____.
   a. densitometer
   b. viscometer
   c. nuclear detector
   d. gravity tester

9. The Jackson turbidimeter uses _____ as a light source.
   a. a laser
   b. a strobe
   c. sunlight
   d. a candle

10. When testing for flash point, a blue halo constitutes a flash point.
    a. True
    b. False

11. A closed-cup method of flash point testing results in a lower flash point (in degrees) than an open-cup test.
    a. True
    b. False

12. _____ is defined as the negative logarithm of the hydrogen ion activity.
    a. ORP
    b. fusion
    c. pH
    d. hydrogen sulfide

13. The reciprocal of resistance is _____.
    a. pH
    b. conductivity
    c. ohm/cm
    d. the letter R

14. The primary element of a high-temperature electrochemical oxygen sensor is a _____ cell.
    a. mercury oxide
    b. uranium lattice
    c. magnetic field
    d. ceramic zirconium oxide

15. Oxygen has a very high magnetic susceptibility compared to other gases.
    a. True
    b. False

16. Carbon monoxide is a fuel used in combustion.
    a. True
    b. False

17. At a temperature of –110°F, at atmospheric pressure, carbon dioxide _____.
    a. turns to liquid
    b. turns to gas
    c. solidifies
    d. forms carbon and oxygen

18. In low concentrations, hydrogen sulfide gas smells like _____.
    a. sweet crude oil
    b. rotten eggs
    c. ammonia
    d. citric fruit

19. Volatile organic materials that make up the THC contain at least one hydrogen atom and at least one atom of _____.
    a. oxygen
    b. nitrogen
    c. helium
    d. carbon

20. The most common THC analyzer operates on the principle of _____.
    a. displacement
    b. infrared absorption
    c. FID
    d. ORP

21. Clean rooms are typically large users of energy.
    a. True
    b. False

22. Gas chromatography is a technique commonly used to separate and analyze _____ organic compounds.
    a. inert
    b. volatile
    c. solid
    d. non-volatile

23. Infrared radiation is visible in the dark to the human eye.
    a. True
    b. False

24. The UV spectrophotometer is used to determine wavelength.
    a. True
    b. False

25. The primary standard method of calibration for UV absorption analyzers is _____ calibration.
    a. absorbance
    b. transmittance
    c. electronic
    d. chemical

Trade Terms Introduced in This Module

Absolute zero: The temperature at which substances have no thermal energy. Generally accepted as –273.15°C, or –459.67°F; it is 0° on the Kelvin scale.

Absorbance: The ratio of absorbed to incident radiation.

Acidity: The property of a sample that has a pH less than 7.0.

Alkalinity: The property of a sample that has a pH greater than 7.0.

Anaerobic decay: Organism decay that is not dependent upon oxygen for occurrence.

Aqueous solution: Any solution in which water is the solvent.

Buffer solution: A solution that is resistant to changes in pH that contains either a weak acid and its salt or a weak base and its salt.

Butanol: An alcohol with the chemical formula $CH_3(CH_2)_3OH$. It is formed during anaerobic fermentation, using bacteria to convert the sugars to butanol and carbon dioxide.

Chromatography: A technique used to separate components in a sample mixture by the differential distribution of the components between a mobile phase and a stationary phase.

Collimated: Directed in a straight line.

Coulometric: The principle by which an analytical sensor functions by receiving a measured amount of current from a constant current source, often generated by internal reaction.

Diffraction: The modification of light or deflection of its rays.

Dynamic viscosity: Equal to density times kinematic viscosity. For the laminar flow of a fluid, it is the ratio of the shear stress to the velocity gradient perpendicular to the plane of the shear.

Electrolyte: Liquid that has the properties required to conduct electricity.

Emissivity: The ratio of the radiation emitted by a surface to that emitted by a black body at the same temperature.

Greenhouse effect: A natural warming process of the earth.

Hemoglobin: A red pigment in red blood cells that can bind with oxygen and is largely responsible for the blood's oxygen-carrying capacity.

Inert: A substance that is not reactive.

In-situ: An on-line or in-line analyzer that is in direct contact with the process, such as a flue gas oxygen analyzer.

Isothermal: Set at one temperature; unchangeable.

Kinematic viscosity: Determined by measuring the time required for a fixed amount of liquid to flow through a capillary tube under the force of gravity.

Laboratory principle: Comparison of a sample to a standard.

Logarithm: A mathematical term that indicates the power to which a number must be raised to arrive at a given number.

OSHA ceiling: Maximum concentration for a short period (usually between 5 and 30 minutes, but each gas is different). Four such exposures are usually allowed per day, and average exposures must still be within the time-weighted average (TWA) or permissible exposure limit. This is the cumulative average concentration over an 8-hour day, 40-hour week to which a worker can safely be exposed.

Oxidation: A reaction in which the atoms in an element lose electrons and the valence of the element is correspondingly increased.

Paramagnetic: Relating to or being a substance in which an induced magnetic field is parallel and proportional to the intensity of the magnetizing field but is much weaker than in ferromagnetic materials.

Reciprocal: The inverse of a number; for example, the reciprocal of 5 is 1/5.

Reduction: The absorption of electrons by one molecule.

Reflectivity: The ratio of the intensity of the total radiation reflected from a surface to the incident radiation.

Refraction: An oblique direction given to a ray of light passing into a different medium or through a medium of varying density.

Repeatability: The ability of an instrument to repeatedly produce the same measurement signal from the same sample. The measurement of repeatability is typically specified as the expected deviation in the signal—for example, a repeatability of 1 part in 100,000 or 1:100,000.

Rotameter: A vertically oriented tapered glass tube with the large end at the top and a metering float that is free to move within the tube. Fluid flow causes the float to begin to rise in the tube as the upward pressure differential and buoyancy of the fluid overcome the effect of gravity.

Sequestered: Emissions that are absorbed or taken out of the atmosphere and stored in a terrestrial or oceanic reservoir.

Shearing force: Structural strain in a substance or body caused by parallel layers shifting against one another in opposite directions.

Siemens/cm: Reciprocal of resistance, and it is the unit of measurement for conductivity, replacing the former unit, mho/cm.

Spectrometry: The science dealing with analysis of the optical spectrum of an object in order to determine the chemical makeup.

SUS: Saybolt universal seconds as determined by the Standard Method of Test for Saybolt Viscosity (*ASTM D88-56*). It may be determined by use of the SUS conversion tables specified in *ASTM Method D2161-66* following determination of viscosity in accordance with the procedures specified in the Standard Method of Test for Viscosity of Transparent and Opaque Liquids (*ASTM D445-65*).

Transmittance: The ratio of the radiant energy transmitted to the total radiant energy incident on a given body.

Yttrium: A malleable silvery metal element that belongs to a group of high-quality, rare-earth elements. A mixture of yttrium and europium (another rare-earth element) oxides is widely used as the red phosphor on television screens. In its oxide form, it is used in conjunction with zirconium oxide in oxygen analyzer sensing cells.

Additional Resources

This module is intended to be a thorough resource for task training. The following reference works are suggested for further study. These are optional materials for continued education rather than for task training.

www.isa.org. The website of the Instrument Society of America.

The Condensed Handbook of Measurement and Control, 1997. N.E. Battikha. Research Triangle Park, NC: Instrument Society of America.

Industrial Pressure, Level & Density Measurement, 1995. Bela G. Liptak. Boca Raton, FL: CRC Press.

Instrumentation Reference Book, 2002. Walter Boyes. 3rd ed. Research Triangle Park, NC: Instrument Society of America.

Measurement & Instrumentation Principles, 2001. Alan S. Morris. Boston: Butterworth-Heinemann.

Figure Credits

Beckman Coulter, Inc.	408F15, 408F16
Bacharach, Inc.	408F22
American Gas & Chemical Company, Ltd.	408F23
TSI, Inc.	408F25

NCCER CURRICULA — USER UPDATE

NCCER makes every effort to keep its textbooks up-to-date and free of technical errors. We appreciate your help in this process. If you find an error, a typographical mistake, or an inaccuracy in NCCER's curricula, please fill out this form (or a photocopy), or complete the online form at **www.nccer.org/olf**. Be sure to include the exact module ID number, page number, a detailed description, and your recommended correction. Your input will be brought to the attention of the Authoring Team. Thank you for your assistance.

Instructors – If you have an idea for improving this textbook, or have found that additional materials were necessary to teach this module effectively, please let us know so that we may present your suggestions to the Authoring Team.

NCCER Product Development and Revision
13614 Progress Blvd., Alachua, FL 32615

Email: curriculum@nccer.org
Online: www.nccer.org/olf

❏ Trainee Guide   ❏ AIG   ❏ Exam   ❏ PowerPoints   Other _____

Craft / Level: _____   Copyright Date: _____

Module ID Number / Title: _____

Section Number(s): _____

Description: _____

_____

_____

_____

_____

Recommended Correction: _____

_____

_____

_____

_____

Your Name: _____

Address: _____

_____

Email: _____   Phone: _____

Index

Bit(s)
  and binary codes, 6.8, 6.11–6.12
  definition, 6.8, 6.33
  least significant, 1.21, 6.8
  most significant, 1.21, 6.8
  overflow, 6.24
  and PLC memory, 6.18–6.19
  stop, 6.16, 6.33
Bit distribute (BTD), 6.25, 6.26
Black body, 8.25
Blowdown, 3.3
Blow-out, 3.2
Boolean logic and programming language, 1.2, 1.5, 6.20, 6.21
BTD. *See* Bit distribute
Bubbler, 3.7, 8.2
Buffer, of a fiber-optic cable, 3.7, 3.15
Buffering, 1.23
Buffer solution, 8.12, 8.33
Bus topology, 7.9
Butanol, 8.22, 8.33

Cable, fiber-optic, 3.3, 3.7–3.8
Calculators, electronic, 1.13, 1.38, 1.42
Calibration
  analog, 2.1, 2.9–2.15
  of analyzers, 8.1
  bench, 2.18
  definition, 2.1, 2.2, 2.25
  of differential pressure transmitters, 2.5–2.8
  of discrete particle counters, 8.22
  five-point method for, 2.3–2.4, 2.25
  of HART® devices, 2.16–2.18, 3.12
  improper, 2.18, 3.12
  of inductive conductivity cells, 8.15
  isolation of control loop during, 2.6
  of a loop, 3.10–3.13
  of oxidation-reduction probe, 8.11
  of pH-sensitive electrodes, 8.13
  pneumatic, 2.1, 2.4–2.9
  process, 2.2–2.3, 2.23
  reiterative nature of, 2.3
  schedule, 7.16
  test equipment for, 2.1–2.2, 2.3, 2.4–2.18
  three-point method for, 2.3, 2.4, 2.25
  three signal groups for, 2.1
  using smart instruments, 2.1, 2.15–2.18, 2.23, 3.12
  UV spectrometers, 8.28–8.29
  of valve positioners, 2.21–2.22
  vs. proving, 3.8
  vs. tuning, 2.4, 3.10
Capillary system, 2.9
Carbon dioxide, 8.18–8.19
Carbon monoxide, 8.17–8.18
Cascade control, 5.16–5.17
Cathode ray tube (CRT), 8.5
Cavitation, 4.5
Chemical components, 8.22–8.23
Chromatography, 8.1, 8.22–8.23, 8.33
Chronic, 4.2, 4.15
Cladding, of fiber-optic cable, 3.7, 3.15
Clean room, 8.21
Clear signal, 1.14
Client, 7.11, 7.12, 7.19
Clocked RS latch, 1.13–1.14
Clock signal
  in binary coded decimal counter, 1.37
  in data latch, 1.14, 1.15
  in D flip-flop, 1.15

  in down counter, 1.33–1.34
  in four-bit binary counter, 1.31
  in JK flip-flop, 1.16, 1.18
  in Johnson counter, 1.40
  in ring counter, 1.39
  in shift register, 1.19–1.20
  in up counter, 1.32–1.33
  in up/down counter, 1.34–1.36
Code
  ASCII, 6.8, 6.11, 6.16
  binary, 1.32, 6.11–6.12. *See also* Binary coded decimal
  Gray, 6.12
  illegal, 1.42
  weighted, 1.21, 1.30
Collimated, 8.22, 8.33
Colorimeter, 8.6–8.7
Combination logic, 1.10, 1.47
Commissioning a loop, 3.1, 3.15, 4.10–4.11
Communication
  between control boxes. *See* Data highway
  data transfer between devices. *See* Conversions
  HART® protocol, 2.15–2.16, 6.17
  parameters for data transfer, 6.16
  protocol standards, 7.1, 7.10–7.11, 7.12
  remote, 7.12–7.14
  with the system operator, 4.2, 5.15
Complete response (CR), 5.4–5.5
Computers. *See also* Microprocessor
  central, 7.4, 7.6
  circuits, 1.42
  communication between. *See* Communication, remote; Data highway; Local area network; Protocol standards
  and Hart® devices, 2.16
  historical background, 7.1, 7.2
  inspection, 3.3
  mainframe, 7.12
  to monitor valve controllers, 2.22
Condensate, 2.11, 3.3
Condensation particle counter (CPC), 8.22, 8.23
Conductivity of a liquid, 8.13–8.15
Conduit, 3.2, 3.3, 4.10
Construction Specification Institute (CSI), 4.11
Contaminants, 3.3
Contention, 7.11
Continuity test, 3.4–3.8, 3.15
Control box, 7.6, 7.7, 7.8, 7.12
Controller
  definition, 7.19
  digital, 1.5, 1.6, 2.22, 3.4
  and the field transmitter, 3.2
  historical background, 7.1
  proportional, 5.2
  temperature, 2.14
  troubleshooting, 4.4
  tuning the, 2.4, 5.1
  used in distributed control system, 7.2
Control room
  centralization with DCS, 7.2, 7.3–7.4
  diagrams and drawings, 4.10
  inspection of components, 3.3
  panel, 4.7, 4.8, 4.10, 4.11
Control systems, distributed. *See* Distributed control systems
Conversions
  AC/DC, 6.13–6.14
  analog to/from digital, 2.14, 2.16–2.17, 3.11, 6.15–6.17
  analog to pneumatic, 4.7
  within Hart® devices, 2.16–2.17, 3.11–3.12
  temperature to millivolt, 2.14

to/from BCD, 6.12
to/from binary numbering system, 6.9
to/from octal numbering system, 6.10–6.11
within transducers, 2.18–2.19
Cooking temperature, 2.14
Core, of fiber-optic cable, 3.7, 3.15
Coulometric, 8.17, 8.33
Counters
binary coded decimal, 1.37–1.38, 1.42, 1.43, 1.44
down, 1.33–1.34, 1.44, 6.24
four-bit binary, 1.31–1.32
instructions in a PLC, 6.22–6.24
in I/O modules, 6.17
Johnson, 1.39–1.40, 1.44
overview, 1.18, 1.30
particle, 8.22, 8.23
programmable, 1.40–1.42, 1.44
ring, 1.38–1.39, 1.44
ripple (asynchronous), 1.30, 1.32–1.33
ripple carry (synchronous), 1.30, 1.37, 1.44
switch-tail. See Counters, Johnson
transition from shift register, 1.38
up, 1.32–1.33, 1.44, 6.24
up/down, 1.34–1.36, 1.44
Counter/timer done (DN), 6.23, 6.24
Coupler, opto (optical), 6.6
CPC. See Condensation particle counter
CR. See Complete response
Critical frequency, 5.1, 5.8
CRT. See Cathode ray tube
CSI. See Construction Specification Institute
CTD. See Counters, down
CTU. See Counters, up
Current loop trim, 2.17, 2.18, 3.12

Dampened-oscillation method, 5.10, 5.19
Damping, 2.18
Data
comparison, 6.25
false or erroneous, 1.38, 1.42–1.43
from HART® devices, 2.16
loading, 1.20–1.21
manipulation. See Conversions
movements in a shift register, 1.19, 1.20
numerical, in I/O modules, 6.15
parallel, 1.21, 1.44
parameters for transfer between devices, 1.23, 6.16
serial, 1.20, 1.44
storage
and counters, 1.30
in PLC, 6.19
by shift registers, 1.18, 1.20, 1.28, 1.44
in storage bits, 6.18–6.19
transfer instructions in ladder diagram, 6.25–6.26
Data field, 7.10, 7.19
Data highway, 7.6, 7.7, 7.8, 7.19
Data table, 6.18
DCS. See Distributed control systems
Deadtime, 5.3, 5.11, 5.12, 5.13, 5.16
Decay ratio, 5.10, 5.19
Decimal numbering system, 6.8, 6.33. See also Binary coded decimal
Decoders, 1.42–1.43
Decrement, 6.24, 6.33
Densitometer, 8.3
Density, 8.2–8.3
Derivative, 5.2, 5.19

Derivative action, 5.16
Derivative time, 5.6–5.7
Deutsches Institut für Normung. See DIN
D flip-flop, 1.15, 1.19
Differential pressure transmitter (DP). See Transmitter, differential pressure
Differentiator, 5.6, 5.19
Diffraction, 8.22, 8.33
Digital logic circuits
ANSI/ASQC standards, 1.43–144
arithmetic elements, 1.42
counters. See Counters
decoders, 1.42–1.43
flip-flops (latches), 1.10–1.18
number systems used in, 6.8–6.12
overview. See Gates
shift registers. See Shift registers
Digital signals
in control valve positioners, 2.22
conversion to/from analog, 2.14, 2.16–2.17, 3.11, 6.15–6.17
definition, 2.9, 2.25
DIN, 6.2, 6.33, 8.9
Discrete, 6.1, 6.33, 7.2, 7.19
Discrete particle counter (DPC), 8.21–8.22
Displacer, 8.2–8.3
Distributed control systems (DCS)
definition, 7.2, 7.19
evolution of, 7.2–7.8
with human interface, 6.2, 6.6, 7.8–7.14, 7.15
maintenance, 7.14–7.17
for manufacturing processes, 7.1–7.2
overview, 7.1
Dividers, 1.18
D latch, 1.14
DN. See Counter/timer done
Documentation
for commissioning a loop, 4.10–4.11
improvement with DCS, 7.2, 7.4
maintenance history files, 7.15–7.16
troubleshooting a loop, 4.2
Downtime, 7.14, 7.16
DP. See Transmitter, differential pressure
DPC. See Discrete particle counter
Drivers. See Decoders
Dry leg, 2.12, 2.25
Dry process, 7.2, 7.19

Edge triggered, 1.15, 1.16, 1.28, 1.47
Electrochemical sensors, 8.20
Electrodes, 8.12–8.13, 8.14, 8.17, 8.20
Electrolyte, 8.12, 8.17, 8.20, 8.33
Electromagnetic spectrum, 8.24
Element
arithmetic, 1.42
control, 2.14, 5.2
final, 7.2, 7.19
memory, 1.10, 1.11
primary, 3.2, 7.2, 7.3, 7.19
typical, 7.3
Emissions, 8.18–8.19, 8.20, 8.28
Emissivity, 8.25, 8.33
Empirical equations, 5.9, 5.19
Encoder, 6.17
Energy balance, 5.3–5.4
English statement programming language, 6.20, 6.21
Ethernet, 7.12

Factory characterization, 2.15, 2.25
Feedback, 1.11, 2.14, 2.22
Fiber-optics. *See* Cable, fiber-optic
FID. *See* Flame ionizer detector
Fieldbus, 2.1, 7.7, 7.19
Field process area, 4.7
Fire point, 8.8
First-order lag plus deadtime (FOLPDT), 5.2
Fisher® valve controller, 2.22
Fisher® valve positioners, 2.20, 2.21–2.22
Flame detector, UV, 8.29
Flame ionizer detector (FID), 8.20–8.21
Flash point, 8.8–8.9
Flip-flops. *See also* Latches
 D, 1.15, 1.19
 definition, 1.10, 1.44, 1.47
 JK. *See* JK flip-flop
 race condition with, 1.12–1.13
 relationship to shift register, 1.18–1.19
 T, 1.18
 three basic control signals, 1.13
 trailing edge triggered, 1.28
Flow, 2.7, 2.10, 5.4
Flowchart, troubleshooting, 4.5–4.6
Flow meters, magnetic, 2.16, 2.19
Fluid, Newtonian, 8.5
Fluke® calibrators, 2.9–2.10, 2.14–2.15, 3.5, 3.9, 3.10
Fluke® pressure modules, 3.9, 3.10
Fluke® 789 ProcessMeter™, 3.5
FOLPDT. *See* First-order lag plus deadtime
Food processing, 2.14
Force bar, 2.5, 2.6
Foxboro® pneumatic temperature transmitter, 2.9
Frequency shift key technology (FSK), 2.16
Fulcrum, 2.5, 2.6, 2.25
Full adder, 1.42, 1.43
Functional block programming language, 6.20, 6.21
Furnace burner, 1.5, 1.6
Fuse, 6.14

Gain of process, 5.1, 5.5, 5.14, 5.15–5.16, 5.19
Galvanic sensors, 8.17
Gas analyzer, 2.16
Gas chromatography (GC), 8.22–8.23
Gates
 AND. *See* AND gate
 NAND, 1.5, 1.7–1.8, 1.26–1.27, 1.47
 NOR, 1.8, 1.13, 1.28, 1.30, 1.47
 OR. *See* OR gate
 in a series, additive propagation delay in, 1.2
 three basic control signals, 1.13
 XOR, 1.8–1.9, 1.42, 1.47
GC. *See* Gas chromatography
Greenhouse effect, 8.19, 8.33
Grounding, 3.3, 3.15, 6.29

Half adder, 1.42
Hardwired systems, 6.2, 6.6
HART® communication devices
 in calibrating a loop, 3.10–3.13
 connection to transmitter, 3.12–3.13, 6.17
 and digital valve controller, 2.22
 in loop continuity test, 3.5
 overview, 2.16–2.18, 2.23, 3.15
HART® Communication Foundation, 2.15
HART® communications protocol, 2.15, 2.16, 2.22, 2.23, 6.17
Helium, 2.9

Hemoglobin, 8.17
Hexadecimal numbering system, 6.11, 6.33
High-temperature electrochemical sensor, 8.15–8.16
Highway addressable remote transducer. *See* Hart® *entries*
Historians, 7.11, 7.19
HMI. *See* Human/machine interface
Human/machine interface (HMI), 6.2, 6.6, 7.8–7.14, 7.15
Hydrogen sulfide, 8.19–8.20

IEEE. *See* Institute of Electrical and Electronics Engineers
Increment, 6.24, 6.33
Incremental changes, 5.14–5.15
Indicators, 2.2
Inert, 8.18, 8.33
Infrared radiation, 8.24–8.27
Input
 auto-step or auto-ramp, 2.10, 2.14, 3.9
 display on Fluke® calibrators, 3.9
 energies (measured variables), 2.2
 to Fisher® valve positioners, 2.20, 2.21
 to Hart® devices, 2.17, 2.18
 parallel, 1.30, 1.44
 preset, 1.40, 1.41
 in proportional control, 5.2, 5.5–5.6
 relationship to output states, 1.2
 reset, 1.11
 role in calibration, 2.1–2.2
 set, 1.11
 shift left, 1.28, 1.29
 shift right, 1.28, 1.30
 to a transducer, 4.9
 true, 6.23
 zero, 22
Input/output systems. *See* I/O modules
Input table, 6.18
In-situ, 8.19
Inspections, 1.44, 3.1–3.3, 3.13, 4.2, 8.26
Instability, apparent, 5.15
Installation
 digital logic circuits, 1.43–1.44
 dry leg, 2.12
 flow transmitter, 2.10
 loop. *See* Commissioning a loop; Proving a loop
 mechanical, verification of, 3.1–3.3
 PLC, 6.27–6.30
 wet leg, 2.12
Institute of Electrical and Electronics Engineers (IEEE), 7.10
Instrumentation
 calibration of. *See* Calibration
 drawing of location, 4.10
 hardware, 4.11. *See also* Programmable logic controller, hardware
 index of tag numbers, 4.11
 loop checks, 3.1–3.13
 maintenance and repair, 7.14
 manufacturer's literature on, 7.17
 operating range, 2.2
 pneumatic, continuity tests for, 3.6–3.7
 range *vs.* span, 2.2
 troubleshooting in the field, 4.4–4.5
 tuning, 2.4
 zeroing, 2.2, 2.4, 2.18, 2.21, 2.25, 3.12
Integral, 5.2, 5.19
Integral time, 5.6, 5.16
Integrator, 5.6, 5.19
International Standards Organization (ISO), 1.44
Internet, 7.12, 7.14, 7.15

Inverter, 1.5, 1.6, 1.47
I/O modules
    communication with HART® devices, 2.16
    five classes of, 6.13–6.17
    overview, 6.1, 6.2, 6.3–6.5, 7.5
    wiring, 6.29–6.30
Isothermal, 8.22, 8.33

JK flip-flop
    in counters
        binary, 1.31, 1.32
        binary coded decimal, 1.37–1.38
        down, 1.34
        Johnson, 1.39, 1.40
        ripple carry, 1.37
        up, 1.32, 1.33
    overview, 1.15, 1.16–1.18
    in parallel in–serial out shift register, 1.26, 1.27
    in ring, 1.38
JMP. *See* Jump
JSR. *See* Jump to subroutine
Jump (JMP), 6.26
Jump to subroutine (JSR), 6.26, 6.27

Label (LBL), 6.26–6.27
Laboratory analyses. *See* Analyzers
Laboratory principle, 8.1, 8.33
LAD. *See* Ladder diagram programming language
Ladder diagram programming language (LAD), 6.2, 6.19, 6.20, 6.21–6.27, 6.28
Lag, 5.3
LAN. *See* Local area network
Laser light, 8.22
Latches. *See also* Flip-flops
    clocked RS, 1.13–1.14
    data (D), 1.14
    in ladder diagram instructions, 6.22, 6.23
    RS, 1.10–1.11, 1.17
    RS NAND, 1.12–1.13
    RS NOR, 1.11–1.12
    slave, 1.17
LBL. *See* Label
Least significant bit (LSB), 1.21, 6.8
Level transmitter (LT), 2.11–2.13
Light, backscattered, 3.8, 3.15
Light spectrum, 8.27
Liquid analyzer, 2.16
Listening skills, 4.2
Local area network (LAN), 7.8–7.10, 7.11–7.12, 7.19
Lockout and tagout procedures, 3.4, 3.5, 4.8
Logarithm, 8.12, 8.33
Logs, 7.2
Loop
    calibration of a, 3.10–3.13
    commissioning a, 3.1, 3.15, 4.10–4.11
    continuity tests, 3.4–3.8
    control, 7.2, 7.3, 7.19
    conventional 4–20mA instrument, 3.10
    current loop trim, 2.17, 2.18, 3.12
    ground, 3.3, 3.15
    history file of a, 7.16
    multi-function, 2.2, 2.4, 2.25, 3.9, 7.4, 7.5
    proving a, 3.1, 3.8–3.10, 3.13, 3.15, 4.7–4.10, 4.15
    troubleshooting a, 4.1–4.3, 4.8–4.10
    tuning a
        basic equations, 5.3–5.8
        closed loop methods, 5.8–5.10
        open loop methods, 5.10–5.14

        overview, 5.1, 5.17
        review of proportional control, 5.2–5.3
        terms and definitions, 5.3
        visual, 5.14–5.17
    uncommissioned, 4.7, 4.15
Loop diagram, 4.2, 4.3, 4.4, 4.5, 4.11
Loop sheet, 3.4, 3.8, 4.7–4.8
LSB. *See* Least significant bit
LT. *See* Level transmitter

Machine stage programming language, 6.20, 6.21
Machine tool station, 1.38
Magnetic field, 8.14, 8.16
Maintenance and repair
    acquiring expertise for, 7.16–7.17
    diagnostic alerts, 2.16
    digital logic circuits, 1.43–1.44
    distributed control systems, 7.14–7.17
    electrical equipment and circuits, 8.26
    fiber-optic cable, 3.3
    instrumentation, 7.14–7.16
    on-site, 7.16
    oxidation-reduction probe, 8.11
    record keeping on, 4.2
    use of default values during, 7.6
Malfunction, 1.38
Manual mode, 4.4, 5.10, 5.15
Manufacturing processes, 2.14, 7.1–7.2, 8.15, 8.18–8.19
Masked move (MVM), 6.25, 6.26
Master control relay (MCR), 6.26, 6.27
Master-slave principle, 1.17–1.18
Material safety data sheet (MSDS), 8.8, 8.9
MCR. *See* Master control relay
Memory capacity
    and counters, 1.30
    in PLCs, 6.18–6.19
    random access memory, 1.42, 6.16
Memory device. *See* Shift registers
Memory elements, 1.10, 1.11. *See also* Flip-flops; Latches
Metal oxide semiconductor field-effect transistor. *See* MOSFET devices
Mho/cm, 8.13, 8.14
Microprocessor(s)
    control systems based on, 7.1
    within the DCS, 7.5
    definition, 7.19
    within Hart® devices, 2.16, 2.17, 3.11–3.12
    within the PLC, 6.1, 6.16, 6.17–6.19, 7.5
Microscopy, optical, 8.21
Modulus, 1.30, 1.40, 1.41–1.42
Monitoring. *See also* Analyzers
    air quality, 8.19, 8.21
    central, 7.3–7.4, 7.6
    storage of information by historians, 7.11
MOSFET devices, 1.18, 1.47
Most significant bit (MSB), 1.21, 6.8
MOV. *See* Move
Move (MOV), 6.25, 6.26
MSB. *See* Most significant bit
MSDS. *See* Material safety data sheet
Multi-drop topology, 7.9
MVM. *See* Masked move

NAND gate, 1.5, 1.7–1.8, 1.26–1.27, 1.47. *See also* RS NAND latch
Nanosecond, 1.2, 1.47
National Bureau of Standards, 7.16
National Electrical Code®, 6.29

National Fire Protection Association (NFPA), 1.5
Natural gas, 8.20
NC. *See* Normally closed
NEC®. *See National Electrical Code*®
Negative-going edge, 1.2, 1.16, 1.47
NFPA. *See* National Fire Protection Association
NMHC. *See* Non-methane hydrocarbons
NO. *See* Normally open
Noise, 1.38, 3.3, 6.14, 6.15
Non-methane hydrocarbons (NMHC), 8.20
NOR gate, 1.8, 1.13, 1.28, 1.30, 1.47. *See also* RS NOR latch
Normally closed (NC), 6.21–6.22, 6.23, 6.24
Normally open (NO), 6.21–6.22, 6.23, 6.24
NOT AND gate. *See* NAND gate
NOT function, 1.5. *See also* Inverter
Nuclear detector, 8.3
Number sequence. *See* Counters
Number systems used in digital operation, 6.8–6.12

Occupational Safety and Health Administration (OSHA), 8.9, 8.19
Octal numbering system, 6.9–6.11, 6.33
Ohm's law, 8.14
One-quarter dampened wave, 5.9, 5.19
OPC. *See* Optical particle counter
Operating system, 7.11, 7.19
Operators
    communication with, 4.2, 5.15
    responsibilities, 7.2, 7.11, 7.16
Optical particle counter (OPC), 8.22
Optical time domain reflectometer (OTDR), 3.8
OR gate
    application to furnace burner, 1.5, 1.6
    in a combination logic circuit, 1.10
    exclusive, 1.8–1.9
    memory capability, 1.10
    overview, 1.4–1.5, 1.6, 1.47
    relationship to NAND gate, 1.7
    relationship to NOR gate, 1.8
    in up/down counters, 1.34
Orifice plate, 3.2, 3.8
ORP. *See* Oxidation-reduction potential
Oscillation
    definition, 5.19
    due to a sticking valve, 5.17
    and loop tuning, 5.8, 5.9, 5.10, 5.15, 5.16
    and speed of system response, 5.1
    troubleshooting, 4.3–4.6
Oscilloscope, 1.4
OSHA. *See* Occupational Safety and Health Administration
OSHA ceiling, 8.19
OTDR. *See* Optical time domain reflectometer
Output
    derivative, 5.6–5.7
    display on Fluke® calibrators, 3.9
    energies (signals), 2.2–2.3
    fixed, for test mode, 2.16, 2.17, 3.12
    within Hart® devices, 2.17, 2.18
    PID, 5.8
    proportional, 5.2, 5.19
    relationship to input, 1.2
    role in calibration, 2.1–2.2
    serial or parallel, 1.20, 1.21, 1.44
    with synchronous counters, 1.37
    time-proportioned, 6.17
    true, 6.21, 6.22, 6.23
Output table, 6.18
OV. *See* Overflow bit

Overflow bit (OV), 6.24
Over the hump, 5.16, 5.19
Oxidation, 8.10, 8.33
Oxidation-reduction potential (ORP), 8.10–8.11, 8.17
Oxygen, 8.15–8.17, 8.21

P. *See* Proportional control
Panel drawings, 4.3, 4.11
Panel junction zone, 4.7, 4.8
Parallel counters, 1.30
Parallel enable line (PE), 1.27
Parallel loading of data, 1.20–1.21
Paramagnetic, 8.34
Paramagnetic oxygen analyzer, 8.16–8.17
Parity, 6.11, 6.33
Particulates, 8.21–8.22
PE. *See* Parallel enable line
Personnel. *See* Operators; Technicians
pH, 8.12–8.13
Phase shift, 5.1, 5.8, 5.15
Photocell, 8.7, 8.8, 8.16
PI. *See* Proportional/integral control
PID. *See* Proportional/integral/derivative control
P&ID. *See* Piping and installation drawing
Piping and installation drawing (P&ID), 3.2, 3.8, 4.2, 4.5, 4.10
Plant management, 7.2
PLC. *See* Programmable logic controller
Pneumatic instrumentation
    continuity tests for, 3.6–3.7
    control valve positioners, 2.20–2.22
    differential pressure transmitters, 2.5–2.8
    temperature transmitters, 2.9
    tubing, 3.2, 3.3, 3.7
Pneumatic signal
    for calibration, 2.1, 2.4–2.9
    in control valve positioners, 2.20–2.22
    conversion of analog signal to, 4.7
    conversion of voltage signal to, 2.18, 2.19, 2.22
    in differential pressure transmitters, 2.5–2.8
    in temperature transmitters, 2.9
Polling, 7.11
Positioners, control valve, 2.1, 2.19–2.22
Positive-going edge, 1.2, 1.47
Power supply
    for a loop during troubleshooting/proving, 4.7, 4.8–4.9
    for a PLC, 6.12–6.13
    relationship to control loop, 4.9
    for test equipment, 3.9
    for a Wally Box®, 2.4
Preset signal, 1.14, 1.40, 1.41
Pressure gauge, 2.2
Pressure head, hydrostatic, 2.11, 2.25
Pressure module on test equipment, 3.9
Pressure transmitter. *See* Transmitter, differential pressure
Probes, 8.11, 8.14–8.15
Process gain, 5.1, 5.5
Processor, in PLC, 6.17–6.19
Program, definition, 7.5, 7.19
Programmable counter, 1.40–1.42, 1.44
Programmable logic controller (PLC)
    and the field transmitter, 3.2
    hardware, 6.10–6.17, 6.27
    hardwired and PLC systems, 6.2–6.8
    introduction, 6.1–6.2, 7.4–7.5
    isolation during loop continuity test, 3.4
    numbered systems used in, 6.8–6.12
    processors, 6.17–6.19, 7.5

programming and installation, 6.27–6.30
in proving a loop, 3.8
with smart instruments, 2.1
software, 6.2, 6.18, 6.19–6.27
Programmable read only memory (PROM), 6.18, 6.33
Programming languages, 6.19–6.21
PROM. *See* Programmable read only memory
Propagation delay, 1.2–1.3, 1.23, 1.37, 1.47
Proportional band, 5.5–5.6
Proportional control (P), 5.2
Proportional/integral control (PI), 5.2
Proportional/integral/derivative control (PID), 5.2–5.3, 5.7–5.8, 5.19, 6.17
Protocol standards, 7.1, 7.10–7.11, 7.12
Proving a loop, 3.1, 3.8–3.10, 3.13, 3.15, 4.7–4.10, 4.15

Quality assurance/quality control, 1.43–1.44, 7.2

Race condition, 1.12–1.13, 1.14, 1.16, 1.47
Radiation
    electromagnetic, 8.24, 8.25
    infrared, 8.24–8.27
    light spectrum, 8.27
    nuclear, 8.3
    ultraviolet, 8.27–8.29
Radioactivity, 8.3
RAM. *See* Random access memory
Random access memory (RAM), 1.42, 6.16, 6.33
Range, of an instrument, 2.2, 2.18, 3.12
Rate control, 5.2
Ratio turbidity analyzer, 8.7–8.8
Reaction rate method, 5.13–5.14, 5.19
Read only memory (ROM), 6.18, 6.33
Reciprocal, 5.2, 5.19, 8.13, 8.34
Record keeping. *See* Documentation
Reduction, 8.10, 8.34
Redundancy, 7.3, 7.4, 7.7–7.8, 7.16
Reflection turbidity analyzer, 8.7
Reflectivity, 8.25, 8.34
Refraction, 8.22
Refrigeration system, 2.14
Relay instructions, 6.21–6.22, 6.23
Remote I/O module, 6.17
Remote server unit (RSU), 7.14
Repeatability, 8.1, 8.34
Repeats per minute (RPM), 5.6, 5.19
Reset
    latches, 1.10–1.13, 1.16, 1.18
    in loop tuning, 5.2, 5.9, 5.10, 5.12, 5.13, 5.14
Resistance, 3.3, 3.5, 3.12, 4.9, 8.13, 8.14
Resistance temperature detector (RTD), 2.14, 3.2, 3.9
Response, 5.1, 5.3, 5.15–5.17
RET. *See* Return
Retentive time on (RTO), 6.23
Return (RET), 6.27
Ring topology, 7.8, 7.10
ROM. *See* Read only memory
Rosemount transmitters, 2.10–2.11, 2.13–2.14, 3.11
Rotameter, 8.2, 8.11, 8.34
RPM. *See* Repeats per minute
RS latch, 1.10–1.11, 1.16
RS NAND latch, 1.12–1.13
RS NOR latch, 1.11–1.12, 1.18
RSU. *See* Remote server unit
RTD. *See* Resistance temperature detector
RTO. *See* Retentive time on

Safety
    carbon monoxide, 8.17
    fiber-optic sources, 3.7
    with flash point tester, 8.9
    Fluke® pressure modules, 3.10
    lockout and tagout procedures, 3.4, 3.5, 4.8
    during maintenance or repair, 7.16
    operating range limit, 2.2
    personnel protection indicators, 8.19–8.20
    PLC installation, 6.29
    troubleshooting a loop, 4.3
Saybolt universal seconds (SUS), 8.9, 8.34
SBR. *See* Subroutine
SCADA. *See* Supervisory control and data acquisition
Scan, 6.16, 6.18, 6.25, 6.33
SCR. *See* Silicon-controlled rectifier
Semiconductor sensors, 8.20
Sensor trim, 2.16
Sequential function chart (SFC), 6.2
Sequester, 8.19, 8.34
Serial loading of data, 1.20
Series, 1.2
Server, 7.8, 7.12
Setpoint
    and distributed control systems, 7.1, 7.2, 7.3
    and loop tuning, 5.1, 5.2, 5.3, 5.9, 5.15, 5.16
Settling time, 2.18
SFC. *See* Sequential function chart
Shearing force, 8.3, 8.34
Shift registers
    basic, 1.18–1.21, 1.44
    disabled, 1.30
    five-bit, 1.21
    four-bit, 1.20–1.21, 1.28
    parallel in–parallel out, 1.28–1.30
    parallel in–serial out, 1.25–1.28
    serial in–parallel out, 1.23–1.25, 1.26
    serial in–serial out, 1.22–1.23, 1.24, 1.25
    transition to counter, 1.38
    universal, 1.28–1.30
Siemens/cm, 8.13, 8.14, 8.34
Siemens® pneumatic differential pressure transmitter, 2.7
Signal(s)
    analog. *See* Analog signal
    clear, 1.14
    clock, 1.13, 1.14
    digital. *See* Digital signals
    distortion. *See* Noise
    error, 5.2
    output energies from instruments, 2.2–2.3
    overview, 2.2–2.3
    pneumatic. *See* Pneumatic signal
    preset, 1.14, 1.40, 1.41
Signal conditioner (transducer), 2.18–2.19, 3.9
Silicon-controlled rectifier (SCR), 6.14
Simulation test, of a loop's operation, 3.8
Sinking operation, 6.13, 6.33
Smart instruments, 2.15–2.18, 3.2, 3.3, 3.5, 3.11, 3.15, 6.17
Software
    for DCS, 7.11, 7.19
    for PLC, 6.2, 6.18, 6.19–6.27
SONET. *See* Synchronous optical network
Sourcing operation, 6.13, 6.33
Span, of an instrument, 2.2, 2.3, 2.18, 2.21–2.22, 2.25, 3.12
Special state, use of term, 1.38
Specification sheet, 4.11
Specific gravity, 2.11, 2.12, 8.2–8.3

Spectrometry, 8.26–8.27, 8.34
Speed of circuit, 1.2
ST. *See* Structured text language
Stability of a system, 5.1, 5.3
Standards
    for calibration, 7.16
    for digital logic circuits, 1.43–144
    DIN, 6.2, 6.33, 8.9
    for fire prevention, 1.5
    for flash point testing, 8.9
    for particulates, 8.21
Star topology, 7.8, 7.10
Stepper control module, 6.17
Stiction, 5.17, 5.19
Stop bit, 6.16, 6.33
Storage bit, 6.18–6.19
Strobe line, 1.43
Stroke, 4.9, 4.15
Structured text language (ST), 6.2
Subroutine (SBR), 6.27
Supervisory control and data acquisition (SCADA), 7.12, 7.14
SUS. *See* Saybolt universal seconds
Symbols
    AND, 1.2
    amplifier, 1.6
    exclusive OR, 1.8, 1.9
    inverter, 1.6
    in ladder diagram programming language, 6.21–6.27
    on a loop diagram, 4.3, 4.4
    NAND, 1.7
    NOR, 1.9
    NOT, 1.7
    RS latch, 1.11
    on troubleshooting flowchart, 4.5–4.6
Synchronization, 1.13, 2.3
Synchronous counter, 1.30, 1.36–1.37, 1.44
Synchronous optical network (SONET), 7.12
System checks, dynamic, 6.30

Tag number, 3.4, 4.11
Tanks, 2.5, 2.11–2.13
TC. *See* Thermocouple; Time constant
TCP/IP. *See* Transmission control protocol/Internet protocol
TCV. *See* Temperature control valve
Technicians
    responsibilities, 7.15
    verbal feedback from system operator to the, 4.2
Teflon®, 8.17
Telephone communication, 3.6, 7.12
Temperature control valve (TCV), 2.14
Temperature monitoring/control module, 6.17
Temperature transmitters, 2.9, 2.14–2.15
Test equipment
    calibration of, 2.1–2.2, 2.3, 2.4–2.18
    for proving a loop, 3.9–3.10, 4.8
THC. *See* Total hydrocarbon
Thermocouple (TC), 2.14, 3.2, 3.9
Thermography, 8.24, 8.25–8.26
Time
    deadtime, 5.3
    derivative, 5.6–5.7
    integral, 5.6, 5.15, 5.16
    over the hump, 5.16, 5.19
    settling, 2.18
Time constant (TC), 5.3, 5.4, 5.12
Time constant method, 5.11–5.13, 5.19

Time-proportioned output (TPO), 6.17
Timer instructions, 6.22–6.24
Timer off delay (TOF), 6.23
Timer on delay (TON), 6.23
Timing diagram
    counters, 1.31, 1.33, 1.34, 1.35
    D flip-flop, 1.15
    D latch, 1.14
    down counter, 1.34
    exclusive OR, 1.9
    AND gate, 1.4
    JK flip-flop, 1.17
    JK master-slave, 1.17
    NAND gate, 1.8
    NOR gate, 1.9
    OR gate, 1.4–1.5
    overview, 1.47
    RS latch, 1.11
    shift registers, 1.19, 1.25, 1.26, 1.28
TOF. *See* Timer off delay
Toggle flip-flop (T), 1.18
Toggling, 1.16, 1.31
Token passing, 7.11
TON. *See* Timer on delay
Tools, repair, 7.17
Topology of local area network, 7.8–7.10
Total hydrocarbon (THC), 8.20–8.21
tpHL. *See* Turn-on delay
tpLH. *See* Turn-off delay
TPO. *See* Time-proportioned output
Tracer, fiber-optic, 3.8
Trailing edge triggered, 1.28, 1.47
Transducer
    calibration, 2.18–2.19
    E/P, 2.18
    I/P, 2.18, 2.19, 4.7, 4.8, 4.9, 4.15
    test equipment for loops with a, 3.9
Transfer frame, 7.10
Transformer, 2.14
Transistor-transistor logic (TTL), 1.2–1.3, 1.17, 1.47, 6.17
Transmission control protocol/Internet protocol (TCP/IP), 7.12, 7.14
Transmission turbidity analyzer, 8.6–8.7
Transmittance, 8.25, 8.34
Transmitter(s)
    analog, 2.10–2.15
    connection to HART® communication device, 3.12–3.13
    definition, 2.1
    differential pressure, 2.5–2.8, 2.10–2.14, 3.2–3.3, 4.4–4.5
    field, 3.2–3.3, 3.5
    level, 2.11–2.13
    pneumatic, 2.5–2.8, 2.9
    smart, 2.15–2.18, 3.2, 3.3, 6.17
    temperature, 2.9, 2.14–2.15
    troubleshooting in the field, 4.4–4.5
Trim, 2.16, 2.17, 2.18, 3.12, 5.9
Troubleshooting
    an oscillating process, 4.3–4.6
    flowchart, 4.5–4.6
    a loop, 4.1–4.3, 4.8–4.10
Truth table
    amplifier, 1.6
    binary coded decimal counter, 1.38
    definition, 1.2, 1.47
    D latch, 1.14
    down counter, 1.34
    exclusive OR, 1.9
    four-bit binary counter, 1.32

AND gate, 1.2
  inverter, 1.6
  JK flip-flop, 1.16
  Johnson counter, 1.40
  NAND gate, 1.5, 1.7
  NOR gate, 1.8, 1.9
  OR gate, 1.5
  ring counter, 1.39
  RS NOR gate, 1.12
  three-input AND gate, 1.3
TTL. *See* Transistor-transistor logic
Tubing, 3.2, 3.3, 3.7
Tuning. *See* Controller, tuning; Instrumentation, tuning;
  Loop, tuning
Turbidimeter, Jackson, 8.5–8.6
Turbidity, 8.5–8.8
Turn-off delay (tpLH), 1.2
Turn-on delay (tpHL), 1.2
Twisted ring counter. *See* Counters, Johnson

Ultimate period method, 5.8–5.10, 5.19
Ultraviolet radiation, 8.27–8.29
Universal shift registers, 1.28–1.30
U.S. Federal Code of Regulations, 8.21

Valve(s)
  block, 2.6, 3.3
  control (final loop element), 4.7, 4.8, 5.17
  controllers, 2.22
  effects of partially blocked, 4.5
  positioners, 2.1, 2.19–2.22
  pressure relief, 3.6
  safety, 1.5, 1.6
  spring-loaded, 2.5
  sticking, 5.17
  stroke, 4.9
  temperature control, 2.14
Valve actuator, 2.19
Viscometers, 8.4–8.5
Viscosity
  definition, 8.3
  dynamic, 8.4, 8.33
  kinematic, 8.4, 8.33
VOC. *See* Volatile organic contaminants
Volatile organic contaminants (VOC), 8.20

Voltage
  conversion of temperature to, 2.14
  conversion to pneumatic signals, 2.18, 2.19, 2.22
  and Ohm's law, 8.14
  and oxidation-reduction potential, 8.10

Wallace & Tiernan® Company, 2.4, 3.6
Wally Box®
  how to use, 2.7–2.8
  for monitoring output signal from temperature
    transmitters, 2.9
  overview, 2.4, 2.5, 3.6–3.7
  upgrades, 2.5, 3.6
WAN. *See* Wide area network
Warehouse, cold storage, 2.14
Water, inches of, 2.2, 2.25
Weighted code, 1.21, 1.30
Wet leg, 2.12
Wet process, 7.2
Wide area network (WAN), 7.12, 7.13
Wika Instrument Corporation, 3.6
Windup, reset, 5.2
Wiring
  continuity test, 3.5–3.6
  inspection, 3.3
  I/O, 6.29–6.30
Wiring diagram, 4.7, 4.10
Workstation, 7.8, 7.11

XOR (exclusive OR) gate, 1.8–1.9, 1.42, 1.47

Yttrium, 8.15, 8.34

ZCL. *See* Zone control
Zero
  absolute, 8.24, 8.33
  checking for, 4.5
  elevated, 2.7, 2.13
  suppressed, 2.11
  trim, 2.17, 2.18, 3.12
  true, 2.2, 2.4
Zeroing an instrument, 2.2, 2.4, 2.18, 2.21, 2.25, 3.12
Ziegler-Nichols method and equations, 5.8–5.10, 5.12–5.13,
  5.13–5.14, 5.16, 5.17
Zirconium oxide, 8.15–8.16
Zone control (ZCL), 6.26